Macromolecular Engineering
Recent Advances

Macromolecular Engineering

Recent Advances

Edited by

Munmaya K. Mishra

Texaco Inc., R&D
Beacon, New York

Oskar Nuyken

Technical University of Munich
Garching, Germany

Shiro Kobayashi

Tohoku University
Aoba, Sendai, Japan

Yusuf Yağci

Istanbul Technical University
Maslak, Istanbul, Turkey

and

Bidulata Sar

PFI, Inc.
Hopewell Junction, New York

Springer Science+Business Media, LLC

Library of Congress Cataloging-in-Publication Data
On file

Proceedings of the International Conference on Advanced Polymers via Macromolecular Engineering, held June 24–28, 1995, at Poughkeepsie, New York

ISBN 978-1-4613-5778-0   ISBN 978-1-4615-1905-8 (eBook)
DOI 10.1007/978-1-4615-1905-8

© 1995 Springer Science+Business Media New York
Originally published by Plenum Press, New York in 1995
Softcover reprint of the hardcover 1st edition 1995

10 9 8 7 6 5 4 3 2 1

PREFACE

This volume *Macromolecular Engineering: Recent Advances* has been developed based on the 1st International Conference on *"Advanced Polymers Via Macromolecular Engineering"* (APME '95), June 24–29, 1995 at the Vassar College campus, Poughkeepsie, New York. In APME '95, 100 oral and over 50 poster presentations are to be delivered from scientists around the globe. The scientific program covers recent advances in macromolecular engineering. It is our vision that the knowledge of the past and the promise of the future are blended together in APME '95 to enrich and stimulate the scientists, which will bring about the progress of *macromolecular engineering*. Scientists from over 30 countries will be joining together to share this vision.

Although over 150 papers are to be presented in APME '95 conference, we could not include all the papers in this book for a variety of reasons, most importantly the authors willingness to contribute to this volume in time to meet the deadline. However, the 24 comprehensive chapters included in this volume are a true reflection of some of the important themes of *macromolecular engineering* that are part of the APME '95 conference. We believe *macromolecular engineering* is the key to developing new polymeric materials and, to this end, it is hoped this volume will aid in this introspection.

An editorial task of this volume and the international conference of this magnitude would not have been possible without the help of many of our co-workers, international scientists, and support of several institutions. We would like to particularly thank to Akzo Nobel Research (Netherlands), DuPont Company, Marcel Dekker, Inc., 3M Company, Procter & Gamble Co., Texaco Inc., and Viscotek Corp. for their generous financial support. Also, we would like to express our thanks to the Polymer Frontiers International, organizer and sponsor of the APME '95 conference.

M. K. Mishra, USA
O. Nuyken, GERMANY
S. Kobayashi, JAPAN
Y. Yaǧci, TURKEY
B. Sar, USA

CONTENTS

GROUP TRANSFER POLYMERIZATION AND ITS RELATIONSHIP TO OTHER LIVING SYSTEMS

Owen W. Webster

DuPont Central Research and Development
Wilmington, Delaware 19880-0328

INTRODUCTION

The large number of organic function groups that can be attached to the ester function of methacrylate and acrylate monomers makes this series of monomers attractive for synthesis of polymers with diverse properties. When this diversity of functionality is coupled with living polymerization techniques the range of new product possibilities is staggering. Polymer synthesis chemists have therefore been searching for living polymer systems for methacrylates and acrylates that operate at above ambient temperature, use reasonably low cost initiators, tolerate moderate amounts of impurities, control chain tacticity and allow the functionality that is to be introduced to survive the polymerization process.

Davis, Haddleton, and Richards have recently reviewed the controlled polymerization of acrylates and methacrylates, (Ref. 1). Müller has compared Group Transfer Polymerization (GTP) to anionic polymerization (Ref. 2). In this review the major living methacrylate polymerization techniques will be compared to GTP with an eye to possible industrial utilization.

The term "living" will be used whenever all polymer chains in a system grow at the same time, low polydispersity results and the molecular weight is controlled by the monomer/initiator ratio. To classify initiators, enolate species on the propagating chain ends will be identified, although in practice these enolate ends may have been formed by addition of other reagents to the MMA, (metal hydride, alkyl lithium etc.).

SILICON ENOLATES AS INITIATORS FOR METHACRYLATE POLYMERIZATION - GROUP TRANSFER POLYMERIZATION

In 1983 DuPont announced a new procedure for polymerization of acrylic monomers (Ref. 3). In this procedure trimethylsilyl capped chain ends inserted monomer at temperatures as high as 100°C. High enough so that river water can be used to cool a batch reactor under reflux. The polymerization was living and thus providing routes to

Macromolecular Engineering, Edited by M.K. Mishra et al.
Plenum Press, New York, 1995

1

low polydispersity-block, star and telechelic polymers of predetermined molecular weight. The name given to the procedure was Group Transfer Polymerization (GTP), a name that unfortunately, to some among us, implies direct transfer of the silyl group to incoming monomer (Eq. 1).

PMMA   (1)

In the more widely accepted mechanism the silyl group (10^{-3}M), which protects chain ends from termination, does not directly transfer to incoming monomer but rapidly exchanges with a small amount of enolate ended polymer, (10^{-5}M) generated by the catalyst (Eq. 2). On this basis dropping the term group transfer polymerization has been suggested (Ref. 4,5). However, in view of the large body of literature that uses the term GTP and the fact that in the final product and all intermediate stages transfer of the silyl group has indeed occurred, we recommend that the name group transfer polymerization should continue to be used.

(2)

Although bifluoride was first used as the nucleophilic catalysts for GTP it proved to be too active for operation at above room temperatures. Tetrabutylamomnium biacetate or bibenzoate are now the catalysts of choice. The extra mole of acid in the catalyst reacts rapidly with the trimethylsilyl ketene acetal initiator to generate trimethylsilyl carboxylate, a livingness enhancing agent. Since the catalyst is used at levels less than 1% based on initiator the destruction of this small amount of ketene acetal has no noticeable effect on the molecular weight of the polymer. M_n = [Monomer]/(Initiator) x m. w. of monomer.

GTP requires dry monomer with pendent functional groups free of active hydrogen (Ref. 6). The initiator is expensive enough so that polymer prepared by GTP cannot compete economically with free radical initiated product. Molecular weights over 50,000 are difficult to obtain. GTP is used by DuPont to prepare block polymer dispersing agents for pigments.

When one uses a block polymer from nucleophilic catalyzed GTP as a dispersing agent, the trimethylsilyl groups on the chain ends hydrolyze to form hexamethyldisiloxane. This innocuous material can be left in the polymer solution along with the small amount of catalyst. On the other hand, Lewis acid catalyzed GTP requires large amounts of Lewis acid, $ZnBr_2$, $ZnCl_2$ or alkyl aluminum dichloride (up to 30% based on monomer, Ref. 7). Since these acids would have to be removed before the resins could be used, the cost would be prohibitive.

TRANSITION METAL ENOLATES AS INITIATORS FOR METHACRYLATE POLYMERIZATION

In significant work that introduces its use of transition metal enolates for living methacrylate polymerization, Yasuda has shown that the polymerization of methyl methacrylate can be initiated by $(Cp*_2SmH)_2$ addition to MMA to give enolate **1**, (Ref. 8, Eq. 3). The structure of **1** was confirmed by X-ray.

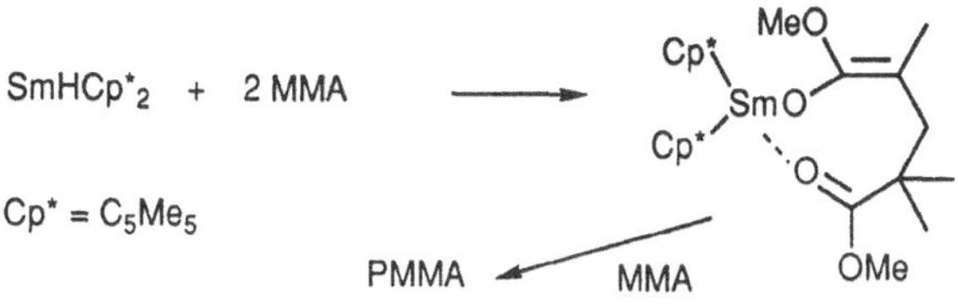

$$SmHCp*_2 \;+\; 2\,MMA \longrightarrow \mathbf{1}$$

$$Cp* = C_5Me_5$$

$$PMMA \longleftarrow MMA$$

(3)

No catalyst is needed for this group transfer like process. Polymers in the 100,000 molecular weight range with polydispersities as low as 1.05 have been obtained. In addition block polymers with ethylene can be made (Ref. 9).

Novak (Ref. 10) has suggested that the polymerization of MMA by $(Sm.\ Cp*_2)_2$ first produces a bisenolate by election transfer and dimerization.

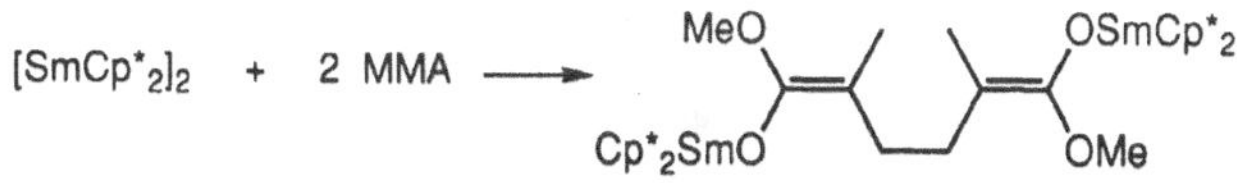

$$[SmCp*_2]_2 \;+\; 2\,MMA \longrightarrow$$

(4)

For comparison he prepared the bisalkyl initiator $(Cp*_2Sm(\mu\text{-}h^3\text{-}CH_2CHCH)]_2$. It initiated MMA in a living fashion giving quantitative yields of PMMA with polydispersities of ~1.1 after 2 h at 0°C. A plot of Mn vs. monomer/initiator ratio shows the expected linear relationship. Although the lanthanide initiators are unparalleled in their ability to polymerize MMA to high molecular weight, low polydispersity PMMA. The cost of preparing lanthanide Cp* derivatives and difficulty in handling these air sensitive materials, may hold back industrial development.

In work with more industrial potential Collins (Ref. 11) has shown that ester enolates of Zirconium metallocenes initiate living polymerization of MMA in the presence of the corresponding cationic methyl complex (eq. 5).

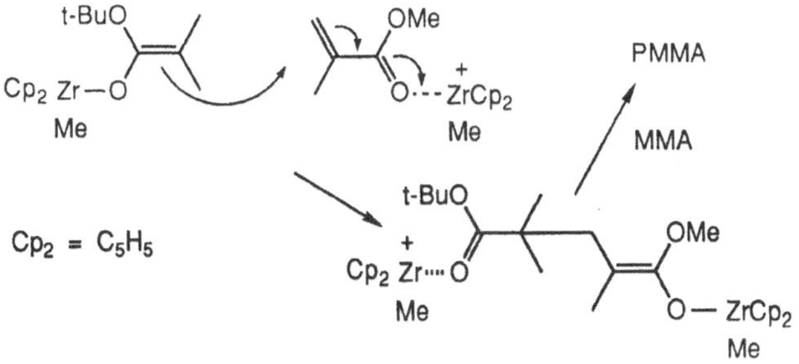

$$Cp_2 = C_5H_5$$

(5)

Molecular weights in the 100,000 range were obtained with polydispersities ~1.15. When a chiral initiator system was used isotactic PMMA was formed. Since the metallocene derivatives were made from cyclopentadiene they would be more readily available than the Sm metallocene derivatives made with pentamethylcyclopentadiene. Collins' polymerizations were conducted at 0°C. Can variations be found that will operate at 80°C?

ALUMINUM ENOLATES AS INITIATORS FOR METHACRYLATE POLYMERIZATION

Immortal Polymerization (IP)

Inoue (Ref. 12) has shown that an ester enolates attached to an aluminum porphyrin (**2**) polymerizes MMA slowly at ambient temperature (100% yield in 8 h) but more rapidly at 80°C (80% yield in 2 h). The ester enolates are generated by insertion of MMA in the Al-X bond of the TPP Al-X. When X=Me irradiation from a Xenon arc (<420 nm) is required, for X = -SPr no light is needed. The polymerization rates increase by a factor of 45,000 in the presence of hindered bulky Lewis acids for example aluminum phenoxide, **3** (Eq. 6). Once the enolate is formed the polymerization meets the requirements for living polymerization with molecular weight increasing monotonously with conversion and production of low polydispersity polymer. Inoue has obtained PMMA with molecular weights in excess of 1 million with a polydispersity of 1.2. The technique was named immortal polymerization since when TPPAl X was used to polymerize epoxides, it was a living system proton sources would not kill.

TPPAIX

2

PMMA   (6)

Although the polymerization operates at industrial process temperatures, the porphyrin initiators are deeply colored and expensive. Inoue has reported that less colored and lower cost planar aluminum complexes such as **4** also operate in for immortal polymerization (Ref. 13). Methacrylate polymerization with these complexes will be of keen interest.

Screened Anionic Polymerization (SAP)

Haddleton and Ballard have reported an aluminum mediated living polymerization system for MMA possibly related to Inoue's IP process (Ref. 14). In this process, named screened anionic polymerization (SAP), a large cationic component screens the propagation terminus from side reactions. A mixture of t-butyl lithium/diisobutyl aluminum o-di-t-butylphenoxide is used as the initiator. The growing chain end is probably an aluminum enolate (Eq. 7, Ref. 1). The molecular weights are higher than theory based on the monomer/t-butyl Li ratio. The structure of the cation associated with the growing enolate end is vague. SAP is a robust process operating at near ambient temperature and gives polymer up to 60,000 m.w. with polydispersities close to 1.1. The polymerization proceeds in toluene but not in THF. SAP is industrially feasible, however, river water cooling would not be practical at room temperature.

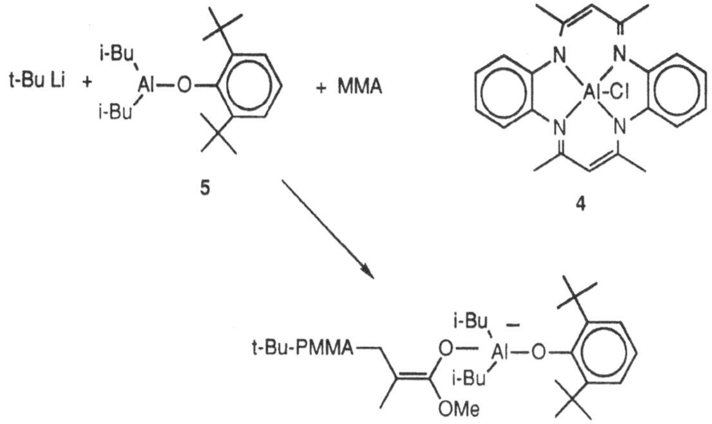

$$Li^+ \text{ Complex with}$$
$$\text{t-Bu Li} + \mathbf{5} \tag{7}$$

Along similar lines Teyssie (Ref. 15) very recently demonstrated that two-bulky phenolates on the aluminum are better than one and were able to polymerize MMA at 40°C in the present of **6** and an anionic initiator.

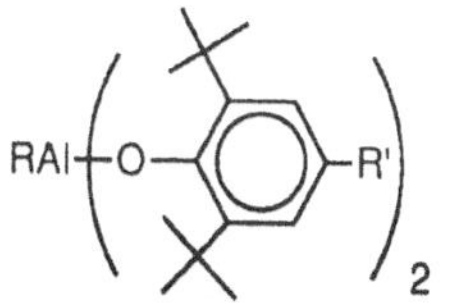

R = Me, Et, i-Bu
R' = H, Me, t-Bu

6

t-Bu Li/trialkyl aluminum can also be used for SAP and proceeds well in toluene at 20° C (Ref. 16).

LITHIUM ENOLATES AS INITIATORS FOR METHACRYLATE POLYMERIZATION

Lithium ester enolates generated by addition of hindered lithium alkyls to methacrylate esters have been used for over 30 years as initiators for living polymerization of methacrylates (Ref. 17). The main problem with the systems is that at temperatures above -60° C , chain termination occurs by the well known back biting cyclization of the chain ends (Eq. 8). This limits the method to research synthesis since in an industrial plant the refrigeration required to remove the heat of polymerization would cost too much. The lowest workable temperatures for an industrial process are in the 80° C range.

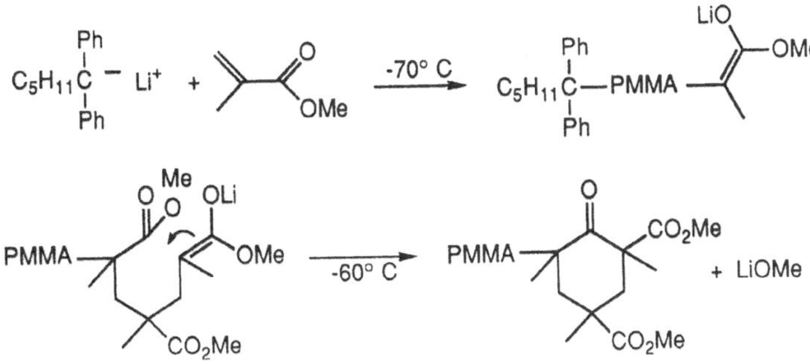

$$\tag{8}$$

In the 70's Lochmann showed that the addition of bulky metal alkoxides to a living anionic polymerization of MMA permitted one to operate at up to 20° C (Ref. 18). Complexation of the metal alkoxide to the enolate chain ends no doubt hinders the back biting termination (Eq. 9).

$$Na_6(O\text{-}t\text{-}Bu)_5^+$$

(9)

In recent work Teyssie has shown that lithium 2-(2-methoxyethoxy)ethoxide is a better livingness enhancing agent than lithium t-butoxide. With this reagent present, polymerizations can be conducted at 0° C (Ref. 19). In NMR studies Teyssie showed that the lithium enolate from methyl isobutyrate forms a 1/2 complex with added lithium alkoxide (Ref. 20). This complex retards chain termination.

Lithium chloride has a similar stabilizing affect on chain ends. However operation temperatures were only raised to -20° C (Ref. 21).

TETRABUTYLAMMONIUM AND OTHER LARGE ION ENOLATES FOR INITIATION OF METHACRYLATES AND ACRYLATES

Reetz (Ref. 22) and later Sivaram (Ref. 23) have shown that a growing enolate ended methacrylate or acrylate polymer chain can be stabilized by the use of a tetrabutylammonium (TBA) counter ion. Although the first work used TBA thiolates as initiators, it appears TBA dialkyl malonate enolates work better. Polymers with polydispersities in the 1.2-1.4 range and some degree of molecular weight control were obtained. Reetz proposes that the α protons on TBA hydrogen-bond to the enolate oxygen. In addition elimination of TBA alkoxide in chain termination reactions (Eq. 8) would be thermodynamically and kinetically unfavorable due to the high degree of charge separation in this ion-pair substance. In spite of the initial excitement over non-metallic initiation it now appears that industrial use will be hampered by the serious decomposition of the TBA by reaction with the strongly basic enolate polymer ends.

The addition of crown ethers to growing enolate chain ends having sodium or potassium gegenions but not lithium, permits one to operate living MMA polymerization at near room temperature. Dibenzo-18-crown-6 works better than 18-crown-6 for sodium ion (Ref. 24). At 20-60° C 18-crown-6/ potassium dialkyl malonate gave polymerswith polydispersities in the 1.5-1.9 range and molecular weights which increased with conversion (Ref. 25). The molecular weights were however 30% lower than theory based on the monomer /initiator ratio. Since potassium t-butoxide was used to generate the malonate enolate the t-butanol generated may be lowering the molecular weight by chain transfer.

It has been known for some time that alkoxides will initiate the polymerization of MMA at above room temperature in alcohol solvent (Ref. 26). The molecular weight is controlled to some degree by the amount of alcohol. Since ester interchange occurs between the alcohol and ester groups, block polymers can not be made. Suzuki (Ref. 27) showed that dicycloheyl-18-crown-6 would activate disodium PEO (M_n 980) for initiation of MMA to give an ABA block polymer. We have found that potassium t-butoxide/18-crown-6 in toluene initiates MMA at 60-80° C. to give PMMA with polydispersities in the 1.8 range (Ref. 25). Since both chain termination reactions for anionic methacrylate polymerization, ketene formation and dimerization (Eq. 10) and end group backbiting (Eq. 8), generate alkoxide, chain transfer will be a major cause of molecular weight dispersity broadening. A test for this kind of problem would be to see if block polymers can be made.

Crown ethers are expensive. A low cost large cation stable to strong bases would make alkoxide initiation industrially viable.

Possibly the most unusual large ion system is the initiation of MMA with P4-base/ethyl acetate (Ref. 28). the strongly basic P4-base generates an enolate ion from the acetate which initiates MMA at 0-50° C to give PMMA with a polydispersity of 1.1 (Eq. 11). Unlike other living MMA polymerization this one does not work well at temperatures below 0° C.

At this point P4-phosphazene base is too expensive to be considered for large scale industrial use. However, it is made for low cost ingredients, PCl_5, Me_2NH and t-butylamine, and cost would no doubt drop drastically if the synthesis were scaled up.

SUMMARY

The use of large gegenions and end group complexation have moved living anionic methacrylate polymerization from the Dry Ice^R to the wet ice temperature range. However GTP still reigns as the method of choice for large scale synthesis of block and star methacrylate polymers.

REFERENCES

1. Davis, T.P., Haddleton, D. M., and Richards, S. N., 1994, Controlled polymerization of acrylates and methacrylates, *J. Macromol. Sci.-Chem. Phys.* C34(2): 243-324.
2. Müller, A. H. E., 1990, Group transfer and anionic polymerization: a critical comparison, *Makromol. Chem., Macromol. Symp.* 32: 87-104.
3. Webster, O. W., Hertler, W. R., Sogah, D. Y., Farnham, W. B., and RajanBabu, T. V. J., 1983, Group transfer polymerization. 1. A new concept for addition polymerization with organosilicon initiators, *J. Am. Chem. Soc.* 105: 5706-5707.

4. Quirk, R. P., and Bidinger, G. P., 1989, Mechanistic role of enolate ions in "Group Transfer Polymerization", *Polymer Bulletin* 22: 63-70.

5. Jenkins, A. D., 1993, Anionic polymerization at ambient temperature, *Eur. Polym. J.* 29: 449-450.

6. Sogah, D. Y., Hertler, W. R., Webster, O. W., and Cohen, G. M., 1987, Group transfer polymerization. Polymerization of acrylic monomers, *J. Am. Chem. Soc.* 20: 1473-1488.

7. Hertler, W. R., Sogah, D. Y., Webster, O. W., and Trost, B. M., 1984, Group transfer polymerization. 3 Lewis acid catatysis, *Macromolecules* 17: 1415-1417.

8. Yasuda, H., Yamamoto, H., Yamashita, M., Yokota, K., Nakamura, A., Miyake, S., Kai, Y., and Kanehisa, N., 1993, Synthesis of high molecular weight poly(methyl methacrylate) with extremely low polydispersity by the unique function of organolanthanide (III) complexes, *Macromolecules* 26: 7134-7143.

9. Yasuda, H., Furo, M., Yamamoto, H., Nakamura, A., Miyake, and S., Kibino, N., 1992, New approach to block copolymerization of ethylene with alkyl methacrylates and lactones by unique catalysis with organolanthanide complexes, *Macromolecules* 25: 5115-5116.

10. Boffa, L. S., and Novak, B. M., 1994, Bimetallic Samarium (III) initiators for the living polymerization of methacrylates and lactones. A new route into tetechelic, triblock, and "link-functionalized" polymers, *Macromolecules* 27: 6993-6995.

11. Collins, S., Ward, D. G., and Suddaby, K. H., 1994, Group transfer polymerization using metallocene catalysts: Propagation mechanisms and control of polymer stereochemistry, *Macromolecules*, 27: 7222-7224.

12. Adachi, T., Sugimoto, H., Aida, T., and Inoue, S., 1993, Aluminum thiolate complexes of porphyrin as excellent initiators for Lewis acid-assisted high-speed living polymerization of methyl methacrylate, *Macromolecules* 26: 1238-1243.

13. Inoue, S., Aida, T., Sugimoto, H., Kawamura, C., and Kuroki, M., 1993, Novel catalyst systems for the synthesis of poly(alkylene oxide) with controlled molecular weight, *Proc. ACS Div. Polym. Mater. Sci., Engl.* 69: 428-

14. Ballard, D. G. H., Bowles, R. J., Haddleton, D. M., Richards, S. N., Sellens, R., and Twose, D. L., 1992, Controlled polymerization of methyl methacrylate using lithium aluminum alkyls, *Macromolecules*, 25: 5907-5913.

15. Wang, J. S., Teyssie', Ph., Heim, Ph., Vuillemin, B., 1994, French Patent 94, 06891.

16. Haddleton, D. M., Muir, A. V. G., O'Donnell, J. P., and Twose, D. L., 1993, Synthesis of block copolymers and homopolymers of methacrylates using a mixed Al/Li alkyl initiator, *Polym. Prepr. Am. Chem. Soc. Div. Polym. Chem.* 34 (2): 564-565.

17. Anderson, B. C., Andrews, G. D., Arthur, P., Jacobson, H. W., Melby, L. R., Playtis, A. J., and Sharkey, W. H., 1981, Anionic Polymerization of Methacrylates. Novel functional polymers and copolymers, *Macromolecules*, 14: 1599-1601.

18. Lockmann, L., Kolarik, J., Doskocilova, D., Voska, S., and Trekoval, J., 1979, Metallo esters. VII - Stabilizing effects of sodium tert-butoxide on the growth center in anionic polymerization of methacrylate esters, *J. Polym. Sci., Polym. Chem. Ed.*, 17: 1727-1737.

19. Wang, J.-S., Jerome, R., and Teyssie, Ph., 1994, Anionic polymerization of acrylic monomers 19. Effect of various types of ligands other than lithium chloride on the stereochemistry of anionic polymerization of methyl methacrylate, *Macromolecules*, 27: 4902-4907.

20. Wang, J.-S, Jerome, R., and Teyssie, Ph., 1994, Anionic polymerization of acrylic monomers. 18. NMR characterization of a unique complex between lithium 2-(2-methoxyethoxy) ethoride and methyl α-lithioisobutyrate, *Macromolecules*, 27: 4896-4901.

21. Fayte, R., Forte, R., Jacobs, C., Jerome, R., Ouhadi, T., Teyssie, Ph., and Varshney, S. K., 1987, New initiator system for the "living" anionic polymerization of tert-alkyl acrylates, *Macromolecules*, 20: 1442-1444.

22. Reetz, M. T., Knauf, T., Minet, U., and Bingel, C., 1988, Metal-free carbanion salts as initiators for the anionic polymerization of acrylic and Methacrylic acid esters, *Angew. Chem. Int. Ed. Engl.*, 27: 1373-1374.

23. Sivaram, S., Dhal, P. K., Kashikai, S. P., Khisti, R. S., Shinde, B. M., and Baskaran, D., 1991, Approaches to controlled polymerization of methyl acrylate through functional anionic initiators, *Polym. Bull,* 25: 77-81.

24. Wang, J.-S., Jerome, R., Bayard, Ph., Baylac, L., Patin, M., and Teyssie, Ph., 1994, Anionic polymerization of acrylic monomers. 15. Living anionic copolymerization of mixtures of methyl methacrylate and tert-butyl acrylate as promoted by dibenzo-18-crown 6, *Macromolecules*, 27: 4615-4620.

25. Webster, O. W., (1994) Does anion-catalyzed group-transfer polymerization proceed via an endate intermediate, *J. Macromol. Sc., Rure App. Chem.*, A31: 927-0935.

26. Haggard, R. A., and Lewis, S. N., 1984, Methacrylate oligomers via alkoxide-initiated polymerization, *Prog. Org. Coatings,*
    12: 1-26.
27. Suzuki, T., Murakami, Y., and Takegami, Y., 1980, Synthesis and characterization of block copolymers of poly(ethylene oxide) and poly(methyl methacrylate), *Polym. J.* 12: 183-192.
28. Pietzonka, T., and Seebach, P., 1993, the P_4-phosphazene base as part of a new metal-free initiator system for the anionic polymerization of methyl methacrylate, *Angew. Chem. Int. Ed. Engl.,* 32: 716-717.

FUNDAMENTALS AND PRACTICAL ASPECTS OF "LIVING" RADICAL POLYMERIZATION

Krzysztof Matyjaszewski

Department of Chemistry
Carnegie Mellon University
4400 Fifth Avenue
Pittsburgh, Pennsylvania 15213

INTRODUCTION

Living radical polymerization is very difficult if not impossible to realize. Complete elimination of chain breaking reactions, especially termination, is not possible in radical processes. Therefore quotation marks are used for a term of "living" radical polymerization. However, well-defined polymers with predetermined molecular weights, low polydispersities and terminal functionalities as well as block copolymers can be prepared via radical polymerization. Thus, although complete suppression of termination is not possible, preparation of controlled polymers with relatively low molecular weights has been successful. In this article fundamentals and some practical synthetic aspects of controlled radical polymerization will be discussed.

EFFECT OF SLOW INITIATION, TERMINATION, AND TRANSFER ON KINETICS, MOLECULAR WEIGHTS AND POLYDISPERSITIES

Well-defined polymers and block copolymers with the polymerization degrees predetermined by the ratio of concentrations of reacted monomer to the introduced initiator, $DP = \Delta[M]/[I]_o$, and with low polydispersities, $M_w/M_n < 1.1$, are usually prepared in living systems, i.e. in chain growth processes without chain breaking reactions. However, the absence of chain breaking reactions is not a sufficient condition for the synthesis of well defined polymers. For example, slow initiation may lead to molecular weights much higher than expected and also to larger polydispersities; slow exchange between species of different reactivities and different lifetimes may result in polymodal molecular weight distributions, even if no irreversible chain breaking reactions are present. Two graphs shown below (Figure 1 and Figure 2) present the quantitative effect of slow initiation, unimolecular termination and transfer to monomer on kinetics and evolution of molecular weights with conversion.

Macromolecular Engineering, Edited by M.K. Mishra et al.
Plenum Press, New York, 1995

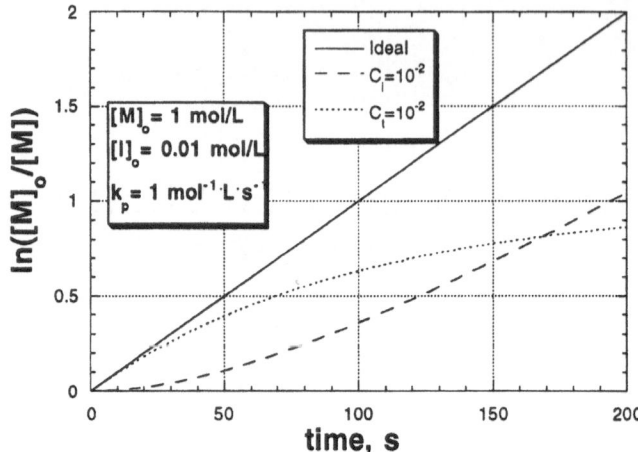

Figure 1. Effect of slow initiation and termination on kinetics of chain polymerization.

In the ideal living system, with one type of active center which is formed by fast initiation and without chain breaking reactions, straight semilogarithmic kinetic plots and linear evolution of molecular weight with conversion is expected. However, an increase in the slope of the semilogarithmic anamorphoses and initially higher than predicted molecular weights are observed with slow initiation.

This is also accompanied by enhanced polydispersities and can be assigned to the continuous increase of the number of growing chains. On the other hand, termination reduces the number of growing chains and decreases the polymerization rate, sometimes even to limited conversions, depending on k_p/k_t and $[M]_o/[I]_o$ ratios.

The total number of chains in systems with unimolecular termination and with termination by disproportionation in radical polymerization remains constant and therefore a linear increase of molecular weights with conversion is preserved provided that initiation

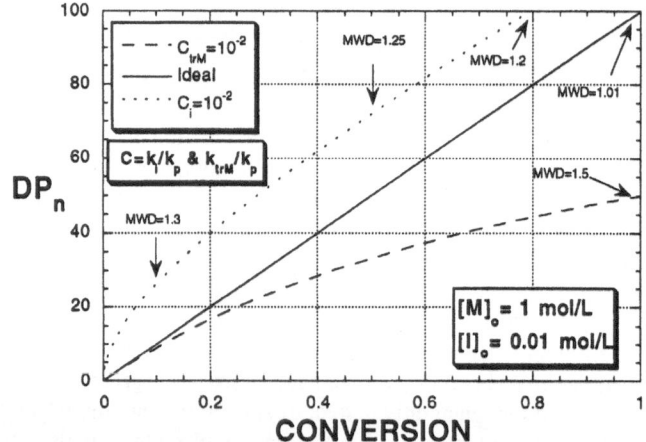

Figure 2. Effect of slow initiation and transfer to monomer on molecular weights.

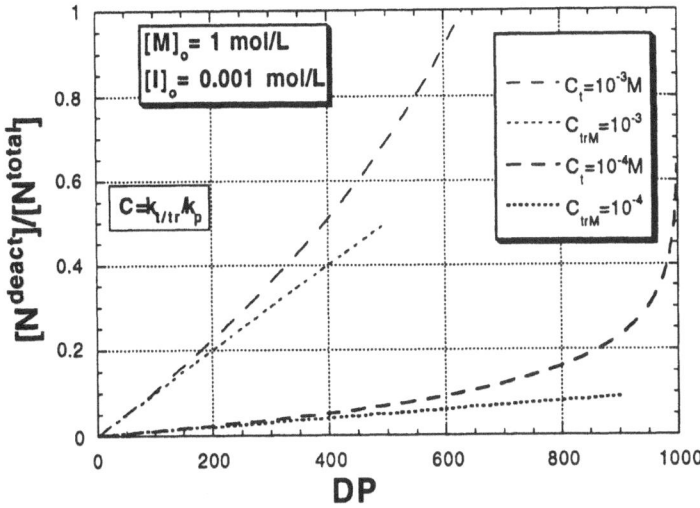

Figure 3. Proportion of chains deactivated by termination and transfer to monomer as a function of DP.

is fast. Transfer usually has no effect on kinetics but it generates an increasing number of dead chains leading to lower than expected molecular weights.

Figure 3 shows the dependence of the ratio of chains which have been irreversibly deactivated by either termination or transfer to monomer, to the total number of chains. If this ratio equals zero, all chains are active and potentially functionalizable; if the ratio equals 1, then all chains are deactivated. In order to prepare well-defined block copolymers and end-functionalized polymers, the proportion of deactivated chains must be very small, <5%. As shown in Figure 3, the proportion of deactivated chains increases with chain length and also with the ratio $k_{t/tr}/k_p$. The proportion of deactivated chains can be significantly reduced for the same transfer/termination coefficient, if low molecular weight polymers are synthesized.

In the case of transfer to monomer, the proportion of deactivated chains increases monotonously with conversion up to a certain value at the final conversion. For the case of termination, it is necessary to stop the reaction at an appropriate stage; otherwise all chains will become deactivated.

Polydispersities are also reduced for shorter chains even for the same ratio of rate constants of propagation and transfer. Fig. 4 shows the reduction of polydispersities in the hypothetical systems with fast initiation, no termination and transfer to monomer, $k_p/k_{trM} = 10^{-3}$, as a function of $[M]_o/[I]_o$ at complete conversion. Polydispersities can be reduced from $M_w/M_n = 2$ for the ratio $[M]_o/[I]_o = 10^4$, to $M_w/M_n = 1.5$ for the ratio 10^{-3} and down to $M_w/M_n = 1.1$ for the ratio $[M]_o/[I]_o = 10^2$.

Both figures demonstrate that polymers with a low proportion of deactivated chains and with very narrow polydispersities can be prepared in the presence of chain breaking reactions such as transfer and termination, if molecular weights are limited to a relatively low range. Many well-defined polymers prepared recently have rather low molecular weights, indicating that control is due to a careful selection of the polymerization conditions rather than to the substantial increase in chemoselectivities and new polymerization mechanisms.

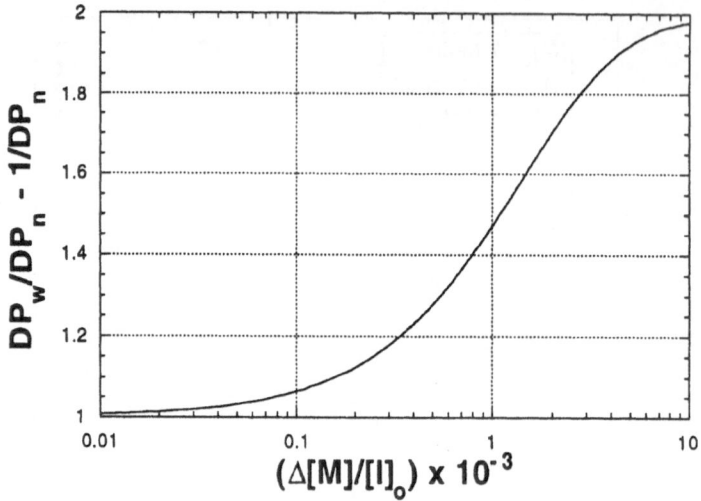

Figure 4. Dependence of polydispersities at complete conversion on $[M]_o/[I]_o$ ratios for systems with fast initiation and transfer to monomer, $k_p/k_{trM} = 10^{-3}$.

BASIS OF RADICAL POLYMERIZATION

Radical polymerization includes five elementary reactions:

- slow homolytic cleavage of a peroxide, diazo, or other similar compounds; $k_d < 10^{-5 \pm 1}$ s^{-1},

$$I\text{-}I \xrightarrow{k_d} 2\,I^\bullet \tag{1}$$

-relatively fast reaction of primary radicals with monomer to generate the first growing species; because $k_d < k_o[M]$, the decomposition is the rate determining step in the initiation process:

$$I^\bullet + M \xrightarrow{k_o} P_1^\bullet \tag{2}$$

- fast propagation ; $k_p \approx 10^{3 \pm 1}$ mol^{-1}·L·s^{-1},

$$P_n^\bullet + M \xrightarrow{k_p} P_{n+1}^\bullet \tag{3}$$

- very fast termination between growing radicals via coupling or disproportionation; $k_t \approx 10^{7 \pm 1}$ mol^{-1}·L·s^{-1},

$$P_n^\bullet + P_m^\bullet \xrightarrow{k_t} P_{n+m} \,/\, (P_n^{=} + P_m\text{-H}) \tag{4}$$

- transfer reactions including transfer to monomer, initiator, solvent, polymer, and specially added transfer agents.

$$P_n^\bullet + X \xrightarrow{k_{tr}} P_n + X^\bullet \tag{5}$$

Usually high molecular weight polymers are formed because contribution of transfer is small, initiation is slow and produces a low stationary concentration of growing radicals ($[P^\bullet] \approx 10^{-7\pm1}$ mol/L) which terminate in a bimolecular way. Ratio of the rate of propagation to that of termination ("livingness") decreases with $[P^\bullet]$, because propagation is first order but termination is second order in respect to $[P^\bullet]$. However, as demonstrated in Fig. 3, the proportion of chains marked by termination increases with chain length. Therefore, well defined polymers may be formed in radical polymerization only if chains are relatively short and concentration of free radicals is low enough. There is an apparent contradiction between these two requirements because usually a decrease of the concentration of radicals leads to higher molecular weights. However, the two conditions can be accommodated in systems with reversible deactivation of growing radicals.

REVERSIBLE DEACTIVATION OF GROWING RADICALS IN THE CONTROLLED RADICAL POLYMERIZATION

As discussed in the previous sections, controlled polymerization requires a low proportion of deactivated chains, which can be achieved by keeping molecular weights sufficiently low. This necessitates a relatively high concentration of the initiator or, in other words, low $[M]_o/[I]_o$ ratios. However, because termination is bimolecular, the contribution of termination becomes more significant at high $[I]_o$, when a large concentration of radicals, $[P^\bullet]$, is generated.

In order to solve the discrepancy between high $[I]_0$ and low $[P^\bullet]$, it is necessary to establish an exchange between dormant and active species. The concentration of the dormant species can be equal to $[I]_0$ and the concentration of momentarily growing species to $[P^\bullet]$. The total number of growing chains will be equal to $[I]_0$ ($\approx [P-R]_o + [P^\bullet]$), and radicals would be present at a very low stationary concentration, $[P^\bullet]$, and therefore the contribution of termination and the proportion of irreversibly terminated chains should be very low. There are three possibilities to realize the concept of controlled radical polymerization.

1. Reversible Homolytic Cleavage of Covalent Species

$$P\text{-}R \underset{k_{deact}}{\overset{k_{act}}{\rightleftarrows}} P^\bullet + R^\bullet \tag{6}$$

The covalent species P-R can reversibly and homolytically cleave to produce the growing radical $P^\bullet$, capable of propagation, and the dormant radical $R^\bullet$, which, ideally, should react only with $P^\bullet$ but not with the monomer or with itself to form inactive dimers. $P^\bullet$ can react not only with M and $R^\bullet$ but also with another $P^\bullet$, leading to termination. Because termination rate is proportional to $[P^\bullet]^2$, and propagation rate to $[P^\bullet]$, the contribution of termination and the proportion of deactivated chains increases with $[P^\bullet]$.

This case is probably most frequently postulated in controlled radical polymerizations. As examples of $R^\bullet$, dithiocarbamate radicals,[1, 2] nitroxyl radicals,[3, 4, 5, 6] and also bulky organic radicals such as triarylmethyl and substituted diarylmethyl species can be used [7, 8]. The problems with most of them, except nitroxyl radicals, is that the scavenging radicals can initiate the polymerization themselves and that they participate in side reactions leading to the degradative transfer. Some systems involving nitroxyl radicals will be discussed in detail later. The degenerative transfer (cf. infra) may be also present in some of these systems.

2. Reversible Homolytic Cleavage of Persistent Radicals

$$P^\bullet + X \xrightleftharpoons{} \{P\text{-}X\}^\bullet \tag{7}$$

The persistent radical $\{P\text{-}X\}^\bullet$ should only cleave homolytically to form $P^\bullet$ and the species X, but it should not react with monomer. X should be an inert compound capable only of reacting with $P^\bullet$. $P^\bullet$, as in the previous case, is a typical growing radical which can react with X, with M, and with $P^\bullet$. As before, if $[P^\bullet]$ is very low, the proportion of deactivated chains is low as well.

The role of X can be played by some elementoorganic or organometallic species with an even number of electrons. Some success has been reported with group XIII and XV elements such as aluminum [9] and phosphorus [10] as well as with organometallic derivatives of Co,[11, 12] Cr,[13, 14, 15] and other transition metals. In some cases, metal with an odd number of electrons acts as a scavenger. In some cases not only radical but also ionic and / or coordinative polymerization may occur, and it is necessary to carefully evaluate the mechanism of the polymerization. In some systems, it may happen that both mechanisms can operate simultaneously. That is, not only homolytic but also heterolytic bond cleavage may take place. The proportion of each pathway will depend on the nature of the metal or element, ligands and medium effects.

3. Degenerative Transfer

$$P_n^\bullet + P_1\text{-}R \xrightleftharpoons{k} P_1^\bullet + P_n\text{-}R \tag{8}$$

The concept of degenerative transfer is based on a thermodynamically neutral exchange reaction between active and dormant species. This can be visualized as transfer of the R moiety between all chains in such a way that the concentration of $P^\bullet$ remains very low and number of chains roughly corresponds to [P-R]. P-R can not react directly with monomer and P-R can not react with one another; it can react only with $P^\bullet$. $P^\bullet$ would be generated by a classic radical initiating system such as AIBN, BPO, redox, etc. $P^\bullet$ can react with monomer, for propagation, with P-R, for degenerative transfer, and can also react one with another, for termination. The latter reaction can not be completely avoided, although its contribution will depend on the concentration of $P^\bullet$, as described previously. Nevertheless, the maximum amount of deactivated chains would be equal to that of the introduced initiator, $[I]_o$. If the concentration of the transfer agent is much higher than that of the initiator, $[P\text{-}R]_o \gg [I]_o$, then the proportion of deactivated chains will be low enough and controlled polymerization can be accomplished. This additionally requires fast exchange between growing radicals, $P^\bullet$, and the dormant species, P-R. Fast initiation is not required in this case. There are only a few examples of such reactions, and one of them, based on alkyl iodides, will be discussed later.

4. Mixed Systems

Degenerative transfer may operate simultaneously with the reversible cleavage of covalent species and persistent radicals. This means that in addition to unimolecular exchange, the bimolecular process based on the degenerative transfer may take place. For example, the use of alkoxyamines together with classic radical initiators pose the possibility of both reactions 6 and 8.

REVERSIBLE CLEAVAGE OF COVALENT SPECIES

The successful unimolecular exchange requires the equilibrium (eq. 6) to be strongly shifted to the side of the dormant species in order to maintain a low stationary concentration of radicals. In addition, the scavenger should not initiate polymerization and should not be involved in side reactions.

Some potential scavengers such as dithiocarbamyl and trityl radicals react with alkenes and also decompose, e.g. $R_2NC(S)S^{\bullet}$ to $R_2N^{\bullet}$ and CS_2 [16]. On the other hand, nitroxyl radicals do not react directly with alkenes (or they react extremely slowly). They are relatively stable even at elevated temperatures, but they react very rapidly with most organic radicals [17] to form alkoxyamines.

Alkoxyamines have been used in the polymerization of methacrylates and acrylates as initiators, either alone or together with peroxy initiators at temperatures <100 °C [3]. More recently, this system has been extended to the polymerization of styrene [4]. Alkoxyamines were prepared in situ from the nitroxyl radical, TEMPO, and benzoyl peroxide. A small excess of benzoyl peroxide was used so that the concentration of the produced radicals would exceed the concentration of the scavenger, and thereby initiate polymerization. However, temperatures above 120 °C were required to break the $C-ONR_2$ bond. At temperatures >100 °C, styrene polymerizes by self-initiation via formation of the Diels-Alder adduct which is further aromatized, generating radicals.

Thus, radicals are slowly formed also during the polymerization. Radicals formed by self-initiation are initially trapped by scavengers and then, when the concentration of scavengers is low enough and the concentration of growing radicals high enough, polymerization starts. This is evidenced by long induction periods [18]. Thus, for the polymerization of styrene moderated by nitroxyl radicals, it is not necessary to use radical initiators.

Kinetics of the polymerization of bulk styrene at 120 °C in the presence of TEMPO initiated by benzoyl peroxide and AIBN as well as the self-initiated system is quite similar and the amount of the added initiator regulates the length of induction periods. Slopes in semilogarithmic coordinates depend on the concentration of added scavenger.

Figures 5 and 6 show typical SEC traces obtained in the polymerization of styrene initiated by an excess of AIBN (0.03 mol/L) over scavenger (0.01 mol/L) and with its equimolar amount (0.03 mol/L). It can be noted that the peak values continuously move to higher molecular weights. M_n corresponds always to the total number of chains produced by the initiating radicals, but M_{peak} values corresponds to the dormant chains which are reversibly activated. In the latter case, both M_n and M_{peak} values are similar because only a very small proportion of chains have been irreversibly deactivated by bimolecular termination between growing / initiating radicals.

This is illustrated in Figs.7 and 8. Fig. 7 shows that M_n values depend on the concentration of the initiator ($\approx 50\%$ efficiency of initiation), if a relatively small concentration of scavenger is used. However, the M_{peak} values, shown in Fig. 8, correlate to the scavenger concentration because it defines the concentration of chains which are in their dormant state and which are capable of growth.

On the other hand, if the scavenger is used in excess, then both M_n and M_{peak} values are defined by the scavenger concentration because polymerization starts only when the concentration of radicals formed from the initiator and by self-initiation becomes comparable to the initial scavenger concentration.

Polydispersities either increase or decrease with conversion. The former trend can be assigned to slow but continuous self-initiation and termination of radicals produced in excess of the scavenger. The decrease of polydispersities with conversion can be explained by the reduction of the proportion of uncontrolled chains terminated at the early stages with

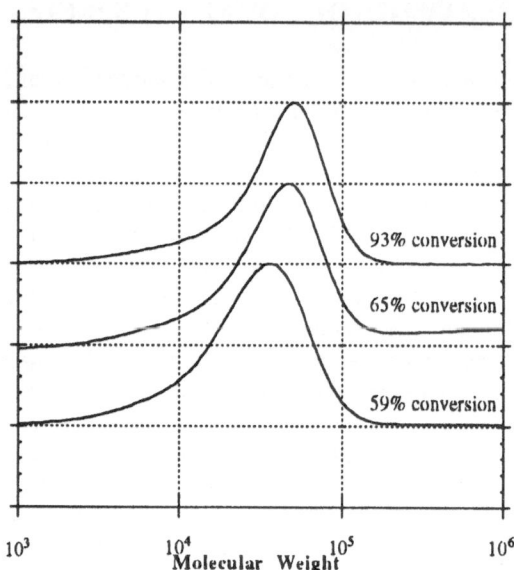

Figure 5. Bulk polymerization of styrene with $[AIBN]_o$=0.03 mol/L and $[TEMPO]_o$=0.01 mol/L at 120 °C.

the excess of radicals. Sometimes, a reduction of polydispersities may also happen if the exchange between active and dormant species is slow in comparison with propagation. Nevertheless, most polydispersities at high conversions are in the range of $M_w/M_n \approx 1.4\pm0.1$, which is much less than that obtained in the classic thermal radical polymerization.

Spontaneous self-initiation in the polymerization of styrene starts at elevated temperatures and could be avoided if the rate of the homolytic cleavage is accelerated. It seems

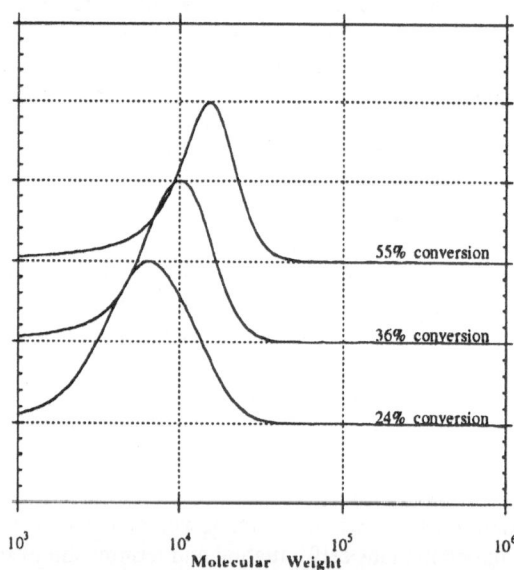

Figure 6. Bulk polymerization of styrene with $[AIBN]_o$=0.03 mol/L and $[TEMPO]_o$= 0.03 mol/L at 120 °C.

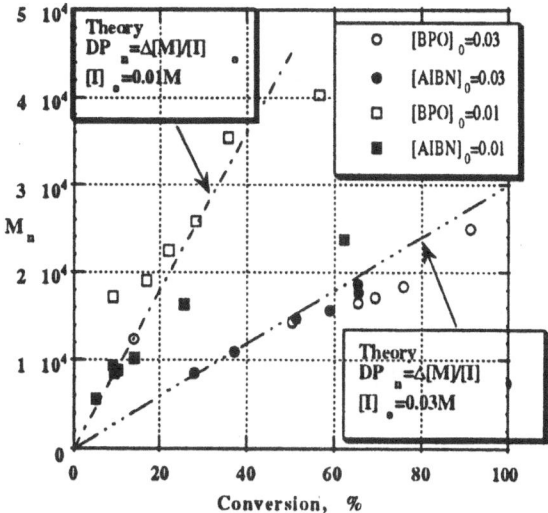

Figure 7. M_n dependence on conversion in bulk polymerization of styrene at 120 °C with BPO/AIBN and [TEMPO]$_o$=0.01 mol/L.

that 4-phosphonooxy-TEMPO provides much faster polymerization than either TEMPO or 4-hydroxy-TEMPO. This is shown in Fig. 9. Higher rates could be due not only to a shift in the equilibrium position, but also due to less efficient scavenging. However, the total number of chains with 4-phosphonooxy-TEMPO is nearly the same as with other nitroxyl radicals. (Fig. 10)

The exact nature of the shift of the equilibrium is not known and it may be due to either faster homolytic cleavage or a slower trapping process. It is possible that the hydrogen

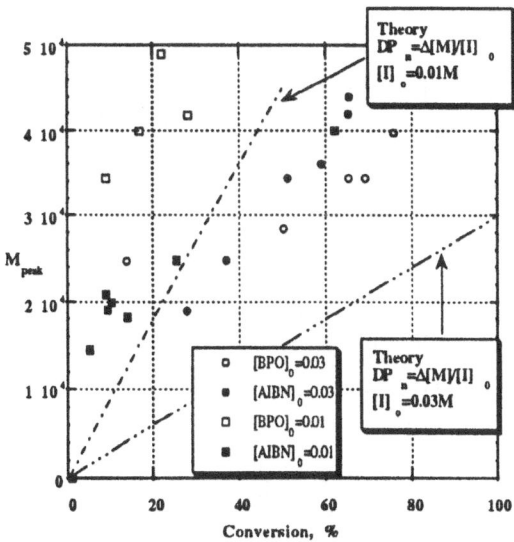

Figure 8. M_{peak} dependence on conversion in bulk polymerization of styrene at 120 °C with BPO/AIBN and [TEMPO]$_o$=0.01 mol/L.

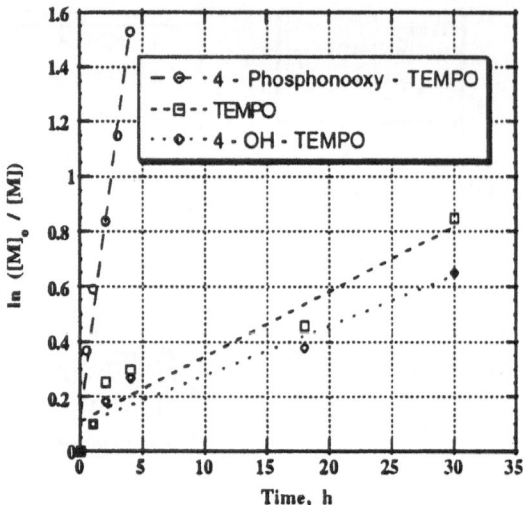

Figure 9. Kinetics of styrene polymerizations at 110 °C with TEMPO derivatives.

bonding from the remote phosphoric acid to the nucleophilic oxygen promotes cleavage and also stabilizes the resulting nitroxyl radical

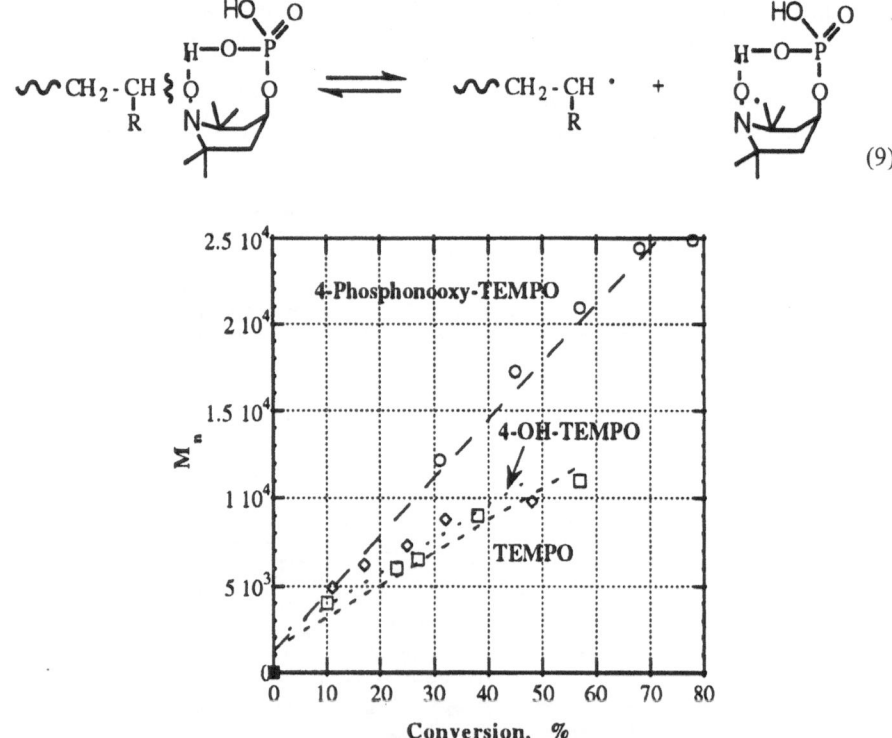

(9)

Figure 10. M_n dependence on conversion in the bulk polymerization of styrene at 120 °C with various TEMPO derivatives.

The fine tuning of the equilibrium constant by structural effects in the scavenger molecule and potentially by medium and additives is one of the most important research directions in this area. It is especially important for the polymerization of monomers which do not self-initiate and can not match-up an excess of scavenger if the initiator is not added in sufficient amount. This is the case of acrylates, vinyl acetate and other alkenes.

DEGENERATIVE TRANSFER

In degenerative transfer, Scheme 1, a transfer agent, R' - X, reacts with a propagating radical ,$P^{\bullet}$, to form a dormant polymer chain, P - X. The new radical, $R'^{\bullet}$, can then reinitiate polymerization. Because of this, the number of polymer chains is equal to the concentration of the transfer agent. The newly formed polymer chain, $P'^{\bullet}$, can then react with the dormant polymer chain, P - X, to form P' - X and $P^{\bullet}$. Optimally, the exchange reaction between the dormant and active radical species should be a thermodynamically neutral reaction. In order for this to be obtained, the transfer agent should resemble the polymer chain end. The compound which was initially employed was 1 - phenylethyl iodide, 1 - PEI.

The basic requirement for the effective degenerative transfer is the sufficient thermal stability of P-R and fast exchange with $P^{\bullet}$ in comparison with propagation. This means that the transfer coefficients should be similar to or larger than 1. This is necessary for the preparation of well defined polymers, especially when $[M]_o \gg [P-R]_o$. Additionally, the reactivity of the P_1-R species in the initially added transfer agent should be similar to or higher than that in the macromolecular species P_n-R. Thus, $P_1^{\bullet}$ should be similar to the growing radical $P_n^{\bullet}$. Inspection of transfer coefficients indicates that large k_{tr}/k_p values are found for compounds with very labile E-Z bonds such as S-S, Si-H, P-H or Br_3C-Br. As mentioned before, the degenerative nature of the process requires that $P^{\bullet}$ must be a carbon centered species, i.e., a 1-phenylethyl derivative for styrene polymerization. Of course, the C-H bond is not sufficiently labile. Apparently, 1-phenylethyl sulfides are not reactive enough either. Thus, in addition to considering organometallic species, the corresponding alkyl iodides were examined by our group for use as the P-R species. Similarly, iodine and selenium compounds have been used successfully in organic synthesis in radical addition reactions to various vinyl compounds [19], [20] Fig. 11 shows the kinetics of the bulk polymerization of butyl acrylate and styrene under various conditions.

The polymerization of butyl acrylate was too fast at 70 °C and although it was carried out at 50 °C was faster than polymerization of styrene at 70 °C. The bulk polymerization of butyl acrylate exhibited a rate acceleration at higher conversions. This may be due to an

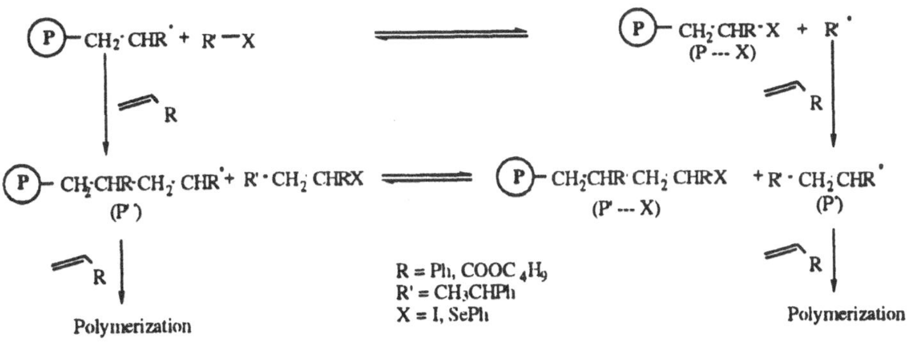

R = Ph, COOC$_4$H$_9$
R' = CH$_3$CHPh
X = I, SePh

Scheme 1

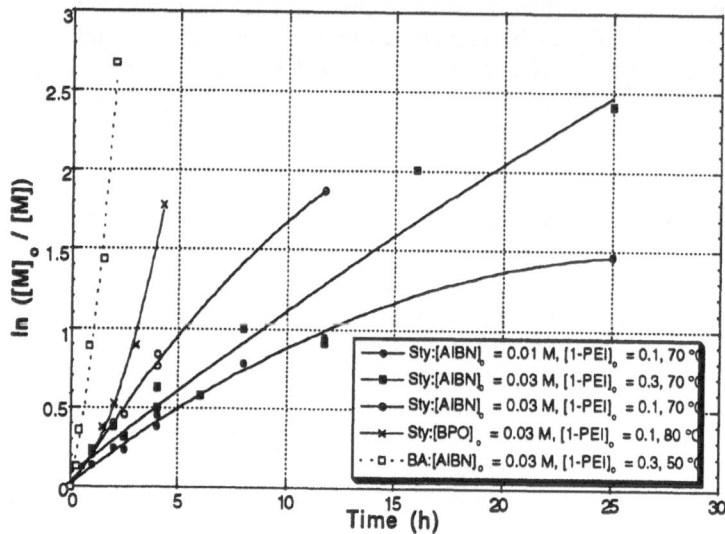

Figure 11. Kinetics of bulk polymerization of styrene and butyl acrylate at various conditions.

increase in the viscosity of the reaction mixture which reduced termination rate like in the Trommsdorff effect. The polymerization of styrene at 80 °C initiated by BPO showed again that the rate of polymerization increased with time as the reaction reached higher conversions.

However, for styrene polymerizations initiated by AIBN semilogarithmic plots were decreasing with time. This could be due to the fact that [AIBN] was substantially reduced when the styrene polymerization had reached high conversions. After 20 hours, the concentration of AIBN would have been ≈ 5% of the original value, resulting in a rate of polymerization ≈ 20% the original. The apparent linearity of the kinetic plots at higher initiator concentrations, may be due to the simultaneous actions of both effects, i. e., rate reduction due to initiator depletion and rate acceleration due to increased viscosity. Increasing the amount of transfer agent, 1 - PEI, resulted in a slower polymerization. As shown in Fig. 11, when $[1 - PEI]_o = 0.1$ M was used, the reaction was 63 % complete in twelve hours, while the reaction using $[1 - PEI]_o = 0.3$ M required sixteen hours to reach similar conversions.

Nevertheless, the main advantage of these systems is the control of molecular weights which increase with conversion and approach the theoretical value of $DP_n = \Delta[M] / [R' - X]_o$, as shown in Fig. 12. The higher molecular weights at the outset than predicted indicate that the rate of exchange of iodine between the propagating radical and the transfer agent is slow in comparison to the rate of propagation. The polydispersities were < 1.5 even at conversions greater than 90%. This is in sharp contrast to the polymerization of styrene without the transfer agent, $M_n / M_w > 2$. When no transfer agent was used, the molecular weight, $M_n \approx 60,000$, did not increase with conversion. Butyl acrylate also provides controlled polymers with AIBN/PEI initiating system [21].

The rates of polymerization increased with temperature. At 70 and 90 °C, there was no significant effect on the evolution of molecular weights with conversion. However, when the polymerization was carried out at 50 °C, the molecular weights and polydispersities were higher than at the higher temperatures indicating that transfer between the propagating radical and the transfer agent was relatively slower.

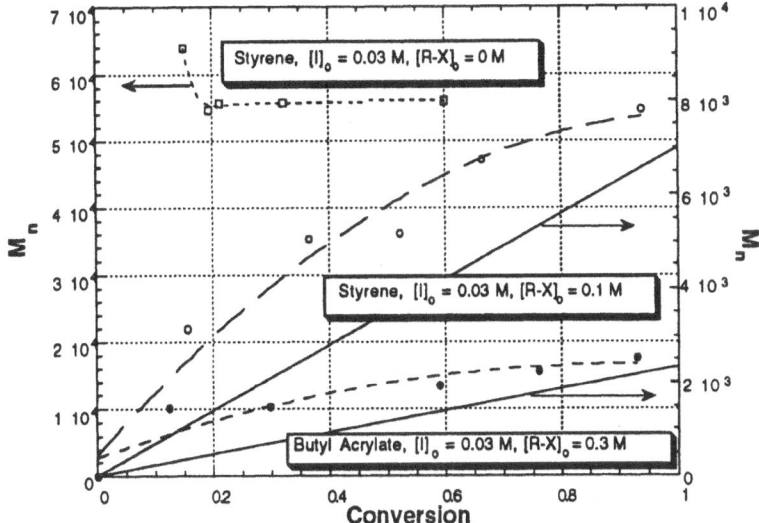

Figure 12. Molecular weight dependence upon conversion for the bulk polymerization of styrene and butyl acrylate at 70 °C. [I] = AIBN, [R-X] = 1-Phenylethyl iodide.

Various transfer agents were used to explore the effect of R' on the control of molecular weights. The following iodine compounds were used in addition to 1 - phenylethyl iodide: perfluorohexyl iodide, perfluoroisopropyl iodide, iodoacetonitrile, isopropyl iodide, n - propyl iodide and phenyl iodide. Of these, only primary alkyl and aryl iodides showed no or very weak effect on molecular weights.

Although the C-I bond can also be cleaved heterolytically, this does not happen for styrene derivatives below 100 °C. Spontaneous, homolytic cleavage does not happen either at reasonable rates for T < 100 °C, as proved in blank experiments. However, addition of 1 - phenylethyl iodide to the polymerization of styrene initiated by AIBN had a tremendous effect on the evolution of molecular weight with conversion and on polydispersities.

Thus, the radical polymerization of styrene and butyl acrylate in the presence of degenerative transfer agents followed typical behavior of controlled polymerization such as low polydispersities and a linear evolution of molecular weight with conversion, although it was initiated by classic radical initiators. Moreover, addition of a new portion of monomer or addition of another monomer extends chain growth leading to block copolymers as in the case of polystyrene and poly(butyl acrylate) [21].

ACKNOWLEDGMENTS

Support from the Office of Naval Research and the National Science Foundation via Presidential Young Investigator Award is gratefully acknowledged.

REFERENCES

1. T. Otsu, M. Yoshida, *Makromol. Chem. Rapid Commun.,* **3**, 127 (1982).
2. P. Sigwalt, P. Lambrinos, M. Rardi, A. Polton, *Eur. Polym. J.,* **26**, 1125 (1990).
3. D. H. Solomon, G. Waverly, E. Rizzardo, W. Hill, P. Cacioli, *U. S. Pat. 4, 581, 429* (1986).

4. M. K. Georges, R. P. N. Veregin, P. M. Kazmaier, G. K. Hamer, *Macromolecules*, **26**, 2987 (1993).

5. M. K. Georges, R. P. N. Veregin, P. M. Kazmaier, G. K. Hamer, *Macromolecules*, **26**, 5316 (1993).

6. D. Mardare, K. Matyjaszewski, *ACS Polymer Preprints*, **35(1)**, 778 (1994).

7. E. Borsig, M. Lazar, M. Capla, S. Florian, *Angew. Makromol. Chem.*, **9**, 89 (1969).

8. A. Bledzki, D. Braun, *Makromol. Chem.*, **184**, 745 (1983).

9. D. Mardare, K. Matyjaszewski, *Macromolecules*, **27**, 645 (1994).

10. D. Greszta, D. Mardare, K. Matyjaszewski, *ACS Polym. Preprints*, **35(1)**, 466 (1994).

11. H. J. Harwood, L. D. Arvanitopoulos, M. P. Greuel, *ACS Polymer Preprints*, **35(2)**, 549 (1994).

12. B. B. Wayland, G. Pszmik, S. L. Mukerjee, M. Fryd, *J. Am. Chem. Soc.*, **116**, 7943 (1994).

13. S. Gaynor, D. Mardare, K. Matyjaszewski, *ACS Polymer Preprints*, **36(1)**, 700 (1994).

14. Y. Doi, S. Ueki, T. Keii, *Macromolecules*, **12**, 814 (1979).

15. Y. Minoura, M. Lee, *J. Chem. Soc. Faraday Trans. 1*, **74**, 1726 (1978).

16. R. S. Turner, R. W. Blevins, *Macromolecules*, **23**, 1856 (1990).

17. E. Rizzardo, *Chem. Aust.*, **54**, 32 (1987).

18. D. Mardare, T. Shigemoto, K. Matyjaszewski, *ACS Polymer Preprints*, **35(2)**, 557 (1994).

19. D. P. Curran, E. Eichenberger, M. Collis, M. G. Roepel, G. Thoma, *J. Am. Chem. Soc.*, **116**, 4279 (1994).

20. C. P. Curran, M. Newcomb, *Acc. Chem. Res.*, **21**, 206 (1988).

21. K. Matyjaszewski, S. Gaywor, J. S. Wang, *Macromolecules*, **28**, 2093 (1995).

LIVING CARBOCATIONIC COPOLYMERIZATIONS.* Part 1

THE CONSTANT COPOLYMER COMPOSITION TECHNIQUE

A. Nagy,[†] I. Országh,[‡] and J.P. Kennedy

Maurice Morton Institute of Polymer Science
The University of Akron
Akron, OH 44325-3909

1. ABSTRACT

The Constant Copolymer Compositon (CCC) technique readily produces copolymers with constant (homogeneous) macro- and microcompositions (microstructures) even from monomer pairs with significantly different reactivities, for example with monomers whose reactivity ratios differ by an order of magnitude. The CCC technique eliminates the compositional drift along copolymer chains which always arise in copolymers made by batch or forced ideal techniques and necessarily lead to inhomogeneous often ill-defined products. In the CCC technique a stream of comonomers is fed continuously to the active copolymerization charge such that the composition of the feed and the rate of feed addition are respectively equal to the composition of the copolymer produced and the rate of copolymerization. The fundamentals of the CCC technique together with a detailed quantitative analysis are presented, and the differences of the conventional and CCC techniques are discussed and illustrated with the industrially important isobutylene- p-methylstyrene copolymerization system.

2. INTRODUCTION

An important objective of copolymerization is the preparation of structurally uniform products. In the copolymerization of two monomers, which necessarily will have unequal reactivities, the relatively more reactive monomer will preferentially enter the copolymer

---

*Part III of the series of publications on living carbocationic copolymerization. For parts I and II see references 9a and 10a.

[†] Visiting Scientist. Permanent address: Central Research Institute for Chemistry of the Hungarian Academy of Sciences, P.O.Box 17, H-1525 Budapest, Hungary.

[‡] Visiting Scientist. Permanent address: Lajos Kossuth University, Department of Physical Chemistry, P.O.Box 7, H-4010 Debrecen, Hungary.

Macromolecular Engineering, Edited by M.K. Mishra et al.
Plenum Press, New York, 1995

which causes a drift in the composition of both the comonomer charge and the copolymer product [1-3]. For the preparation of copolymers with constant macro- and microcompositions, this drift has to be eliminated. (The macrocomposition of a copolymer is the overall ratio of the total amount of comonomers incorporated into a quantity of copolymer, whereas the microcomposition (or microstructure) is defined by the average sequence length distribution of like monomer units; the latter is usually expressed by the number average sequence length [1,2] or by the run number [4,5a,b]).

Macrocompositional heterogeneity and microcompositional heterogeneity are two entirely different concepts. A macrocompositionally homogeneous copolymer may include a variety of isomeric structures (i.e., chains that differ only in regard to comonomer sequences) and is therefore microcompositionally heterogeneous. In the same vein, a homogeneous or constant copolymer microcomposition does not necessarily imply identical copolymer microstructures because the probabilistic nature of the copolymerization process leads to microstructural differences between individual copolymer molecules [6,7]. This statistical structural inhomogeneity, however, is beyond the scope of the discussion that follows.

Both the macro- and microcompositions of a copolymer are determined by statistical laws. For simplicity, only the two parameter (or terminal) model [2] will be considered but our treatment is valid for other copolymerization models as well [2].

In the terminal model, the macrocomposition of a copolymer is governed by the relative rates of two homopropagations and two cross propagations whose rates depend on comonomer concentrations and the reactivity ratios, r_1 and r_2. Depending on the latters, the microstructure will obey different statistical laws; for example, if $r_1 = r_2 = 1$, monomer incorporation will follow Bernoullian statistics [1] and a random copolymer will arise [1]; if $r_1 \neq r_2$ the copolymer will grow according to first order Markov statistics[1], etc. If r_1 and r_2 remain constant during a batch copolymerization, the composition of the comonomer feed, and consequently that of the copolymer, will drift, which in turn will change the macro- and microcomposition of the product.

The consequences of the CCC technique were analyzed with the technologically important isobutylene(IB)/p-methylstyrene(pMeSt) system using the industrially important 97/3 mol%/mol% comonomer feed. Copolymerization fundamentals of the IB/pMeSt pair have been published [9,10] including the derivation of reliable reactivity ratios ($r_{IB} = 0.74 \pm 0.11$, $r_{pMeSt} = 8.0 \pm 3.3$) and the definition of conditions for living copolymerization [9,10]. The next two sections concern a comparative evaluation of various techniques for the synthesis of IB/pMeSt copolymers by the use of 97/3 mol%/mol% comonomer feeds (Section 3) and a quantitative treatment of the CCC technique (Section 4), respectively.

3. COMPARISON OF BATCH-, FORCED IDEAL-, AND CCC TECHNIQUES

The objective of these studies was to elucidate the macro- and microcompositions of IB/pMeSt copolymers prepared with 97/3 mol%/mol% feeds under living copolymerization conditions obtained by various techniques. The needed reactivity ratios and conditions for living copolymerizations have been established earlier [9,10].

3.1. The Batch Technique

Let us consider a batch copolymerization using a comonomer mixture in which $r_1 \neq r_2$. During the initial stages of the processes the composition of the first instantaneous copolymer will be richer in the relatively more reactive monomer (say M_1) than that of the initial charge, so that the composition of M_1 in the charge will gradually decrease with increasing conversion. The rate of this decrease will be governed by the relative monomer reactivity ratios, r_1 and r_2. Necessarily, the instantaneous copolymer composition will reflect the drift in the relative monomer concentrations in the charge and the final or cumulative product will be inhomogeneous because its composition has constantly drifted with that of the charge during the entire copolymerization. Obviously, the cumulative copolymer composition at the end of the copolymerization (at 100% conversion) will be equal to the composition of the initial comonomer charge because all the monomer molecules have been converted to copolymer.

Figure 1A illustrates this scenario with the industrially significant IB/pMeSt system using a 97/3 mol%/mol% initial comonomer charge; the copolymerization conditions and reactivity ratios have been obtained from refs. 9 and 10. As indicated by the computer generated plot, the compositon of the first instantaneous copolymer (at "0% conversion") would be 95.3/4.7 mol%/mol%, and from this point on the instantaneous compositions would strongly drift with increasing conversions. This drift would be particularly strong toward the end of the copolymerization when the more reactive pMeSt has largely disappeared from the charge. The cumulative copolymer composition trace conveys the same message, however, the compositional drift would be less emphasized. The average composition of the end product, of course, would be 97/3 mol%/mol% but the microcomposition of the copolymer would be strongly heterogeneous along the chain.

3.2. The Forced Ideal Copolymerization Technique

The Forced Ideal Copolymerization technique has been developed some time ago [11,12] to overcome the strong compositional drift experienced in batch copolymerizations. By this technique the comonomer feed is added steadily dropwise to the active copolymerization charge making sure that each and every drop of feed is completely consumed before the next drop arrives. Every drop is in fact an individual batch reactor and, as a consequence, the instantaneous copolymer composition will significantly oscillate along the copolymer chains. The length of the repeat segments will be dependent on the ratio of the number of comonomer molecules in one droplet over that of the growing chains. This fluctuation of the instantaneous copolymer composition will decrease and will become less and less noticeable in the cumulative copolymer composition upon every additional drop of comonomer. The cumulative copolymer composition can be characterized mathematically as a nonsymmetrical "damping oscillation".

Figure 1B illustrates the compositions obtained by the forced ideal copolymerization technique with the IB/pMeSt system using a 97/3 mol%/mol% comonomer feed. For simplicity, only 10 drops of comonomer addition are shown. The broken lines indicate the drift in the instantaneous copolymer composition that occur in every drop as a function of conversion (or chain length since the copolymerization is carried out under living conditions). The solid line shows the cumulative copolymer composition vs. conversion. Obviously, again, the average or cumulative composition of the end product will be 97/3 IB/pMeSt, however, the microstructures of the individual chains will not be homogeneous.

3.3. The Constant Copolymer Composition Technique

In contrast to the shortcomings of the above methods, the CCC technique produces macroscopically and microscopically homogeneous copolymers. The CCC technique is essentially a semibatch operation in which the comonomer feed is introduced continuously into an active copolymerization charge and the product accumulates

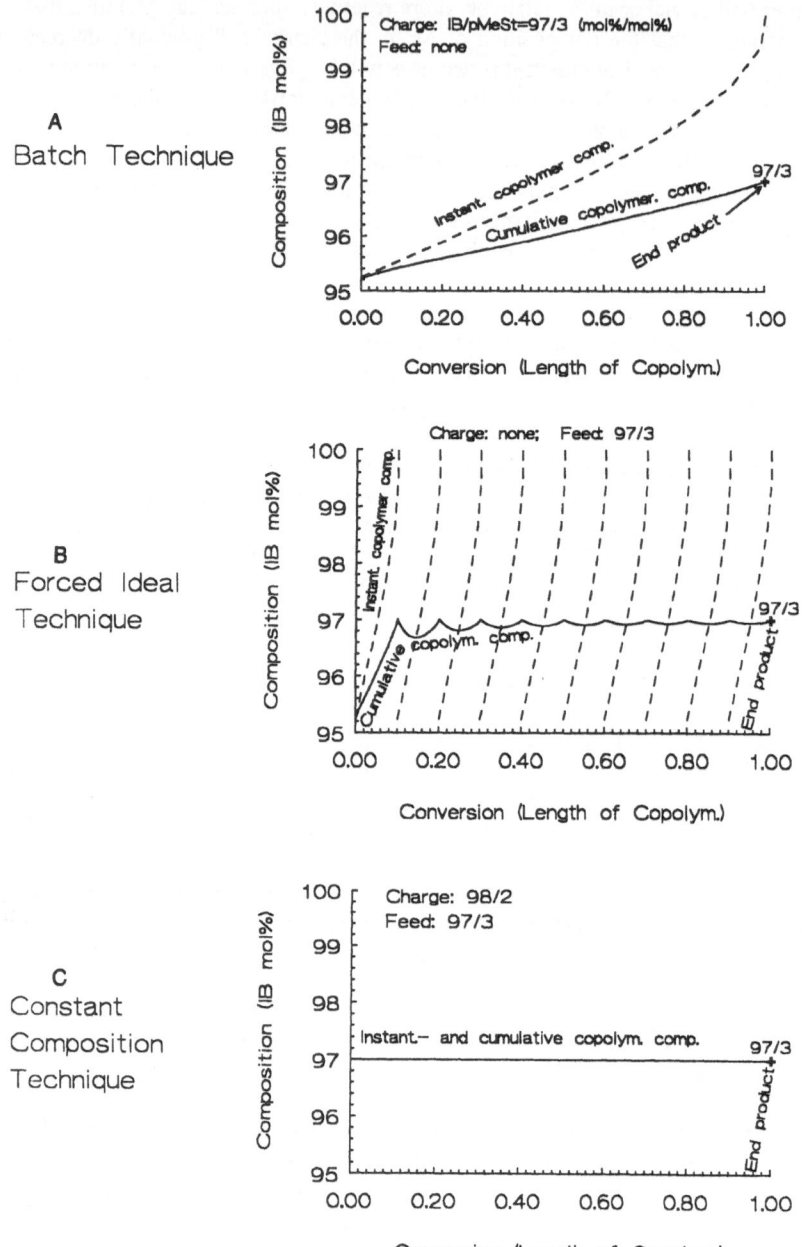

Figure 1. Comparison of various copolymerization techniques in producing of IB/pMeSt copolymers with =
97/3 mol%/mol% overall compositions by living copolymerizations. Broken lines: instantaneous copolymer
compositions, solid lines: cumulative copolymer compositions, as a function of conversion or length of
copolymer chains. A) batch technique, B) forced ideal technique, C) constant copolymer compositon technique.

in the reactor. The composition of the active charge in the reactor and that of the feed are of course not identical: The composition of the former is selected such that it should produce the target copolymer composition, (i.e., it should compensate for the disparity of the monomer reactivity ratios). Importantly, by this technique, the *rates* of comonomer feed introduction and copolymerization are the same, and the *compositions* of the feed and copolymer are also equal. Under these conditions both the macro-and microstructure of the copolymer will be time- or conversion independent. Preliminary experiments have to be carried out to determine the needed feed introduction rate and the two reactivity ratios.

Figure 1C shows the instantaneous copolymer composition profile as a function of conversion in the IB/pMeSt system employing a constant stream of 97/3 mol%/mol% feed. Preliminary experiments provided the two reactivity ratios, (r_{IB} = 0.74±0.11 , r_{pMeSt} = 8.0±3.3), the needed feed introduction rate and indicated that the reactor should initially contain a 98/2 mol%/mol% IB/pMeSt charge which will produce the target 97/3 mol%/mol% copolymer. Since the experiment is carried out under living copolymerization conditions the 97/3 mol%/mol% comonomer feed can be added continuously until the desired molecular weight is reached (the target was M_n > 100,000 g/mol, see ref. 9).

4. QUANTITATIVE TREATMENT OF THE CCC TECHNIQUE

4.1. Material Balances

If a stirred batch reactor is operated by the "semibatch" mode, i.e., the reactor is charged with reactants and additional reactants are added continuously, the material balance is, as follows [13]:

$$\text{Input} = \text{Disappearance by Reaction} + \text{Accumulation} \tag{1}$$

This material balance for the comonomers is simplified by the CCC technique:

$$\text{Input} = \text{Disappearance by Reaction} \tag{2}$$

The material balance for pMeSt:

$$F_s - R_s \cdot V = \frac{d([pMeSt] \cdot V)}{dt} = 0 \tag{3}$$

and for IB:

$$F_B - R_B \cdot V = \frac{d([IB] \cdot V)}{dt} = 0 \tag{4}$$

where F_S and F_B are the input rates of pMeSt and IB in mol/min, R_S and R_B are the rates of reactions consuming pMeSt and IB, [$pMeSt$] and [IB] are comonomer molar concentrations, respectively, V is the total volume of the charge in the reactor and t is the time. Because of continuous comonomer introduction

$$V = V_0 + \int_0^t v \cdot dt \tag{5}$$

where V_0 is the initial volume, and v stands for the input rates of pMeSt and IB in mL/min which may be time dependent, see later. For simplicity, we assume that the volumes are additive. Thus:

$$v = \frac{F_s \cdot M_s}{d_s} + \frac{F_s \cdot M_s}{d_s} \tag{6}$$

where d_S and d_B are the densities- , and M_S and M_B are the molecular weights of pMeSt and IB, respectively.

4.2. Rate of Comonomer Consumption

pMeSt and IB are consumed by propagation:

$$R_s = k_{ss} \cdot \alpha \cdot [I_0] \cdot [pMeSt] + k_{BS} \cdot (1-\alpha) \cdot [I_0] \cdot [pMeSt] \tag{7}$$

$$R_B = k_{SB} \cdot \alpha \cdot [I_0] \cdot [IB] + k_{BB} \cdot (1-\alpha) \cdot [I_0] \cdot [IB] \tag{8}$$

where k_{SS}, k_{BB} and k_{SB}, k_{BS} are the homo- and cross propagation rate constants, $[I_0]$ is the initiator concentration, and α and $1-\alpha$ are the fractions of growing chains with pMeSt$^+$ and IB$^+$ end units, respectively.

The theory of short sequences, or the rate equality of the cross propagation steps should also hold [1,2]:

$$k_{SB} \cdot \alpha \cdot [I_0] \cdot [IB] = k_{BS} \cdot (1-\alpha) \cdot [I_0] \cdot [pMeSt] \tag{9}$$

4.3. Prerequisites for the CCC Technique

According to the instantaneous copolymer composition equation [14a-c], constant copolymer macro- and microcompositions can be sustained only if the comonomer composition in the charge is kept continuously constant:

$$y = \frac{[pMeSt]}{[IB]} \cdot \frac{1 + r_s \cdot \dfrac{[pMeSt]}{[IB]}}{r_B + \dfrac{[pMeSt]}{[IB]}} \tag{10}$$

where y stands for the molar ratio of pMeSt and IB units in the copolymer. Since the reactivity ratios $r_S = k_{SS}/k_{SB}$ and $r_B = k_{BB}/k_{BS}$ are assumed to be constant, $y =$ constant, if

$$\frac{[pMeSt]}{[IB]} = const. \tag{11}$$

Similarly, the prerequisite of the CCC technique can be given by the ratio of the monomer input rates:

$$y = \frac{F_s}{F_B} = const. \tag{12}$$

According to Eqs. 2-12, the individual time functions of the comonomer input rates can be determined and thus constant copolymer compositions can be maintained.

4.4. Determination of the Monomer Input Rates

In order to determine the individual time functions of the monomer input rates (F_S and F_B), the followings have to be considered:

F_S and F_B are interrelated by Equation 12; In our system:

$$\frac{F_s}{F_B} = \frac{3}{97} \tag{13}$$

Combining Eqs.3 and 7 with Eq. 6 yields:

$$v \cdot \left(V_0 + \int_0^t v \cdot dt \right) = \left[k_{ss} \cdot \alpha + k_{BS} \cdot (1 - \alpha) \right] \cdot I_0 \cdot N_{S,0} \cdot \left[\frac{M_S}{d_s} + \frac{M_B}{d_B \cdot \frac{F_s}{F_B}} \right] \tag{14}$$

where I_0 and $N_{S,0}$ are the numbers of moles of initiator and pMeSt initially charged the reactor, respectively. After substituting the right side of Eq. 14 (which contains only constants) for constant A, and rearranging, Eq. 14 becomes

$$V_0 + \int_0^t v \cdot dt = \frac{A}{v} \tag{15}$$

By replacing $\int_0^t v \cdot dt$ with f, and v with $\frac{df}{dt}$, Eq. 15 becomes a separable, differential equation:

$$A \cdot \int_0^t dt = \int_0^t (V_0 + f) df \tag{16}$$

After integration and rearrangement:

$$f^2 + 2 \cdot V_0 \cdot f - 2 \cdot A \cdot t = 0 \tag{17}$$

which yields:

$$f = -V_0 + \sqrt{V_0^2 + 2 \cdot A \cdot t} \tag{18}$$

Since v was replaced with df/dt, thus

$$v = \frac{A}{\sqrt{V_0^2 + 2 \cdot A \cdot t}} \tag{19}$$

Equation 19 is a general expression that describes the time dependence of the combined feeding rates of the two comonomers. As mentioned above, v is time dependent and decreases with increasing time, since the copolymerization rate is constantly decreasing due to dilution by feeding. The individual monomer input rates can easily be generated by substituting the constants in Eqs.6 and 13 into Eq.19.

4.5. A Numerical Calculation of Simulating Equations of the CCC Technique

Equation 19 can be used to calculate v numerically, as a function of time, by substituting the values of constants. Constant A (the right side of Eq. 14) contains α, which can easily be substituted from Eq.9. Based on earlier information [9,10], the following initial conditions were chosen to calculate numerical equations: $k_{SS} = 30$ L.mol^{-1}min^{-1}; $k_{BB} = 3$ L.mol^{-1}min^{-1}; $r_S = 8.0$; $r_B = 0.74$; $F_S/F_B = 3/97$; $[pMeSt]/[IB] = 0.0203$; $I_0 = 2.5 \times 10^{-3}$ mol; $N_{S,0} = 0.1$ mol; $M_S = 118.1$g/mol; $M_{IB} = 56.1$g/mol; $d_S = 0.9$ g/mL; $d_B = 0.7$g/mL; $V_0 = 1000$ mL. The substitution of these quantities into Eq.19 yields

$$v = \frac{3000}{\sqrt{1000^2 + 2 \cdot 3000 \cdot t}}$$

(20)

Figure 2A shows the change in monomer input rate with time so as to maintain constant copolymer composition. The deviation from linearity is due to dilution in the reactor and can be calculated by Eqs. 5 and 20. The total volume in the reactor as a function of time is shown in Figure 2B.

Under CCC conditions rate of the monomer input and copolymer formed must be equal. Since both of these quantities can be readily determined, the diagnostic plot for the

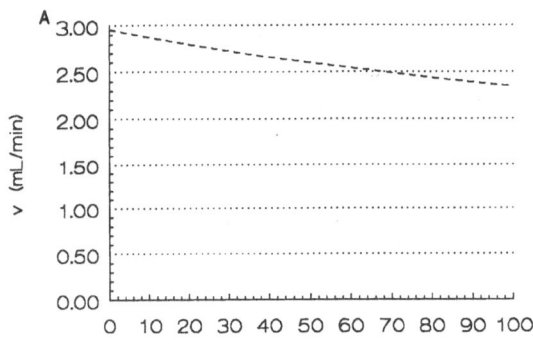

Figure 2. Simulation of the Constant Copolymer Composition Technique. A) total monomer input rate (v) and B) total reaction volume (V) as a function of time in IB/pMeSt living copolymerization. For details, see text and ref. 9.

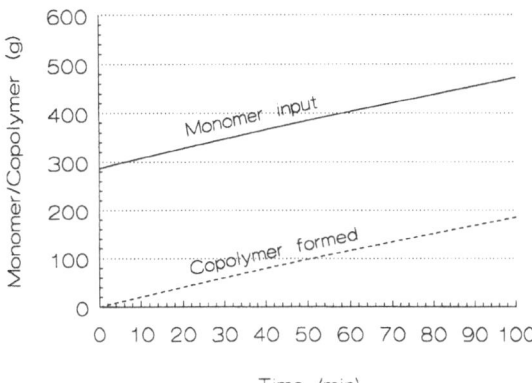

Figure 3. Simulation of the Constant Copolymer Composition Technique. Solid line: total amount of monomer input- (charge + feed), and broken line: total amount of copolymer formed as a function of time. For details, see text and ref. 9.

CCC technique, i.e., the Monomer Input/Copolymer Formed vs Time plot, can be easily constructed. Figure 3 shows such a diagnostic plot for the IB/pMeSt system in which the needed functions were determined by the use of Eqs. 6,13 and 20 in conjunction with the initial conditions. Gratifyingly the traces are parallel indicating that the rates of the monomer input and copolymer formed are equal; the reason for the parallel vertical shift of the monomer input curve relative to the copolymer formed curve is of course due to a difference equal to the comonomer amount initially present in the reactor. Closer inspection of the traces indicate that the rates gradually decrease which is due to the gradual dilution of the charge in the semibatch reactor.

5. CONCLUSIONS

Based on a quantitative comparison of batch-, forced ideal- and constant copolymer composition techniques only the latter one is capable of maintaning conditions needed for producing copolymers with constant macro- and microcompositions. In order to operate a CCC reactor one must determine by preliminary experimentation the reactivity ratios and propagation rates, and the comonomer input rates have to be calculated (see Eqs. 3-19). According to the results of our simulation, the rate of comonomer input must slightly decrease due to dilution. Our next publication will concern a series of CCC experiments and their interpretation [15].

6. ACKNOWLEDGMENT

This material is based upon work partially supported by the National Science Foundation (Grant 94-23202) and the Exxon Chemical Company.

7. REFERENCES

1. G. Odian: Principles of Polymerization, 3rd. Ed., J. Wiley and Sons, 1991.
2. D.A.Tirrell: Encyclopedia of Polymer Sci. and Eng.: 2nd Ed., Vol *4* , 192 (1986), Wiley and Sons.
3. J.P. Kennedy and B. Iván: Designed Polymers by Carbocationic Macromolecular Engineering: Theory and Practice, Hanser, Munich 1991.
4. H.J. Harwood and W.M. Ritchey, J.Polym.Sci. B, *6*, 277 (1968).

5. a. J.P. Kennedy and T. Chou: J. Macromol. Sci. Chem. *A10(7)*, 1357 (1976). b. J.P.Kennedy and T.Chou: Advances in Polym. Sci., *21*, 1 (1976).

6. W.H. Stockmayer: J.Chem.Phys., *13*, 199 (1945).

7. B. Turcsányi: Macromol. Reports, *A30*, (Suppl. 3 and 4), 281 (1993).

8. I.Majoros, A. Nagy and J.P. Kennedy: Advances in Polym. Sci., *112*, 1 (1994).

9. a) I. Országh, A. Nagy, J.P. Kennedy: J. Phys. Org. Chem., Special Issue (1995), in press; b) I. Országh, A. Nagy, J.P.Kennedy: *Abstracts*, MacroAkron '94, 35th Intl. Symp. Macromol., Akron, 7/11-15/1994, p.83.

10. a) A. Nagy, I. Országh, J.P. Kennedy: J. Phys. Org . Chem., Special Issue (1995), in press; b) A. Nagy, I. Országh, J.P.Kennedy: *Abstracts*, MacroAkron '94, 35th Intl. Symp. Macromol., Akron, 7/11-15/1994, p.84.

11. J.Puskás, G. Kaszás, J.P. Kennedy, T. Kelen and F. Tüdös: J. Macromol. Sci. Chem., *A18(9)*, 1315 (1982-83).

12. G. Kaszás, M.Györ, J.P. Kennedy and F. Tüdös: J. Macromol. Sci. Chem., *A18(9)*, 1367 (1982-83).

13. B.L.Crynes and H.S. Fogler: American Institute of Chemical Engineers, Series E, Vol.2, (1981).

14. a. T.Alfrey, Jr. and G.Goldfinger: J. Chem. Phys., *12*, 205 (1944). b. F.R. Mayo and F.M. Lewis: J. Amer. Chem. Soc., *66*, 1594 (1944). c. F.T.Wall: J. Amer. Chem. Soc., *66*, 2050 (1944).

15. I. Országh, A. Nagy, J.P. Kennedy: see subsequent paper in this series.

LIVING CARBOCATIONIC COPOLYMERIZATIONS.[*] Part 2

3

APPLICATION OF THE CONSTANT COPOLYMER COMPOSITION TECHNIQUE FOR THE SYNTHESIS OF ISOBUTYLENE/ p-METHYLSTYRENE COPOLYMERS

I. Országh,[†] A. Nagy,[‡] and J.P. Kennedy

Maurice Morton Institute of Polymer Science
The University of Akron
Akron, OH 44325-3909

ABSTRACT

Compositionally uniform high molecular weight isobutylene (IB)/p-methylstyrene (pMeSt) copolymers have been prepared by using the constant copolymer composition (CCC) technique (see preceding publication) under living carbocationic copolymerization (LC⊕Copzn) conditions. The attainment of the desired copolymer composition of IB/pMeSt = 96.5/3.5 mol%/mol% was demonstrated up to $\overline{M}_n \approx 65,000$ g·mol^{-1}. CCC and LC⊕Copzn conditions have been quantitated by three diagnostic plots, i.e., two plots that prove the attainment of CCC conditions and one that indicate the existence of LC⊕Copzn. Specifically, CCC conditions were demonstrated by a rate plot which shows comonomers input together with copolymer formed as a function of time, and a composition plot which shows feed composition together with copolymer composition as a function of copolymer formed (W_p); finally living copolymerization was demonstrated to proceed by plotting $\overline{M}_n$ (number average molecular weight) as a function of W_p. According to this three prong evidence uniform composition IB/pMeSt copolymers with up to $\overline{M}_n \approx 65,000$ g·mol^{-1} can be conveniently prepared by living copolymerization under readily achievable CCC conditions at -50 °C. This is the first demonstration of the synthesis of a uniform composition binary copolymer by LC⊕Copzn from a comonomer pair whose reactivity ratios differ by about one order of magnitude.

---

[*] Part IV of the series of publications on living carbocationic copolymerization. For part III see preceding section in this book.

[†] Visiting Scientist. Permanent address: Kossuth Lajos University, Department of Physical Chemistry, P.O. Box 7., H-4010 Debrecen, Hungary.

[‡] Visiting Scientist. Permanent address: Central Research Institute for Chemistry of the Hungarian Academy of Sciences, P.O. Box 17., H-1525 Budapest, Hungary.

Macromolecular Engineering, Edited by M.K. Mishra et al.
Plenum Press, New York, 1995

I. INTRODUCTION

The preceding publication [1] described qualitative and quantitative aspects of the CCC technique developed for the synthesis of structurally uniform binary copolymers prepared from monomers having quite dissimilar relative reactivities ($r_1 \approx 10r_2$). The first objective of the present research was to employ the CCC technique, specifically, to demonstrate the synthesis of compositionally homogeneous copolymers of IB and pMeSt [2], a comonomer pair with very different reactivity ratios [3], $r_{IB} = 0.74 \pm 0.11$ and $r_{pMeSt} = 8.0 \pm 3.3$ [4], within the industrially desired $90/10 \geq IB/pMeSt \geq 98/2$ mol%/mol% composition range. The second equally important objective was to demonstrate that the CCC process can be run under $LC^{\oplus}$Copzn conditions and thus to prepare uniform high molecular weight copolymers in the desired composition range of this important monomer pair by a combination of these powerful techniques.

On the basis of earlier research and accumulated experience we have identified the suitable ingredients (initiator, coinitiator, solvent, electron pair donor, proton trap) and their respective concentration ranges for achieving $LC^{\oplus}$Copzn conditions [2]. Under CCC conditions we could not use conversion vs. time plots developed for the diagnosis of irreversible termination [2] because the comonomers had to be added continuously; however, according to earlier investigations [2] irreversible termination is absent under the conditions used in the present research. Electron pair donor and proton trap were used to depress the rate of chain transfer relative to that of propagation and to eliminate impurity induced initiation [2]. The effect of slow initiation on the diagnosis of chain transfer has also been explored and discussed for this system [2,5]. In addition we have developed the needed experimental parameters (reactivity ratios in the composition ranges of interest [2,3,5], initial charge composition and concentration, rates of comonomers input and copolymerization (see later)), so that the stage was set for carrying out IB/pMeSt copolymerizations by the CCC process under $LC^{\oplus}$Copzn conditions.

This paper concerns the first demonstration that high molecular weight (up to $\overline{M_n} \approx 65,000$ g·mol^{-1}) structurally uniform IB/pMeSt copolymers with precisely defined composition can be conveniently synthesized by combining $LC^{\oplus}$Copzn and the CCC technique with the following system: 5-tert-butyl-1,3-dicumyl methyl ether (5-tBu-1,3-DiCumOMe) initiator, TiCl$_4$ coinitiator, ethyl chloride (EtCl) or EtCl/hexanes (77/23% v/v) solvent(s), triethylamine (TEA) electron pair donor, 2,6-tert-butylpyridine (DtBP) proton trap, at -50 °C.

II. EXPERIMENTAL

II.1. Materials

The source and purification of IB/pMeSt comonomers, 5-tBu-1,3-DiCumOMe initiator, TiCl$_4$ coinitiator, EtCl or EtCl/hexanes (77/23% v/v) solvent(s), TEA electron pair donor, DtBP proton trap have been described [2].

II.2. Procedures

Figure 1 outlines the experimental setup and Table I summarizes the conditions used in the 3 experiments described in this report. Copolymerizations were carried out at -50 °C in a glove box under a dry nitrogen atmosphere by using 2L three-neck round bottom flasks equipped with a stirrer immersed in a thermostatted cooling bath. The bath temperature was maintained at 10-14 °C below the target temperature in the reactor because of the relatively

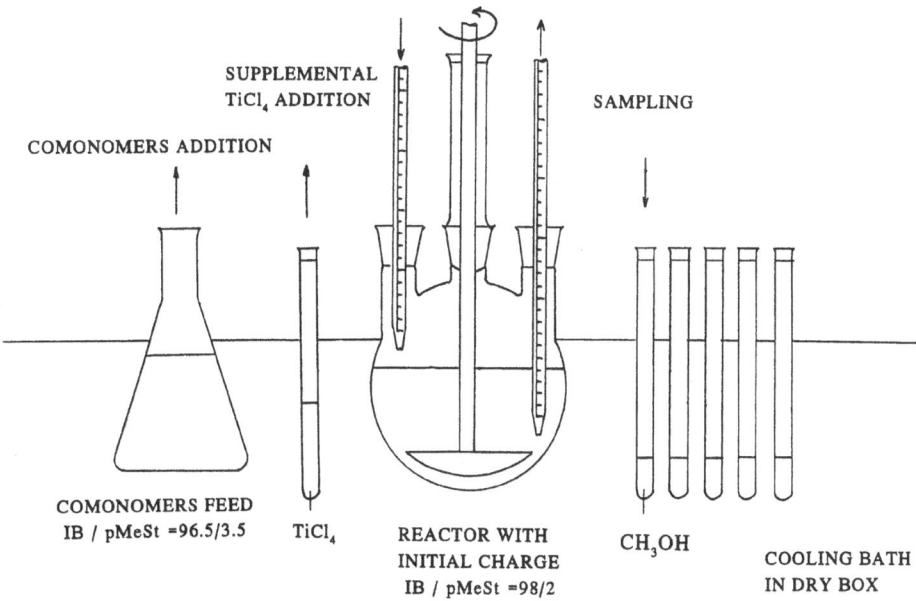

SUPPLEMENTAL
TiCl₄ ADDITION

SAMPLING

COMONOMERS ADDITION

COMONOMERS FEED
IB / pMeSt =96.5/3.5

TiCl₄

REACTOR WITH
INITIAL CHARGE
IB / pMeSt =98/2

CH₃OH

COOLING BATH
IN DRY BOX

Figure 1. Scheme of experimental setup for constant copolymer composition experiments.

slow heat transfer during the rapid exothermic copolymerizations (line 2, Table I). The reactor was charged with the monomers, solvent(s), initiator and additivies and the copolymerization was started by the addition of the prechilled TiCl₄ coinitiator solution (see Initial charge, Table I). The volume of the initial charge was 1000 mL. Immediately upon TiCl₄ addition the copolymerization started as indicated by the appearance of yellow color and temperature increase (line 2, Table I). The copolymerization was continued and maintained by the addition of predetermined amounts and composition of prechilled comonomers to the charge at a predetermined rate (see lines 10 and 11, Table I and Section III. for additional details). To compensate for the decrease of the copolymerization rate caused by dilution due to feed addition, supplemental TiCl₄ (see line 12, Table 1) was also supplied. During the copolymerization a series of aliquots were withdrawn by a prechilled pipette and quenched into methanol placed in a series of test tubes in the cooling bath (see Figure 1). The temperature in the bath and in the reactor was continuously monitored.

Molecular weights were determined by a Waters high pressure GPC instrument. Further procedural details and the characterization method have been described [2,6,7].

III. RESULTS AND DISCUSSION

III.1. Proof for Attaining CCC Conditions: The Diagnostic Plots

Constant copolymer composition (CCC) conditions prevail when the rates of comonomers input and copolymer formation are equal, i.e., **R**=1, *and* the compositions of the feed and the copolymer are also the same, i.e., **C**=1:

$$\frac{Rate_{comonomers\ input}}{Rate_{copolymer\ formed}} = \mathbf{R}$$

(1)

Table 1. Experimental conditions

Experiments	I.	II.	III.
T(cooling bath → charge) (°C)	-60 → -50	-64 → -50	-60 → -50
Initial charge, V_0=1000mL			
Solvent:	EtCl	EtCl/Hexanes=77/23	EtCl/Hexanes=77/23
$10^3[I]_0$ (mol/L)	2.5	2.5	1.25
$10^3[TEA]$ (mol/L)	7.5	7.5	7.5
$10^3[DtBP]$ (mol/L)	5	5	12.5
IB+pMeSt (mol) (IB/pMeSt)*	4.37+0.0892 (98/2)	2.19+0.0446 (98/2)	2.19+0.0446 (98/2)
$10^3[TiCl_4]$ (mol/L)	50	50	50
Comonomers feed (bulk), IB/pMeSt*	96.5/3.5	96.5/3.5	96.5/3.5
Average (actual) feed rate (mL/min)	12.5 (25/2)	4.0 (20/5)	4.0 (20/5)
Supplemental TiCl₄ solution (mol/L)	3.23	3.23	3.23
Average feed rate (mL/min)	0.60	0.39	0.39

* IB/pMeSt is in mol%/mol%

$$\frac{Composition_{comonomers\ feed}}{Composition_{copolymer}} = C \tag{2}$$

These equations lead to diagnostic plots which define **R** and **C**, i.e., the parameters by which CCC conditions can be quantitatively expressed. The diagnostic plots show A) Total comonomers input and the corresponding copolymer formed as a function of time, and B) Feed composition and the corresponding copolymer composition as a function of copolymer formed (conversion). The ratios of the corresponding slopes of these plots yield **R** and **C** (see later). If **R** and **C** are distant from unity, CCC conditions are absent and the experimental conditions have to be modified to approach it.

III.2. Designing the Experiments

Prior to experimentation toward developing CCC conditions one has to establish 1) The concentration and composition of the comonomers in the initial charge in the reactor into which the feed is added, and 2) The input rate of the comonomers feed.

1. Comonomer Composition in the Initial Charge. The pMeSt/IB comonomer composition in the initial charge must be selected so that it, in combination with the incoming comonomers feed, should produce the target copolymer composition (y) [1]. In the present investigations:

$$y = \frac{pMeSt_{in\ copolymer}}{IB_{in\ copolymer}} = \frac{3.5}{96.5} \tag{3}$$

where the numbers indicate mol% compositions. The required *pMeSt/IB* ratio in the initial charge can be obtained by rearranging the instantaneous copolymer composition equation [8]:

$$\frac{pMeSt}{IB} = -\frac{1-y}{2r_S} + \sqrt{\left(\frac{1-y}{2r_S}\right)^2 + y\frac{r_B}{r_S}} = 0.023 \tag{4}$$

where $r_B = 0.74 \pm 0.11$ and $r_S = 8.0 \pm 3.3$, and the subscripts B and S stand for *IB* and *pMeSt*, respectively. Eq. 4 in conjunction with the general (Gaussian) error proliferation rule [9] yields $pMeSt/IB = (2.3 \pm 0.3)/(97.7 \pm 0.3) \approx 2/98$ mol%/mol%. Preliminary experiments have confirmed that the target copolymer composition ($y = 3.5/96.5$) is indeed obtained by the use of the initial charge composition of *pMeSt/IB* = 2/98 under CCC conditions.

2. Input Rate of the Comonomers. As it was shown in the first part [1], the time dependence of the comonomers input rate (in mL·min^{-1}) is:

$$v = \frac{A}{\sqrt{V_0^2 + 2 \cdot A \cdot t}} \tag{5}$$

where V_0 and t are the initial charge volume and time, respectively; the constant A (in mL^2·min^{-1}) can be easily obtained from the kinetic equations of IB/pMeSt copolymerization [1]:

$$A = \frac{r_S + \frac{[IB]}{[pMeSt]}}{\frac{r_S}{k_{SS}} + \frac{r_B \cdot [IB]}{k_{BB} \cdot [pMeSt]}} \cdot I_0 \cdot N_{S,0} \cdot \left[\frac{M_S}{d_S} + \frac{M_B}{d_B} \cdot \frac{F_S}{F_B} \right] \tag{6}$$

where, *[IB]* and *[pMeSt]* indicate concentrations in mol·L^{-1}; k_{SS} and k_{BB} are pMeSt and IB homopropagation rate constants in mL·mol^{-1}·min^{-1}; I_0 and $N_{S,0}$ are the moles of initiator and pMeSt in the initial charge; F_S and F_B are input rates of pMeSt and IB in mol·min^{-1} (under CCC conditions $F_S/F_B = y$); M_S and M_B are the molecular weights of pMeSt (118.1 g·mol^{-1}) and IB (56.1 g·mol^{-1}), and d_S (0.90 g·mL^{-1}) and d_B (0.68 g·mL^{-1}) are the corresponding densities of the monomers at -50 °C. Since the rate of homopolymerization of pMeSt is much higher than that of IB, and *[IB]/[pMeSt]* = 49, r_S/k_{SS} in the denominator of the first term of Eq.6 can be neglected. Based on earlier findings [2,5] and on numerous preliminary experiments, k_{BB} is to be in the range of $1 - 3 \cdot 10^4$ mL·mol^{-1}·min^{-1} under the conditions similar to those used in these experiments. At this point all the necessary information is on hand to perform CCC experiments. The next section describes and analyses three experiments on the road toward CCC conditions under LC$^\oplus$Copzn conditions.

III.3. Synthesis of Uniform High Molecular Weight IB/pMeSt Copolymers Under CCC and LC$^\oplus$Copzn Conditions

Experiment I. Table 1 shows the ingredients and conditions employed. By setting $k_{BB} = 1.5 \cdot 10^4$ mL·mol^{-1}·min^{-1}, we obtained $A = 12.3 \cdot 10^3$ mL^2·min^{-1}, which according to Eq. 5 requires an initial comonomer feeding rate of $v = 12.3$ mL·min^{-1} to reach the target IB/pMeSt=96.5/3.5 mol%/mol% copolymer composition.

Due to a lack of costly measuring and feeding instrumentation we were not able to feed the comonomers into the charge at the continuous and decreasing stream as demanded by Eq. 5, but approached the required 12.3 mL·min^{-1} by adding 25 mL comonomers feed every 2 mins (i.e., average 12.5 mL·min^{-1}) (see line 11, Table I). The addition of the comonomers feed to the charge causes a dilution of the charge. To compensate for this dilution, supplemental quantities of TiCl$_4$ were also added (0.60 mL·min^{-1}) (Section II.2. and lines 12, 13, Table I). The rate of TiCl$_4$ addition had to be established by iterative experiments [10] because the effect of dilution on the rate constants was unknown.

Despite vigorous stirring the charge temperature was 10-14 °C higher than that of the cooling bath (see line 2, Table I). This temperature differential is due to rapid heat evolution coupled with relatively slow heat removal, particularly at the beginning of the copolymerization. Investigations are in progress to eliminate this problem by the use of the highly efficient Leidenfrost reactor [11] and the results will be reported separately[12].

The diagnostic plots shown in Figure 2 summarize the results: As seen by examining the trends of the experimental points in Figure 2A, the comonomers input rate was significantly higher than the rate of copolymer formed, i.e., **R**>1. Consequently, as shown in Figure 2B, the IB content in the copolymer was lower than that of the target, i.e., **C**>1. These results can be easily understood by considering that from the very start of the copolymerization the charge composition has shifted towards lower IB contents because of the overly rapid introduction of the feed whose IB content was lower than that of the initial charge. This compositional drift in the reactor yielded copolymers with IB contents lower than that of the target composition.

As shown by the third diagnostic plot (Figure 2C) the copolymerization exhibits living character up to $\overline{M_n} \approx 35,000$ g·mol^{-1}. Unfortunately, due to the high comonomers concentration in the reactor, the initially homogeneuos charge became heterogeneous after ~20 mins and a suspension formed which became increasingly dense. Because of the insufficient heat transfer in the increasingly heterogeneous system LC$^\oplus$Copzn could not proceed beyond $\overline{M_n} \approx 35,000$ g·mol^{-1} and, as indicated by Figure 2C, the experimental molecular weights started to deviate from the theoretical line.

Experiment II. In Expt. I both **R** and **C** were larger than unity. Thus, in order to attain CCC conditions in Expt. II the comonomer concentration in the initial charge was halved to *[IB]/[pMeSt]* = 2.19/0.0446 (see line 8, Table I). The reduction in the amount of comonomers in the charge was compensated by the addition of the same amount of hexanes so as to maintain the volume and polarity of the charge (see line 4, Table I). In addition, to avoid heterogeneity (suspension formation, see Expt. I), the comonomers input rate was also reduced (see line 11, Table I). Also, it was assumed that $k_{BB} = 1.0 \cdot 10^4$ mL·mol^{-1}·min^{-1}. These values yielded $A = 4.1 \cdot 10^3$ mL^2·min^{-1} which in turn gave for the theoretical comonomers input rate $v = 4.1$ mL·min^{-1}. In actuality we added 20 mL comonomers feed every 5 mins which averages to v=4.0 mL·min^{-1}. Based on preliminary experiments [10] the rate of addition of supplemental TiCl$_4$ solution was 0.39 mL·min^{-1}. The data in Table I summarize the details.

According to the first diagnostic plot, Figure 3A, the comonomers input rate at the start was much lower than that of the copolymerization, i.e., **R**<1. Then **R** slowly increased with time because the rate of copolymerization was faster than that of comonomers input; i.e., the amount of comonomers decreased in the charge and thus the rate of copolymerization gradually diminished. The IB content in the copolymer was higher than that in the feed, i.e., **C**<1. The charge composition drifted toward higher IB (lower pMeSt) contents because the amount of the more reactive pMeSt consumed by the copolymerization was not completely replenished by the feed. Thus the copolymer composition gradually deviated from the target composition.

According to the data shown in the third diagnostic plot, Figure 3C, the copolymerization exhibited living character up to $\overline{M_n} \approx 65,000$ g·mol^{-1}. The copolymerization was homogeneous during the entire experiment.

Experiment III. In Expt. II the rate of the copolymerization was higher than that of the comonomers feed. Thus, in order to attain CCC conditions in Expt. III we decided to reduce the rate of copolymerization by halving the initiator concentration to $[I]_0$=1.25·10^{-3} mol·L^{-1} (see line 4, Table I). The DtBP concentration was increased since the I_o was decreased

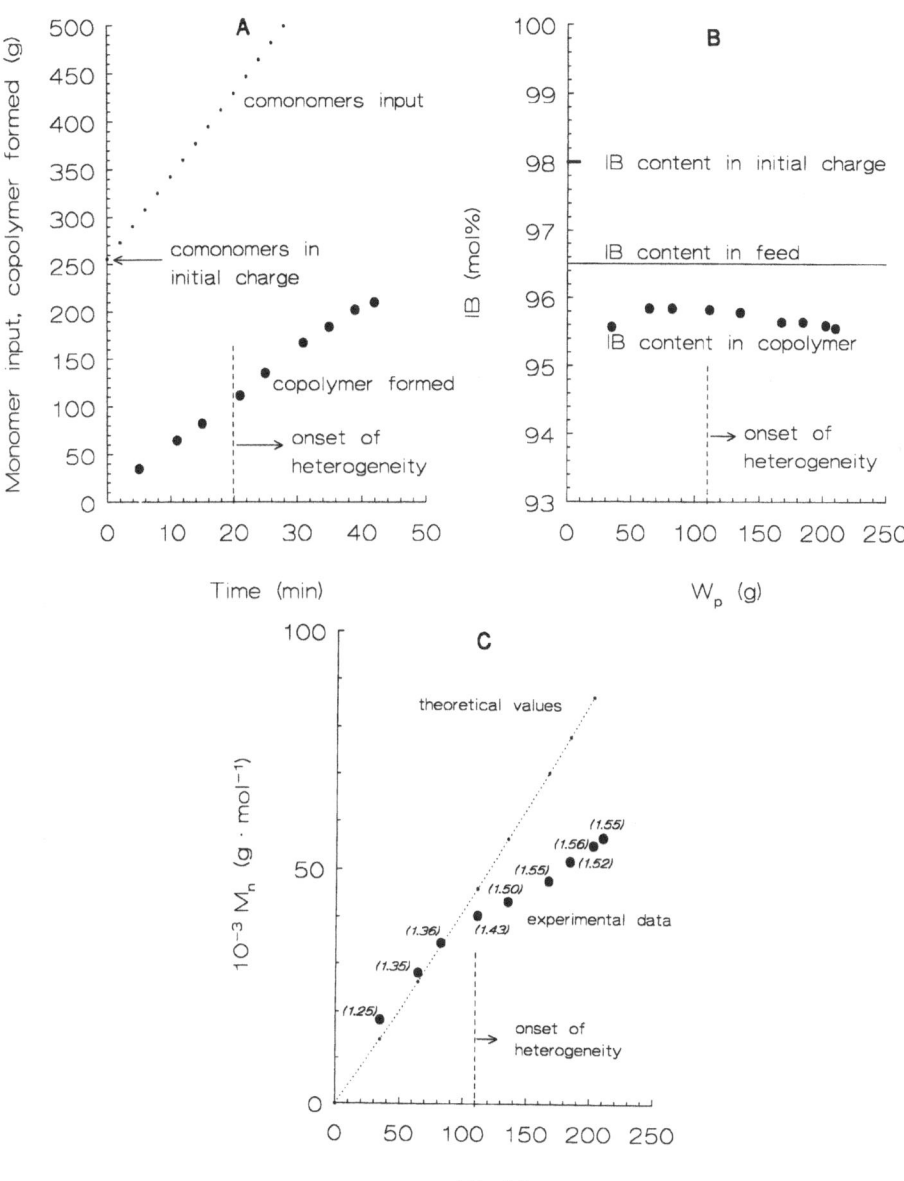

Figure 2. Diagnostic plots for constant copolymer composition condition and living IB/pMeSt copolymerization in Expt. I: A) *comonomers input* and corresponding *copolymer formed* vs. *time*, showing that the rate of comonomers input is higher than that of copolymer formed, i.e., R>1. B) *comonomers composition in feed* and *copolymer composition* vs. W_p, showing that the IB content in the feed is always higher than the IB content in the copolymer, i.e., C>1. C) $\overline{M}_n$ vs. W_p indicates significant chain transfer above $\overline{M}_n \approx 35{,}000$ g·mol^{-1}; the numbers in parantheses indicate $\overline{M}_w/\overline{M}_n$. Experimental conditions: see Table 1, Column I.

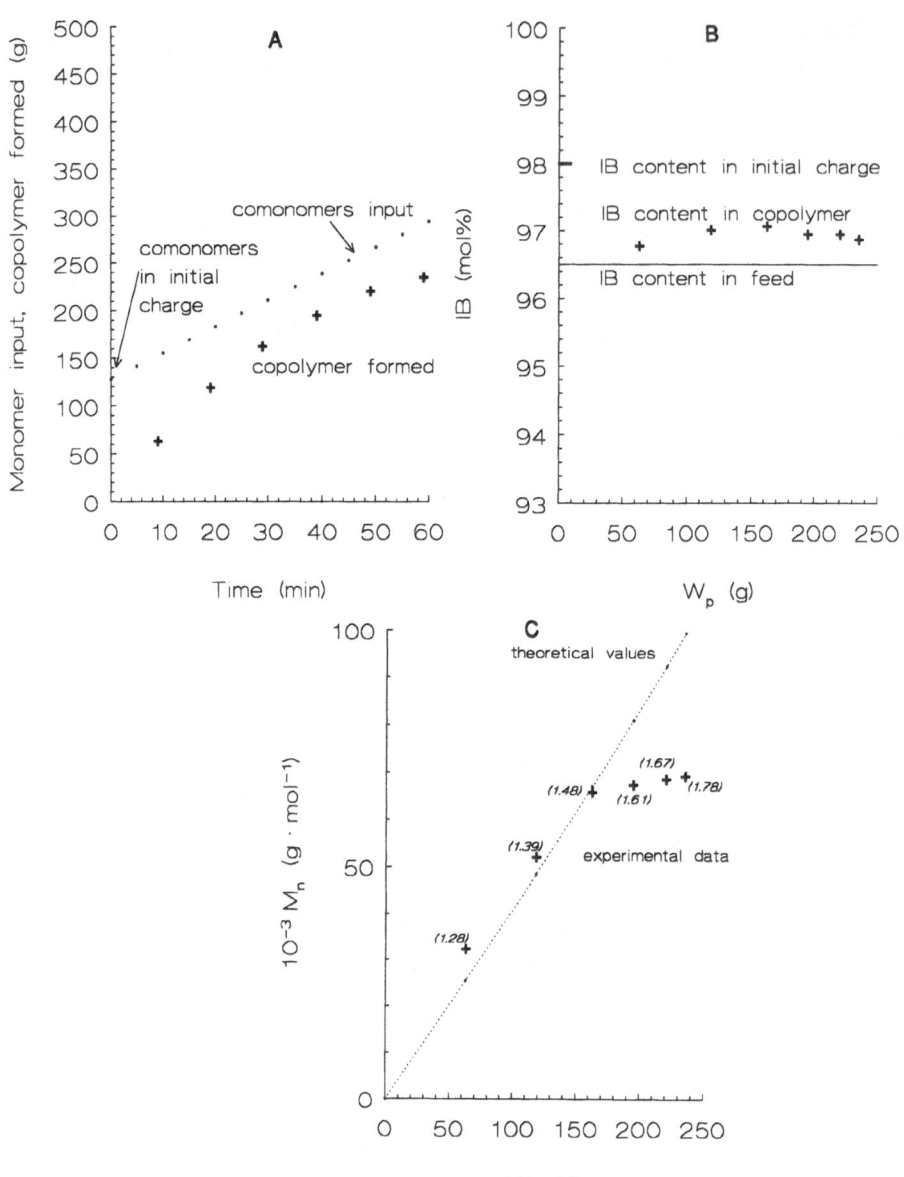

Figure 3. Diagnostic plots for constant copolymer composition condition and living IB/pMeSt copolymeriza-tion in Expt II: A) *comonomers input* and corresponding *copolymer formed* vs. *time*, showing that the rate of comonomers input is initially lower and later becomes higher than that of the copolymer formed, because of the changing rate of copolymer formation, i.e., initially R<1 and later R>1. B) *comonomers composition in feed* and *copolymer composition* vs. W_p, showing that the IB content in the feed is always lower than the IB content in the copolymer, i.e., C<1. C) $\overline{M}_n$ vs. W_p indicates significant chain transfer above $\overline{M}_n \approx 65{,}000$ g·mol⁻¹; the numbers in parentheses indicate $\overline{M}_w/\overline{M}_n$. Experimental conditions: see Table 1, Column II.

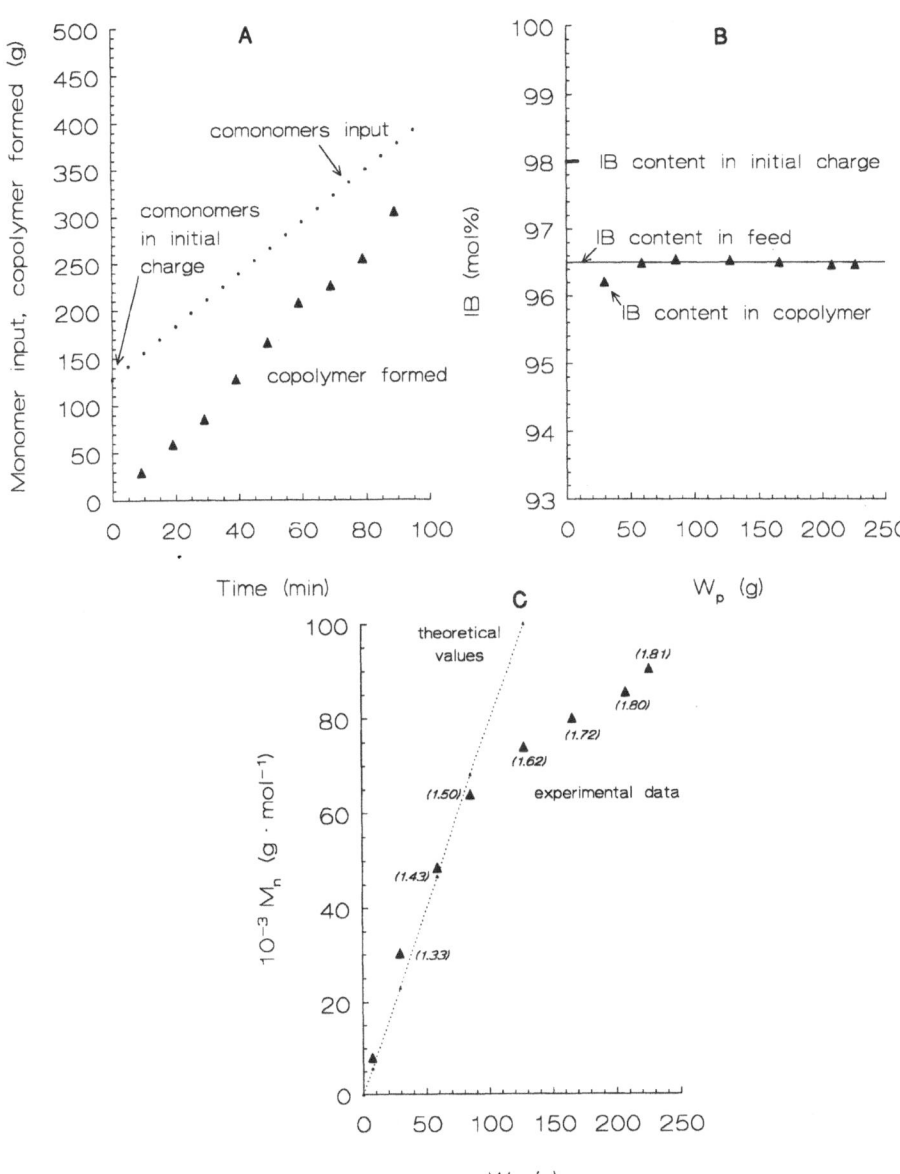

Figure 4. Diagnostic plots for constant copolymer composition condition and living IB/pMeSt copolymeriza-tion in Expt. III: A) *comonomers input* and corresponding *copolymer formed* vs. *time*, showing that at first the rate of comonomers input is slightly lower than that of the copolymer formed, but later they become equal, i.e., R≅1. B) *comonomers composition in feed* and *copolymer composition* vs. W_p, showing that the IB content in the feed is practically equal to the IB content in the copolymer, i.e., C≅1. C) $\overline{M}_n$ vs. W_p indicates significant chain transfer above $\overline{M}_n$ ≈ 65,000 g·mol^{-1}; the numbers in parantheses indicate $\overline{M}_w/\overline{M}_n$. Experimental condi-tions: see Table 1, Column III.

and the system became more sensitive to impurities. The increase in DtBP concentration from $5.0 \cdot 10^{-3}$ to $12.5 \cdot 10^{-3}$ mol·L^{-1} (see line 7, Table I) should not affect copolymerization kinetics [2]. Since we found $v = 4.0$ mL·min^{-1} easy to maintain and since according to Eq. 5 A has to be kept constant, the only parameter in Eq. 6 that can be adjusted was k_{BB}; thus $k_{BB} = 2.0 \cdot 10^4$ mL·mol^{-1}·min^{-1} was chosen. The use of these I_o and k_{BB} values yielded $A = 4.1 \cdot 10^3$ mL^2·min^{-1} which in turn gave for the comonomers input rate $v = 4.1$ mL·min^{-1}. We have approached this requirement by using an average v of 4.0 mL·min^{-1}, in actuality 20 mL/5 min. The data in Table 1 summarize the details.

According to the first diagnostic plot, Figure 4A, the input rate of comonomers was close to equal to that of the copolymerization, i.e., $\mathbf{R} \cong 1$. Toward the end of the experiment $\mathbf{R}$ started slowly to decrease with time because the rate of copolymerization was slightly faster than that of comonomers input. The IB content in the copolymer was equal to that in the feed, i.e., $\mathbf{C}=1$. The composition of the charge did not change, and the copolymer and feed compositions were practically identical during the entire experiment. According to this evidence, then, CCC conditions have been achieved.

The third diagnostic plot, Figure 4C, shows that the copolymerization exhibited living character up to $\overline{M_n} \approx 65,000$ g·mol^{-1}. The copolymerization was homogeneous during the entire experiment. The cumulative composition of the copolymer did not change even at the highest molecular weights (see Figure 4B). As shown by the data in Figure 4C chain transfer becomes noticeable after $\overline{M_n} \approx 65,000$ g·mol^{-1}, i.e., the molecular weights start to deviate from the theoretical line and the MWDs start to broaden.

IV. CONCLUSIONS

Uniform composition high molecular weight ($\overline{M_n} \approx 65,000$ g·mol^{-1}) IB/pMeSt copolymers have been prepared by the use of the constant composition technique under living copolymerization conditions at -50°C. The desired copolymer composition of IB/pMeSt=96.5/3.5 mol%/mol% has been attained in spite of the very large reactivity difference between the comonomers ($r_{IB} = 0.74 \pm 0.11$ and $r_{pMeSt} = 8.0 \pm 3.3$). Diagnostic methods have been developed to express quantitatively the attainment of target uniform copolymer compositions.

V. ACKNOWLEDGMENT

This material is based upon work partially supported by the National Science Foundation (Grant # 94-23202) and the Exxon Chemical Company.

REFERENCES

1. A. Nagy, I. Országh, J.P. Kennedy: the preceding paper in this series and references therein.
2. *a)* I. Országh, A. Nagy, J.P. Kennedy: J. Phys. Org. Chem., Special Issue (1995), in press; *b)* I. Országh, A. Nagy, J.P. Kennedy: *Abstracts*, MacroAkron '94, 35th Intl. Symp. Macromol., Akron, 7/11-15/1994, p.83
3. A. Lubnin, I. Országh, J.P. Kennedy: J.M.S.-Pure Appl. Chem.(1995), in press.
4. *a)* A. Nagy, I. Országh, J.P. Kennedy: J. Phys. Org. Chem., Special Issue (1995), in press; *b)* A. Nagy, I. Országh, J.P. Kennedy: *Abstracts*, MacroAkron '94, 35th Intl. Symp. Macromol., Akron, 7/11-15/1994, p.84.
5. I. Majoros, A. Nagy, and J.P. Kennedy: Adv. in Polym. Sci., *112*, 1 (1994).
6. G. Kaszás, J. Puskás, J.P. Kennedy: Makromol. Chem., Macromol. Symp. *13/14*, 473 (1988).

7. Zs. Fodor, Á. Fodor and J.P. Kennedy: Polym. Bull. *29(6)*, 697 (1992).
8. G. Odian: Principles of Polymerization, 3rd Ed., J.Wiley & Sons, 1991.
9. A. Nagy: Polym. Bull. *14*, 259 (1985).
10. A. Nagy, I. Országh, J.P. Kennedy: unpublished results.
11. M. Zsuga, J.P. Kennedy and T. Kelen: Polym. Bull. *19*, 201 (1988).
12. I. Országh, A. Nagy, J.P. Kennedy: the forthcoming paper in this series.

4

HEXAARMED POLYSTYRENE STARS FROM A NEWLY DESIGNED INITIATOR OF CARBOCATIONIC POLYMERIZATION

Eric Cloutet,[1,2] Jean-Luc Fillaut,[1] Didier Astruc,[1*] Yves Gnanou[2*]

[1] Laboratoire de Chimie Organique et Organométallique, URA CNRS
Université Bordeaux I
351, Cours de la Libération
33405 Talence Cédex, France
[2] Laboratoire de Chimie des Polymères Organiques, URA CNRS, ENSCPB
Université Bordeaux I
Avenue Pey Berland - B.P. 108
33402 Talence Cédex, France

ABSTRACT

A novel hexafunctional initiator $C_6[(CH_2)_2p\text{-}C_6H_4CH(Cl)Me]_6$ has been purposely synthesized to subsequently serve in the cationic polymerization of styrene. This initiator is prepared via $Fe(\eta\text{-}C_5H_5)^+$ mediated perbenzylation of hexamethylbenzene, followed by regiospecific acetylation, reduction and chlorination of outer phenyl rings. The polystyrene stars which are obtained in the presence of this initiator exhibit narrow molecular weight distribution (MWD) and a precise functionality of 6. Because the cationic polymerization of styrene is not totally free of side reactions, such as β-proton elimination of growing carbeniums, the upper limit of the molecular weight control for these star polymers is the range of Mn ~30000 g/mole.

I. INTRODUCTION

From sheer academic topic, star-shaped polymers steadily grow into a field of high potentiality in virtue of the unique properties they exhibit both in solution and in the bulk. Nanoscale-ordered materials, low viscosity paints, thickeners are examples of technological applications that include star polymers.

---

[*] Address correspondence to Drs. Astruc and Gnanou

Macromolecular Engineering, Edited by M.K. Mishra et al.
Plenum Press, New York, 1995

47

The access to star-shaped polymers can be contemplated in two different ways, either through a convergent method using living precursors chains or via a divergent pathway from a plurifunctional initiator.

In convergent methods also called arm-first , the central core of the star either arises from the stoichiometric reaction of living polymers chains with a plurifunctional deactivating agent [1] or results from the polymerization of a difunctional monomer [2] by these living chains. In the latter case, the functionality of the star varies in a rather large range and its average value depends on the experimental conditions used. Stars obtained by deactivation are, in contrast, better defined, provided the latter raction occurs quantitatively and without side reactions. Whatever the pathway followed to build up the central core, none convergent method can afford star polymers that consist of ω-functionalized branches.

To overcome this drawback, repeated attempts have been made to grow branches in a divergent way from a plurifunctional core that is used as initiator. In core-first methods, the central nodule either arises from a true plurifunctional molecule [3] or results from the polymerization of a difunctional monomer [4]. The initiator which is formed in the latter case actually consists of crosslinked polyfunctional nodules of extremely heterogeneous size and functionality. The star-like polymers that are obtained from these polyfunctional species exhibit in turn high polydispersity indices and large fluctuation of their functionality. In spite of these shortcomings, this method of star synthesis has been applied in many instances and particularly with monomers polymerizing anionically. The difficulty to make carbanionic reagents that consist of a precise number of carbanionic centres and therefore the lack of such initiators have led one to use living poly(divinylbenzene) cores instead, as a means to derive structures such as ω- functionalized stars or star block copolymers [5].

In contrast to the situation prevailing in anionic processes, several plurifunctional initiators have been developed for being used in cationic polymerizations. Kennedy and his coworkers have synthesized three [6] and four [7] arms polyisobutylene from initiators containing three and four cumyl functions, respectively. Likewise, Higashimura [8] and Deffieux [9] have prepared three-arm poly(vinyl ether)s from perfectly trifunctional initiators which they purposely designed for the cationic polymerization of vinyl ethers. Lately, Chang et al. [10] have successfully introduced six benzyl halide functions on phosphazene rings and utilized the latter compound as plurifunctional initiator of oxazoline polymerization. Hexaarmed poly(oxazoline)s with different arm sizes have been made available in this way.

The recent discovery by Higashimura [11] that styrene can undergo cationic polymerization under living conditions (with Mn growing linearly vs. the polymer yield) prompted us to elaborate an initiator that could give access to star-shaped polystyrenes of precise functionality via the core-first method. The opportunities granted by $Fe(\eta-C_5H_5)^+$ induced perfunctionalization of polymethylaromatics [12] have been used to devise a novel hexafunctional initiator containing not less than 6 phenylethyl chloride functions (**Scheme 1**).

This article is devoted to the synthesis of the hexaarmed polystyrene (PS) stars which have been obtained from this purposely designed initiator of cationic polymerization of styrene (**Scheme 2**). The experimental conditions that afford living cationic polymerization of styrene have been first accurately examined on linear models before being applied to the synthesis of the PS stars. They were characterized by the classical methods of molecular weight determination.

To evaluate the efficiency of this hexafunctional initiator and determine the functionality of the stars, a controlled amount of phenylethyl chloride -the monofunctional model- has been intentionally added to the reaction medium and used together with the above hexafunctional molecule to cationically polymerize styrene. The actual functionality of the stars has been inferred upon comparing their molecular weight with that of the linear species generated by the monofunctional initiator.

A preliminary communication has recently appeared [13] .

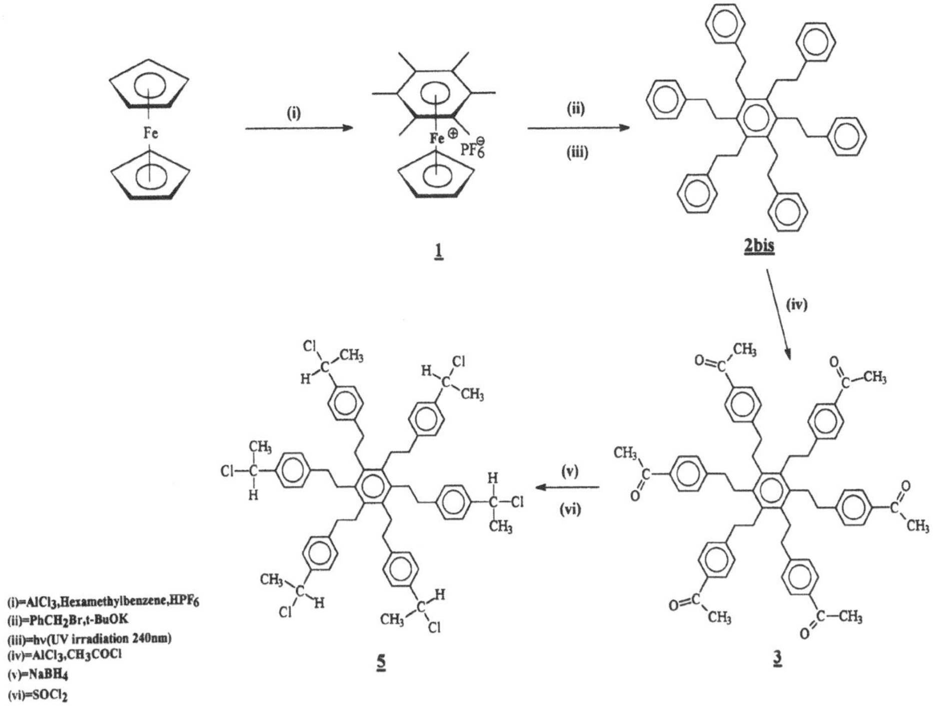

(i)=AlCl₃,Hexamethylbenzene,HPF₆
(ii)=PhCH₂Br,t-BuOK
(iii)=hν(UV irradiation 240nm)
(iv)=AlCl₃,CH₃COCl
(v)=NaBH₄
(vi)=SOCl₂

Scheme 1

II. EXPERIMENTAL PART

a. Development of a Novel Hexafunctional Initiator, 5

Materials. All reactions were carried out under dry Argon atmosphere using Schlenk techniques. Reagent grade 1,2 dimethoxyethane (DME) was predried on Na and distilled from sodium benzophenone. Methylene chloride was stirred overnight with calcium hydride before use. Methanol was refluxed over Mg for 24 hours and distilled. All other reagents were purchased from Aldrich and were used without further purification.

All chemical shifts are reported in parts per million (δ) with reference to tetramethylsilane (Me₄Si) and were measured relative to Me₄Si.

Synthesis of [FeCp(C₆Me₆)]⁺PF₆⁻, 1, (with Cp=C₅H₅). To a 500 mL three-necked flask equipped with a reflux condenser and containing 16.2 g of hexamethylbenzene (100 mmol), 18.6 g of ferrocene (100 mmol), 40 g of powdered anhydrous aluminum chloride (300 mmol) and 2.7 g of aluminum powder (100 mmol), was added through a steel cannula 150 mL of degassed decahydronaphtalene. The mixture was heated to 140°C under N₂ and 2 mL of degassed water (110 mmol) was slowly added. The reaction was stirred at 160°C for 24 hours, during which time the color of the solution turned to dark yellow. The solution was cooled, degassed ice water (250 ml) was carefully added, and the mixture was stirred for a few minutes for hydrolysis. After filtration on celite and extraction with ether, concentrated ammonia was added until reaching pH=9 to precipitate

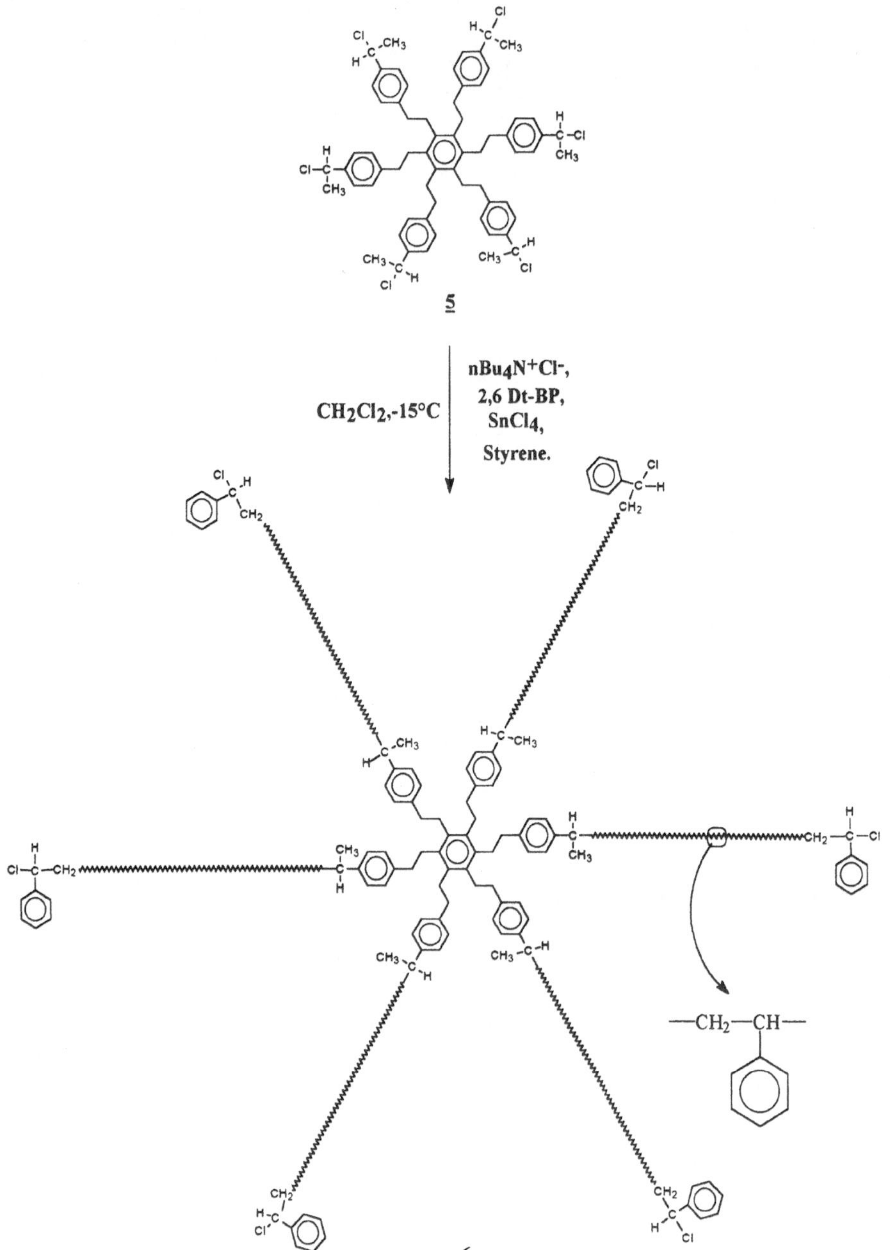

Scheme 2

Al(OH)$_3$. After filtration, a yellow solid was obtained upon addition of an aqueous solution of HPF$_6$. The precipitate was filtered, washed with ether, dissolved in 100 mL of acetonitrile and dried over Na$_2$SO$_4$. A new filtration was then carried out through a small aluminum oxide column. 100 mL of ethanol was added and then slow evaporation of solvent gave a 70% yield of yellow monocrystals. 1: ^{1}H NMR (CD$_3$COCD$_3$, 250 MHz) δ (ppm): 4.7 (s,Cp), 2.52 (s,18H, CH_3); ^{13}C NMR(CD_3CN, 62.90 MHz) δ (ppm): 103.9 (Quaternary C), 77.6 (CH, CP), 15.3 (CH_3).

Synthesis of $[FeCpC_6(CH_2CH_2Ph)_6]^+PF_6^-$, 2, from 1. To a 250 mL three-necked flask containing 3 g (7 mmol) of 1 and 16g (140mmol) of t-BuOK, was added by cannula a solution of 17 mL (140 mmol) of benzyl bromide in 150 mL of DME. The mixture was stirred for 12 hours at 40°C, until its color turned from red to yellowish grey. After the solvent was removed in "*vacuo*", the creamy solid was dissolved in CH_2Cl_2, washed with water, filtered, and dried over Na_2SO_4. Concentration and addition of excess ether provided 3.29 g (67%) of orange powder identified by ^1H and ^{13}C NMR. 2 : ^1H NMR (250 MHz, CD_3COCD_3) δ (ppm) 7.46 (s, C_6H_5, 30H), 5.26 (s, Cp, 5H), 3.50 (unresolved m, CH_2, 24H); ^{13}C NMR δ (ppm) (62.90 MHz, CD_3CN) δ (ppm) 137.1 (Quaternary C of aromatics), 135.1, 130.0 and 129.4 (CH), 99.8 (quaternary C of complexed aromatic ring), 25.3 and 24.7 (CH_2).

Synthesis of $C_6[(CH_2)_2Ph]_6$, 2bis, from 2. 1.42 g (1.47 mmol) of 2 was photolyzed in 200 mL of CH_3CN with UV irradiation (240nm) for 12 hours at room temperature. After the solvent was removed in *vacuo*, the block residue was stirred for 1 hour in concentrated H_2SO_4 to oxidize the iron species formed during the photolysis. The organic compound was extracted from the aqueous layer with ether and dried over Na_2SO_4 ; recrystallization from hexane gave 0.912 g (1.29 mmol) of pure $C_6[(CH_2)_2Ph]_6$. 2bis mp=115-116°C; ^1H NMR CD_3COCD_3, 250MHz) δ (ppm) 7.35 (s, C_6H_5, 30H), 3.0 (unresolved m,CH_2, 24H); ^{13}C NMR ($CDCl_3$, 62.90 MHz) δ (ppm) 142.4 (Quaternary C of aromatics), 133.2, 128.3 and 126.3 (CH), 38.0 and 36.8 (CH_2). Anal. (%) calcd. for $C_{54}H_{54}$: C, 92.25; H, 7.74. Found: C, 92.30; H, 7.70.

Synthesis of $C_6[(CH_2)_2C_6H_4COCH_3]_6$, 3, from 2bis. To 500 mg (0.7 mmol) of 2bis was added by cannula 1 mL (14 mmol) of freshly distilled acetylchloride in 30 mL of CH_2Cl_2. 1.85 g (14mmol) of $AlCl_3$ was added and stirred at room temperature for 16 hours. The product was extracted with CH_2Cl_2-H_2O. The extract was washed with water (3x25mL), dried over Na_2SO_4 and the solvent removed under reduced pressure by a rotary evaporator. Recrystallization from heptane gave 582 mg (0.6 mmol) of white crystals of $C_6[(CH_2)_2C_6H_4COCH_3]_6$, 3, identified and found pure by ^1H and ^{13}C NMR (250MHz) in 86% yield . 3 mp=227-229°C ; ^1H NMR ($CDCl_3$, 250 MHz) δ (ppm) 7.3 (m, AA'BB' syst., C_6H_4, 24H), 3.0 and 2.9 (unresolved m, $(CH_2)_2$, 24H), 2.6 (s, CH_3, 18H); ^{13}C NMR ($CDCl_3$, 62.90 MHz) δ (ppm) 197.7 (C=O), 147.3 (Cq-C=O), 136.4 (C_2-$\underline{C}$q), 135.5 ($\underline{C}$q-CH_2), 128.9 and 128.3 (CH), 37.8 and 32.2 (CH_2), 26.7 (CH_3)

Synthesis of $C_6[(CH_2)_2C_6H_4CH(CH_3)OH]_6$, 4, from 3. The secondary alcohol 4 was prepared by reduction of 3 with sodium boronhydride $NaBH_4$. 0.45 g (12 mmol) of $NaBH_4$ was dissolved in a mixture of solvents $MeOH$-CH_2Cl_2 (20-16mL) for 5 minutes at room temperature. To this, was introduced 0.38 g (0.4 mmol) of 3 dissolved in 4 mL of CH_2Cl_2, and the solution was stirred at room temperature for 16 hours. The reaction mixture was hydrolyzed by addition of 1M NaOH solution, and the product was extracted with water, and dried over Na_2SO_4. After evaporation of the solvent, recrystallization from hexane yielded 279 mg (72%, 0.29 mmol) of the secondary alcohol 4, $C_6[(CH_2)_2C_6H_4CH(CH_3)OH]_6$, in the form of a white powder . 4 ; ^1H NMR (CD_3OD, 250 MHz) δ (ppm) 7.1 (m, AA'BB' syst., C_6H_4, 24H), 4.8 (q, $CH(CH_3)OH$, 6H), 3.3 (s, -OH, 6H), 2.7 and 2.8 (unresolved m, $(CH_2)_2$, 24H), 1.4 (d, $CH(CH_3)$, 18H); ^{13}C NMR (CD_3OD, 62.90 MHz) δ (ppm) 145.4 (Quaternary C of aromatics), 142.3 (Cq-CH(CH_3)OH), 137.7 (Quaternary C of complexed aromatic ring), 129.3 and 126.7 (CH ring), 70.8 (CH(CH$_3$)OH), 38.6 and 33.4 (CH_2)$_2$, 25.8 (CH_3). IR characterization: ν_{OH}=3100–3700 cm^{-1} (NaCl)

Synthesis of $C_6[(CH_2)_2C_6H_4CH(CH_3)Cl]_6$, **5**, *from* **4**. The reaction of chlorination of **4** has required thionyl chloride, used as solvent : 7 mL of SOCl$_2$ were added in a schlenk tube to 279 mg (0.29 mmol) of **4** at 0°C. The reaction mixture was allowed to warm up and stirred for 16 hours at room temperature.

After removal of SOCl$_2$ *"under vacuum"*, the solid residue was washed several times with dry pentane under nitrogen affording 286 mg (0.29 mmol) of fine white powder identified to **5**, $C_6[(CH_2)_2C_6H_4CH(CH_3)Cl]_6$, and obtained in quantitative yield (98%). **5**; ^{1}H NMR (CDCl$_3$, 250 MHz) δ (ppm) 7.3 (m, AA'BB' syst., C$_6$H$_4$, 24H), 5.2 (q, CHCl, 6H), 6H), 3.0 (unresolved m, (CH_2)$_2$, 24H), 1.9 (d, CH(CH$_3$), 18H); ^{13}C NMR (CDCl$_3$, 62.90 MHz) δ (ppm) 142.2 (CH$_2$-C$_6$H$_4$), 140.9 (C$_6$H$_4$-CH(CH$_3$)Cl), 136.6 (Quaternary C of complexed aromatic ring), 128.5 and 127.0 (CH ring), 58.8 (CqH (CH$_3$)Cl), 37.5 and 32.6 (CH_2)$_2$, 26.6 (CH_3).

b. Living Cationic Polymerization of Styrene

Materials. Commercial styrene, 1-phenylethyl chloride (PhEtCl) were distilled twice over calcium hydride under reduced pressure. 2,6 di-tert-butylpyridine (2,6 Dt-BP) was stirred over KOH for one day and distilled just before use. SnCl$_4$ (Aldrich, purity > 99.9%, 1.0 M solution in CH$_2$Cl$_2$) was used as received and handled under dry nitrogen. Tetra-butylammonium chloride (n-Bu$_4$N$^+$Cl$^-$) in CH$_2$Cl$_2$ solution was dried over CaH$_2$ for two days, filtered under dry nitrogen and after removal of solvent *"in vacuo"*, dried under high vacuum at 90°C. CH$_2$Cl$_2$ was dried over CaH$_2$ overnight and purified by cryodistillation just before use; 1,4 dioxane was predried on Na and distilled from sodium benzophenone.

Polymerizations were performed under dry nitrogen and quenched with prechilled methanol containing a small amount of ammonia. For the synthesis of the various linear polystyrenes (Mn,th=12000g/mol), a similar procedure was followed regardless of the initiating systems (a-c) tried.

Typical conditions with PhEtCl/SnCl$_4$/2,6Dt-BP/CH$_2$Cl$_2$ as initiating system (-a-). To a 500 mL reactor equipped with a magnetic stirrer, were charged under slight nitrogen overpressure, at -15°C, 0.15 mL (1.1mmol) of 1-phenylethyl chloride, 49 mL of CH$_2$Cl$_2$, 0.25 ml (1.1 mmol) of 2,6 di-tert-butylpyridine and 15 mL (131 mmol) of styrene. The polymerization was initiated by adding a solution of SnCl$_4$ in CH$_2$Cl$_2$ (5.5 ml, 5.5 mmol) to the mixture, and was quenched with a solution of methanol and ammonia (10mL) after 75 minutes. The quenched reaction mixture was diluted with CH$_2$Cl$_2$, filtered, concentrated and precipitated twice with methanol.

The polystyrene sample was dried and purified by freezed drying from its 1,4 dioxane solution.

With PhEtCl/SnCl$_4$/nBu$_4$N$^+$Cl$^-$/ CH$_2$Cl$_2$ as Initiating System (-b-). 628 mg (2.3 mmol) of tetra-n-butylammonium chloride was first added into the reactor under dry nitrogen atmosphere (glove-box) instead of 2,6 Dt-BP in -a-. The other components and their respective quantity were the same.

With PhEtCl/SnCl$_4$/nBu$_4$N$^+$Cl$^-$/2,6 Dt-BP/CH$_2$Cl$_2$ as Initiating System (-b-). 628 mg (2.3mmol) of nBu$_4$N$^+$Cl$^-$ were added into the reactor, just before introducing PhEtCl

(0.15 ml), CH$_2$Cl$_2$ (49 ml), 2,6 Dt-BP (0.25 ml and SnCl$_4$ (5.5ml) The rest of the polymerization was conducted in the same conditions as those described above.

c. Hexaarmed Polystyrene Stars

For the synthesis of hexaarmed polystyrene stars (Mn,th = 45000 g/mole), a same procedure was followed regardless of the initiating systems (d-e) tried.

Typical Experimental Conditions with Hexafunctional Initiator, 5/SnCl$_4$/nBu$_4$N$^+$Cl$^-$/CH$_2$Cl$_2$ as Initiating System (-d-). Prior to the polymerization, a mixture of 200 mL (0.2 mmol) of 5 (the hexafunctional initiator) and 611 mg (2.2 mmol) of tetra-n-butylammonium chloride were freezed-dried from their dioxane solution (20 mL). The mixture was dissolved in 49 mL of CH$_2$Cl$_2$ at -15°C, 57 mL (198.2 mmol) of styrene was then added to the solution containing the hexafunctional initiator. The polymerization was initiated by adding a solution of SnCl$_4$ in CH$_2$Cl$_2$ (5.5 ml, 5.5 mmol) and was quenched with prechilled methanol (20 ml) containing a small amount of ammonia.

The quenched reaction mixture was diluted with CH$_2$Cl$_2$, filtered, concentrated and precipitated twice with methanol. The polystyrene sample was dried and purified by freezed-drying from its 1,4 dioxane solution.

With Hexafunctional Initiator 5 /SnCl$_4$/nBu$_4$N$^+$Cl$^-$/2,6 Dt-BP/CH$_2$Cl$_2$ as Initiating System (-e-). 0.25 mL (1.1 mmol) of 2,6 di-tert-butylpyridine was added to the mixture containing the hexafunctional initiator 5 and the same reagents as those mentioned above.

d. Polymer Characterization

-^{1}H and ^{13}C NMR spectra were recorded on a Brücker AC 250 FT apparatus.

-Size exclusion chromatography (SEC) measurements were performed using a JASCO HPLC-pump type 880-PU, a Varian apparatus equipped with refractive index /UV dual detection and TSK Gel columns calibrated with polystyrene standards. Some Mw's (specially those of star polymers) were obtained by SEC coupled with the multiangle laser light scattering (MALLS) device (DAWN-F laser photometer) from Wyatt Technonology. THF was used as eluent.

III. RESULTS AND DISCUSSION

a. Synthesis and Characterization of the Hexafunctional Initiator,5

The hexafunctional reagent, 5, which is designed to serve as cationic initiator of styrene polymerization was obtained via a reaction sequence (**Scheme 1**), whose first step is the one-pot hexabenzylation of [Fe(η–C$_5$H$_5$){C$_6$Me$_6$}]$^+$PF$_6^-$, 1. This reaction occurs quantitatively because of the enhanced acidity of benzylic protons of aromatics that are coordinated to an electron-withdrawing transition metal moiety demonstrated and has been applied to number of hexafunctionalizations of [Fe(η–C$_5$H$_5$){C$_6$Me$_6$}]$^+$PF$_6$ [14].

Such (Fe$^{II}\eta$–C$_5$H$_5$arene)$^+$ complexes can be obtained by ligand exchange of a cyclopentadienyl ring (Cp=C$_5$H$_5$) in ferrocene for an arene [12d]. The compound 2, [Fe(η–C$_5$H$_5$){C$_6$(CH$_2$CH$_2$Ph)$_6$}]$^+$PF$_6^-$, which results from the reaction of 1 with benzyl bromide gives 2bis, C$_6$(CH$_2$CH$_2$Ph)$_6$, upon photodecomplexation (UV Hg lamp), oxydation (H$_2$SO$_4$)

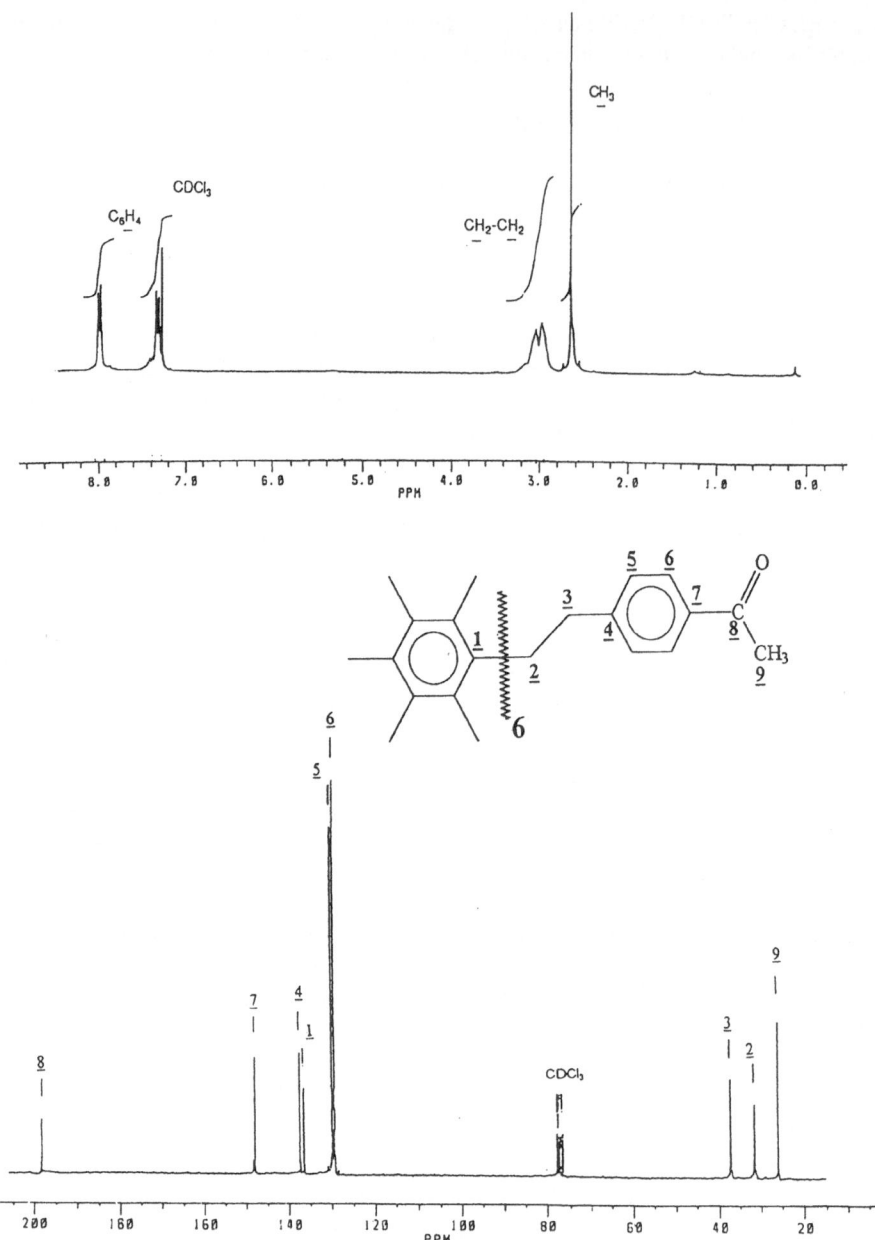

Figure 1. ^{1}H and ^{13}C NMR spectra of $C_6[(CH_2)_2C_6H_4COCH_3]_6$, **3**.

of the ferrocene formed to ferricinium and extraction with ether (86% yield) [14a]. The chemistry concerning the functional derivatives of **2** has been previously developed[14b].

Subsequent to these reactions, regiospecific parafunctionalization of the outer phenyl rings was carried out. This was brought about through reaction of **2** with MeCOCl +Al$_2$Cl$_6$ in CH$_2$Cl$_2$. The resulting hexaketone compound **3**, $C_6(CH_2CH_2p$-$C_6\overline{H}_4COMe)_6$,was obtained in 86% yield (**fig.1**), its subsequent reduction by NaBH$_4$ gave rise to the hexaalcohol **4**,

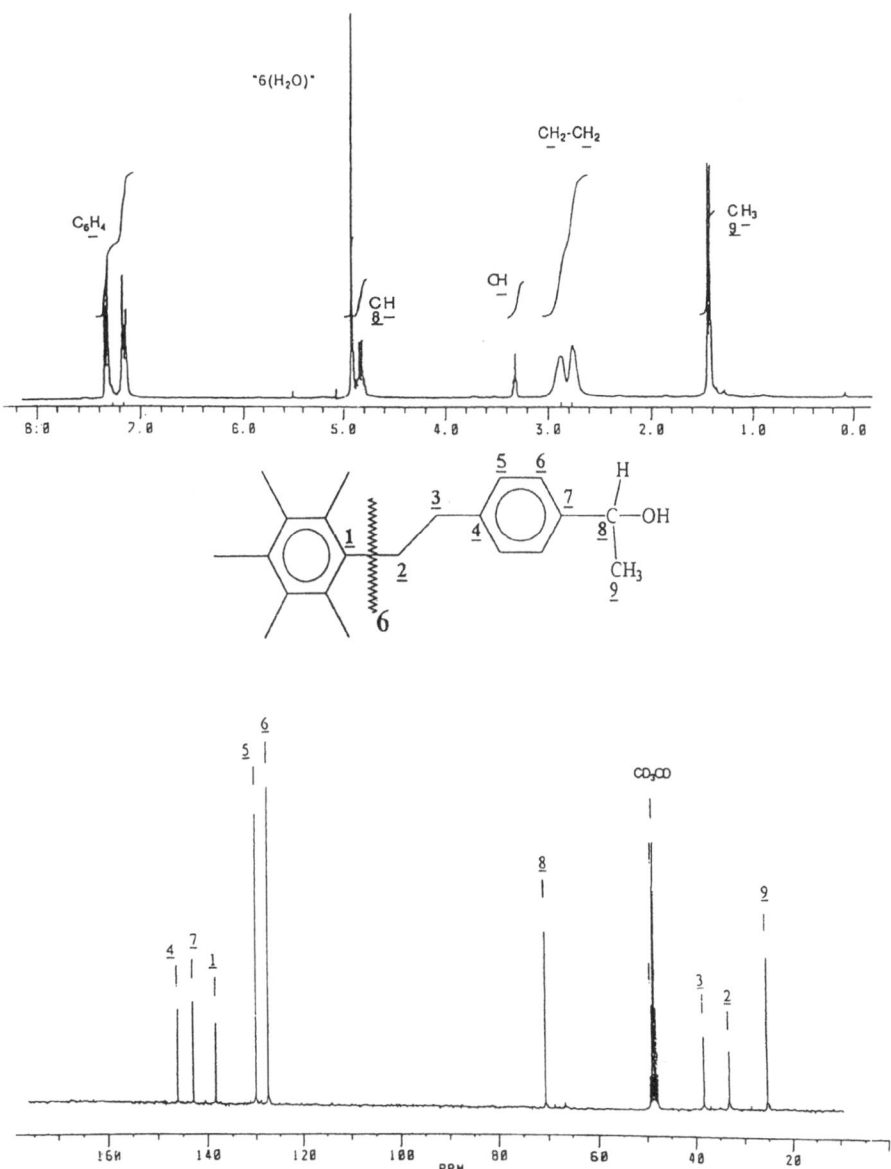

Figure 2. ^{1}H and ^{13}C NMR spectra of $C_6[(CH_2)_2C_6H_4CH(CH_3)OH]_6$, **4**.

$C_6[CH_2CH_2p\text{-}C_6H_4CH(OH)Me]_6$, in 72% yield (**fig.2**). Reaction of the latter compound with $SOCl_2$ afforded the expected initiator, **5**, $C_6[CH_2CH_2p\text{-}C_6H_4CHCl)Me]_6$ in 100% yield (**fig.3**).

All intermediate compounds as well as the ultimate one , 5,have been thoroughly characterized by NMR and elemental analysis after work-up. In all cases, the analytical data collected upon characterization of the isolated products were consistent with the expected structures (**figs.1, 2, 3**). This demonstrates that all the reactions carried out did occur selectively without detectable side reactions.

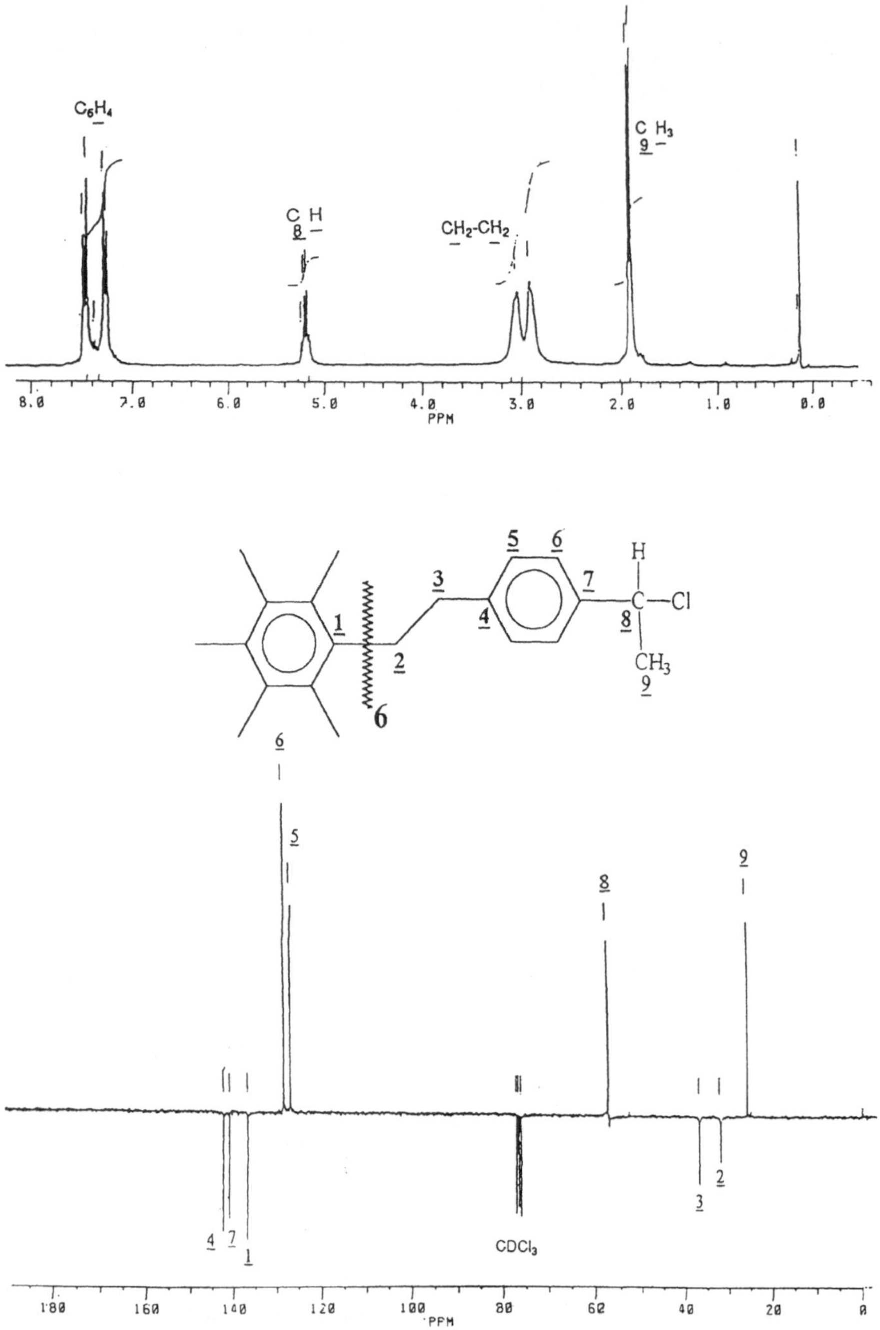

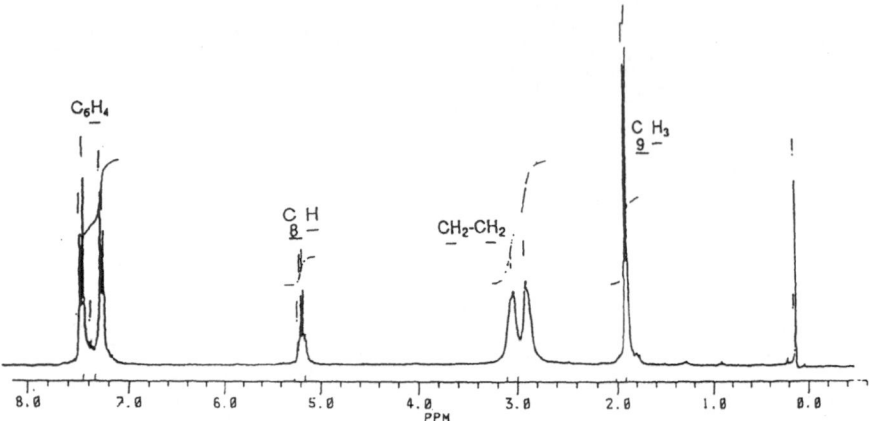

Figure 3. ^{1}H and ^{13}C NMR spectra of $C_6[(CH_2)_2C_6H_4CH(CH_3)Cl]_6$, **5**.

b. Conditions Required For Living Cationic Polymerization of Styrene

Before attempting to make hexaarmed polystyrene stars from 5, we deemed essential to ourselves scrutinize the conditions that afford living cationic polymerization of styrene, on the basis of data already available in the literature [11,15].

Until not so long ago, achieving the cationic polymerization of styrene under living conditions was considered difficult [16], if not impossible. The absence of a strong electron donating substituent on the styrene molecule makes the styryl carbenium ions particularly unstable and thus prone to undergo transfer and termination reactions. In moderately polar solvents, the cationic polymerization of styrene was reported to yield non-living systems and result in bimodal molecular weight distributions [17]. The latter feature was attributed to the coexistence of two growing species that slowly exchange and add monomer at different rates. Of the two populations of polymers chains, the high molecular weight originates from the propagation of dissociated free ions whereas that corresponding to polymers of low molecular weight arises from non-dissociated ion pairs [15]. Because free ions are much more reactive that undissociated species, they get more easily involved in side reactions than ions pairs do and are therefore responsible for the non-living character of styrene polymerization.

Higashimura et al. [11] showed that equilibrium between free ions and ion pairs could be drastically modified in favor of the latter species, provided salts with common ions, such as $nBu_4N^+Y^-$, are added to the reaction medium. These salts not only help to remove free ions, but they also favor the formation of inactive covalent halide species that can be reversibly ionized by the Lewis acid present in the reaction medium. As a result of the use of this salt, the rate of polymerization decreases and the livingness of the system consisting of PhEtCl, $SnCl_4$, $nBu_4N^+Y^-$ dramatically improves. Yet, the different transfer reactions inherent to the cationic polymerization of this monomer do not totally vanish, as pointed out by Matyjaszewski [15b]. The side reaction that predominates when Lewis acid and tetrabuty-lammonium salts are used in CH_2Cl_2 is the transfer to monomer and the concomitant formation of an end-standing unsaturation (**scheme 3**). The proton that is released upon transfer then reinitiates a new polymer chain. The other undesired reactions that generally accompany propagation, such as transfer to counter-anion or Friedel-Craft cyclization reactions were shown to be negligible in a first approximation.

With an initiator that exactly models the growing species, initiation and propagation occur at similar rates so that samples with low polydispersity indices and controlled size can be expected, provided the transfer process is mild enough. The molecular weight of the samples yielded by such systems can be written as :

$1/DPn = [I]_0/x.[M]_0 + k_{tr,M}/k_p$, where x is the conversion, $[I]_0$ the initial concentration of PhEtCl and $[M]_0$ the initial concentration of monomer.

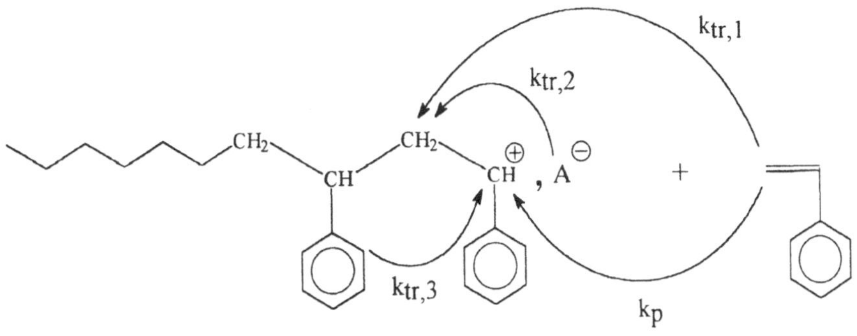

Scheme 3.

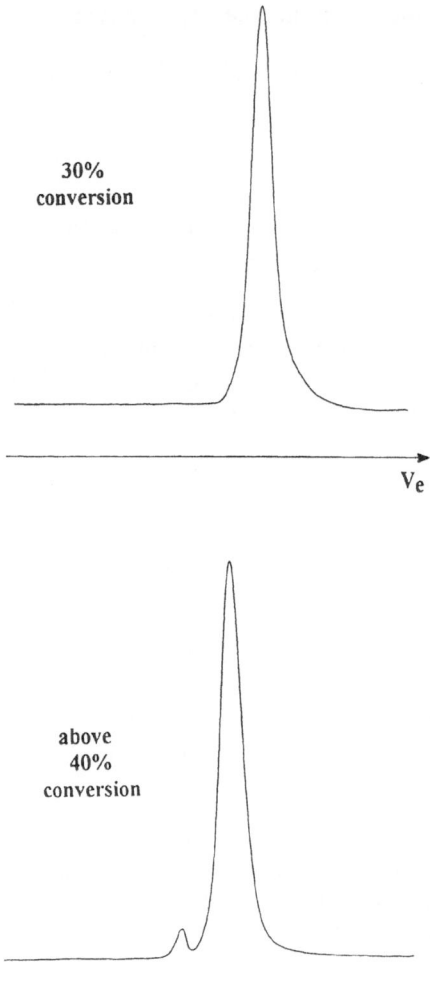

Figure 4. MWD's of the polystyrenes obtained with PhEtCl/SnCl₄/2,6Dt-BP as initiating system at 30% conversion and above 40% in CH₂Cl₂. Weight average molecular weights (Mw(LS)) are drawn from SEC using light scattering detection.

The data available in the literature [15a] for the above system at -20°C show that the ratio $k_{tr,M}/k_p$ is negligible (1-5%) compared to the first term ($[I]_0/x.[M]_0$), as long as low molecular weights are targeted (Mn<5000). Above that range, the consequences of the transfer reactions become detectable by any method of characterization ; the precise control of the molecular weight and the access to samples with narrow molecular weight distribution (MWD) would be then elusive.

With these features in mind, we decided to run our own polymerizations, so as to find out the conditions the best suited to our ultimate objective, that is the synthesis of PS stars of controlled size and functionality. Three slightly different systems have been investigated in this preliminary part devoted to the the cationic polymerization of styrene. In each case, PhEtCl, SnCl₄ and CH₂Cl₂ were respectively chosen as initiator, activator and solvent and 12000 g/mole was the Mn targeted. The variation of Mn and of the MWD has been followed as a function of the conversion of monomer by sampling out aliquots from the reaction medium.

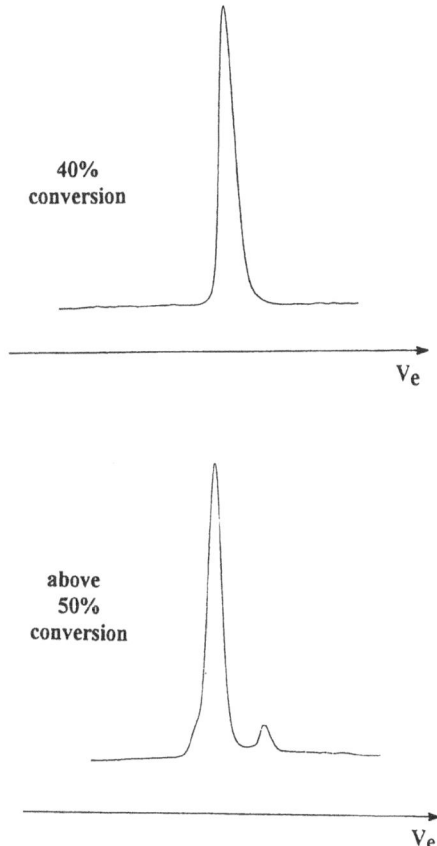

40%
conversion

V_e

above
50%
conversion

Figure 5. MWD's of the polystyrenes obtained with PhEtCl/SnCl$_4$/nBu$_4$N$^+$Cl$^-$ as initiating system at 40% conversion and above 50% in CH$_2$Cl$_2$. Weight average molecular weights (Mw(LS)) are drawn from SEC using light scattering detection.

V_e

In a first experiment -a-, styrene has been polymerized in the presence of 2,6 di-t-butylpyridine, a compound generally used as proton scavenger. When carrying out this polymerization, we had in mind the possibility for this nucleophile to make a reversible complex with the growing carbeniums and form a less reactive or even inactive species such as an onium. In that event, one can expect a lower rate of polymerization and a chain growth easier to control. As long as the monomer conversion stays below 30%, the chains are found to grow linearly with the polymer yield and narrow MWD's (**fig.4**) are obtained. Soon after that critical conversion, a bimodal distribution is seen in SEC traces (**fig.4**), with a small separate peak showing up in the high molecular weight region. The presence of this fraction of high molecular weight chains indicates that a small proportion of free ions still persist in spite of the addition of the above nucleophile (2,6 Dt-BP) to the reaction medium.

In a second experiment -b-, styrene has been polymerized under the same conditions as those put forward by Higashimura [11]. Mn increases proportionally to monomer consumption and slightly deviates from the line corresponding to a true living process, approximately after 50% of conversion is reached. The SEC traces of the samples obtained at 60% conversion display a double peaked MWD. The main peak ends in a shoulder that denotes the presence of a population of high molecular weight species. Besides this major peak, another one of smaller intensity shows up in the low molecular weight region. The formation of polymer chains with contrasted molecular weights next to the main population of macromolecules is indicative of the existence of transfer reactions to monomer (**fig.5**). The latter process occurs by the β-deprotonation of growing carbenium ions (**Scheme 3**) and

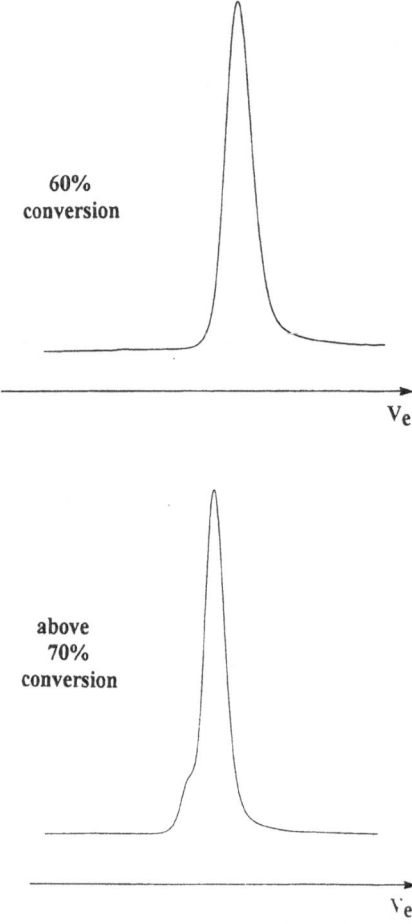

60%
conversion

V_e

above
70%
conversion

V_e

Figure 6. MWD's of the polystyrenes obtained with PhEtCl/SnCl$_4$/nBu$_4$N$^+$Cl$^-$/2,6Dt-BP as initiating system at 60% conversion and above 70% in CH$_2$Cl$_2$. Weight average molecular weights (Mw(LS)) are drawn from SEC using light scattering detection.

concomitant formation of a polymerizable terminal unsaturation. The formation of chains of small size can be attributed to adventitious initiation by the protons released upon β–elimination ,while the presence of high molecular weight species likely results from the attack of β-substituted ω-styryl macromonomer by growing carbeniums.

In a third experiment -c-, a proton trap, -di-t-butylpyridine (2,6 Dt-BP), has been added to the above system of polymerization, that is composed of PhEtCl, SnCl$_4$ and in nBu$_4$N$^+$Cl$^-$. With the use of a proton scavenger, we expected to curb the initiation of new chains by the protons released upon transfer, as it was observed in the previous case. The SEC traces of the different aliquots exhibit narrow and symmetrical MWD's and indicate that Mn grows proportionally to conversion (**fig.6**). Yet, a small shoulder shows up above 70% conversion at lowest elution volume, but no peak due to undesired initiation is detected in the lower molecular weight domain. From these experimental results, it can be inferred that transfer to monomer is suppressed with the use Dt-BP, although the β-proton elimination of growing carbenium ions still occurs. The latter process results in a true termination and

Table 1. Targeted molecular weights ($Mn_{,expected}$) and those obtained from light scattering ($Mn_{,LS}$) and refractometric detections ($Mn_{,RI}$), with $\underline{5}$, $SnCl_4$, and nBu_4N^+,Cl^- as initiating system

$Mn_{,expected}$	$Mn_{,LS}$	$Mn_{,RI}$	$I(M_w/M_n)$
7000	6600	4200	1.06
11000	10400	8900	1.11
16000	14500	11000*	—
37000	35000	30300*	—

*Bimodal distribution

gives rise to unsaturated species that can further polymerize -hence the shoulder- It is likely that in the system $PhEtCl/SnCl_4/nBu_4N^+Cl^-/2,6Dt-BP$, the latter compound not only traps protons but also interacts with the growing carbeniums and decreases their propensity to eliminate β-protons.

This preliminary study of the cationic polymerization of styrene has helped us to define the conditions that afford samples of controlled size and narrow MWD. Polymers with DP~60 and fulfilling the above criteria can be prepared by either of the two last systems described above. The last system, called -c-, affords even upper limits of molecular weight control with DP of about 80. The synthesis of PS stars of precise functionality and well-defined dimensions should be readily accessible, provided the Mn of each their individual arm does not exceed 6000-8000 g/mole.

c. Synthesis of Hexaarmed Polystyrene Stars

The reaction sequence for the synthesis of these polystyrene stars is shown in **scheme 2**. Two systems have been investigated. The first system considered consisted of $C_6[(CH_{2})_2C_6H_4CH(CH_3)Cl]_6$, $\underline{5}$, $SnCl_4$, and $nBu_4N^+Cl^-$. The polymerization was carried out in CH_2Cl_2 at -15°C under similar conditions to those of system -b- which we previously investigated. Being aware that above a targeted DP of 60, the precise control of the sample size and of its MWD is difficult to achieve, we did not endeavour to prepare stars with arms of larger size. Four star samples whose targeted Mn's ranged from 7000 to 37000 -which corresponds to DP of 10 to 60 for each individual arm- have been prepared (**Table 1**).

Characterization by SEC equipped with a light scattering detector was meant to provide both MWD's and actual molecular weights. The values that are drawn from refractometric detection are necessarily misleading, as they are based on a calibration curve established from linear PS samples. The discrepancy which is found between the two sets of data reflects the fact that branched macromolecules exhibit a lower hydrodynamic volume than their linear homologues of same molecular weight. The **left part of fig.7** compiles the SEC traces of the various samples. Narrow MWD's and controlled molecular weights are obtained as long as Mn of the stars does not exceed 10000 g/mole. In contrast, star samples with higher targeted molecular weights -respectively 16000 and 37000- exhibit bimodal distributions and a shoulder in the high molecular weight region. This is obviously due to the already commented on β-elimination of protons which results in the formation of unsaturated species and the initiation of new polymer chains. It is worth mentioning that the effects due to the transfer to monomer are somewhat accentuated on this hexafunctional system as they are detected well before the growing chains reach the critical DP of 50, a value found for linear systems in similar conditions. The probability for two star molecules to get coupled via bimolecular reaction between the terminal

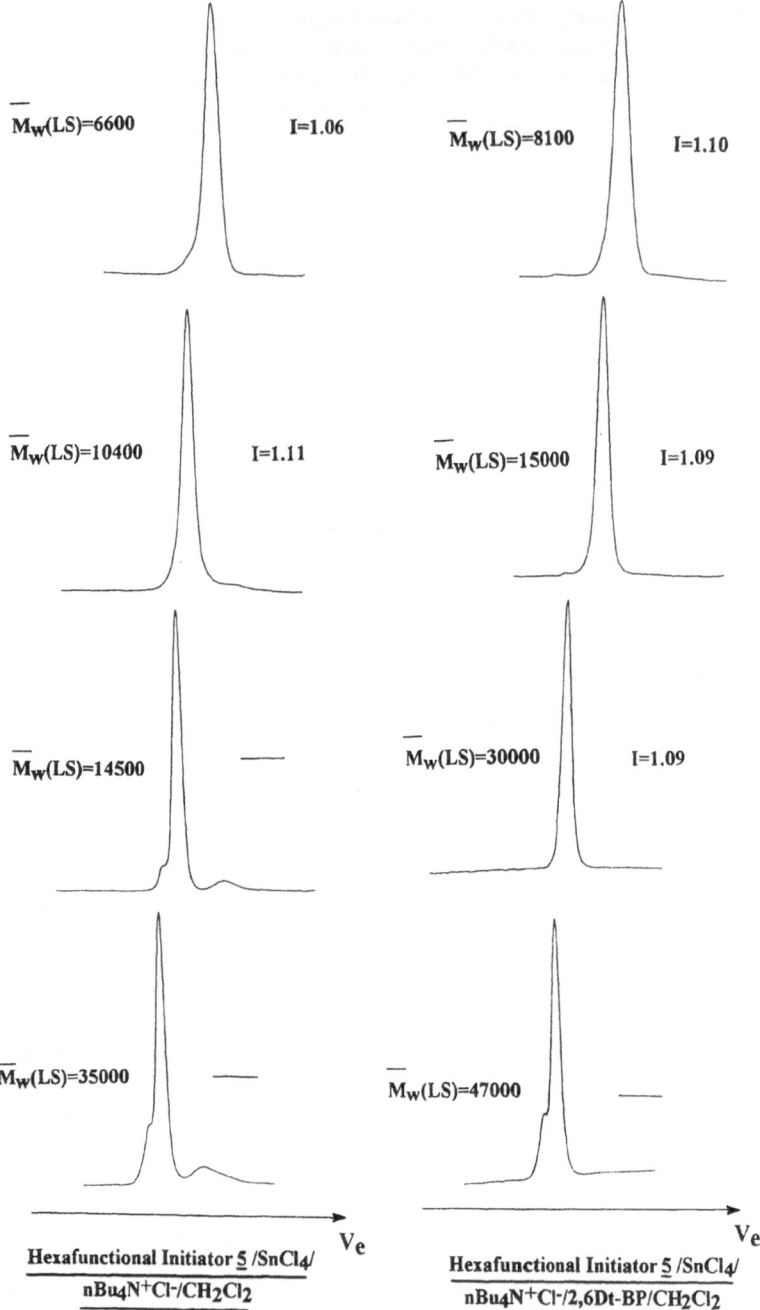

Figure 7. MWD's of the star-shaped polystyrene samples obtained with $\underline{5}$/SnCl$_4$/nBu$_4$N$^+$Cl$^-$ as initiating system (Left part). MWD's of the star-shaped polystyrene samples obtained with $\underline{5}$/SnCl$_4$/nBu$_4$N$^+$Cl$^-$/2,6Dt-BP as initiating system (Right part). Weight average molecular weights (Mw(LS)) are drawn from SEC using light scattering detection.

Table 2. Targeted molecular weights ($Mn_{,expected}$) and those obtained from light scattering ($Mn_{,LS}$) and refractometric ($Mn_{,RI}$) detections with $\underline{5}$, $SnCl_4$, nBu_4N^+,Cl^- and 2,6 Dt-BP as initiating system

$Mn_{,expected}$	$Mn_{,LS}$	$Mn_{,RI}$	I(Mw/Mn)
8000	8100	6700	1.10
15500	15000	10800	1.09
31000	30000	25000	1.09
43000	47000	37000*	—

*Shoulder in high molecular weight region

unsaturation and growing carbeniums is indeed considerably higher than in the homologous monofunctional case.

In a second series of experiments, we added 2,6 Dt-BP to the above system. The presence of this proton scavenger vastly improves the efficacy of the system composed of $\underline{5}$, $SnCl_4$ and $nBu_4N^+Cl^-$. Mn's up to 30000g/mole are obtained with a Poisson type distribution and no side peak is detected (**Fig.7,right part**). A small shoulder is seen in the SEC traces of the sample whose planned molecular weight is equal to 45000g/mole (**Table 2**). These results do confirm our previous observations on the ability of Dt-BP to curb β-proton elimination of growing carbeniums, a process known to trigger two side reactions (initiation of new chains and attack of macromonomer by carbenium).

Star samples that exhibit expected molecular weights and sharp distribution are now accessible up to 30000g/mole upon using 2,6 Dt-BP in conjunction with the system constituted of $\underline{5}$, $SnCl_4$ and $nBu_4N^+Cl^-$.

d. Determination of the Actual Functionality of Polystyrene Stars

To obtain further insight into the structure of our PS stars and in particular determine their actual functionality, another series of experiments has been envisaged. In these polymerizations, the initiating system not only comprises the hexafunctional initiator , $\underline{5}$, $SnCl_4$, $nBu_4N^+Cl^-$, 2,6 Dt-BP, but it also includes a molar amount of PhEtCl, the monofunctional equivalent of $\underline{5}$. Upon addition of styrene, the polymerization that is bound to occur should yield two different populations of macromolecular species : star-shaped polymers arising from the hexafunctional initiator and linear chains because of the presence of PhEtCl. Provided initiation and chain growth occur similarly whether the carbenium is carried by mono- or hexafunctional macromolecular species, the molecular weight of linear chains should be identical to that of the individual arm constituting the stars. Comparison of the average molecular weight of the stars with that exhibited by the linear chains present in the medium and determination of their ratio should then give access to the actual functionality of the branched macromolecules.

Three experiments with different monomer to initiator (both hexa- and monofunctional) ratios have been carried out. Characterization of the samples by SEC/MALLS shows two separate peaks corresponding to the two species present in the reaction medium. The fact that these peaks are narrow and symmetrical regardless of the experiment considered confirm that polymerization had occurred under controlled conditions in each case (**Fig.8**). The actual molecular weights of the respective populations of macromolecules are drawn from light scattering detection. The values obtained for the ratios of the two series of molecular weights demonstrate that $\underline{5}$ has effectively functioned as a true

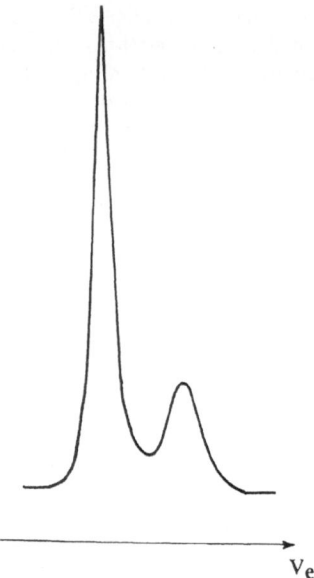

V_e

Figure 8. SEC traces of the sample obtained upon using $\underline{5}$/PhEtCl/SnCl$_4$/nBu$_4$N$^+$Cl$^-$/2,6Dt-BP as initiating system.

hexafunctional initiator and that polymerization has proceeded without detected side reactions. The actual functionality f of the stars falls indeed in close agreement with the expected value of 6 (**Table 3**).

IV. CONCLUDING REMARKS

Although very attractive, the core-first method has never been utilized to prepare star-shaped polystyrenes that exhibit a precise functionality. The lack of reagents fitted with a precise number of functions that would be capable of initiating the polymerization of styrene is the main reason for this situation. We have taken advantage of the possibilities offered by Fe(η-C$_5$H$_5$)$^+$ induced perfunctionalization of polymethylaromatics to develop an entirely original hexafunctional initiator, C$_6$[(CH$_2$)$_2$ p-C$_6$H$_4$H(Cl)Me]$_6$, $\underline{5}$.

When carried out in the presence of $\underline{5}$, 1 equiv. of 2,6 Dt-BP, 5 equiv. of SnCl$_4$, and 2 equiv. of nBu$_4$N$^+$Cl$^-$, the cationic polymerization of styrene affords star-shaped polystyrenes of controlled size and Poisson-type distribution provided the

Table 3. Actual molecular weights of PS stars (Mw$_{,star}$) and of linear PS chains (Mw$_{,linear}$) present in the reaction medium as obtained from light scattering detection. f is obtained upon taking the ratio of Mw$_{,star}$ to Mw$_{,linear}$

Mn$_{,star}$	Mw$_{,linear}$	f
25000	4500	5.6
19000	3300	5.9
15700	2700	5.8

targeted molecular weight does not exceed 30000g/mole. Above that value, the elimination of β-protons from growing carbeniums and its effects on the MWD of the stars become detectable.

ACKNOWLEDGMENTS

The authors are indebted to Alain Rameau from Institut C. Sadron (Strasbourg) for carrying out SEC/MALLS characterizations.

REFERENCES

1. (a) Hadjichristidis, N. ; Guyot, A. ; Fetters, L.J.; *Macromolecules* **1978,** 11, 668 ; **1980,** 13, 191. (b) Alward, D.B. ; Kinning, D.J. ; Thomas, E.L. ; Fetters, L.J. ; *Macromolecules* **1986,** 19, 215.
2. Worsfold, D.J. ; Zilliox, J.G. ; Rempp, P.; *Can. J. Chem.* **1969,** 47, 3379.
3. Gnanou, Y. ; Lutz, P. ; Rempp, P.; *Makromol.Chem.* **1988,** 189, 2885.
4. (a) Eschwey, H. ; Burchard, W. ; *Polymer* **1975,** 16, 180. (b) Lutz, P. ; Rempp, P. ; *Makromol. Chem.* **1988,** 189, 1051.
5. Tsitsilianis, C. ; Lutz, P. ; Graff, S. ; Lamps, J.P. ; Rempp, P. ; *Macromolecules* **1991,** 24, 5897.
6. Mishra, M. ; Wang, B. ; Kennedy, J. ; *Polym.Bull.*. **1987,** 17, 307
7. Huang, K. ; Zsuga, M. ; Kennedy, J. ; *Polym.Bull.*. **1988,** 19, 43.
8. Shohi, H. ; Sawamoto, M. ; Higashimura, I. ; *Makromol.Chem.* **1992,** 13, 2027.
9. Schappacher, M. ; Deffieux, A. ; *Macromolecules* **1992,** 25, 6744
10. Chang, J. ; Ji, Hi ; Han, H. ; Rhee, S. ; Cheong, S. ; Yoon, M. ; *Macromolecules* **1994,** 27, 1376.
11. Higashimura, T. ; Ishihama, Y. ; Sawamoto, M. ; *Macromolecules* **1993,** 26, 744.
12. (a) Astruc, D. ; *Acc.Chem.Res.* **1986,** 19, 377. (b) *Pure Appl. Chem.* **1990,** 62 1165. (c) *Top. Cur. Chem.*, **1991,** 160, 47. (c) *Tetrahedron Report N°157, Tetrahedron,* **1983,**39,4027
13. Cloutet, E.; Fillaut, J.L.; Gnanou, Y.; Astruc, D.; *J. Chem. Soc. Chem. Commun.* **1994,** 2433
14. Hamon, J.R. ; Saillard, J.Y. ; Le Beuze, A. ; Mc Glinchey, M.J. ; Astruc, D. ; *J. Am. Chem. Soc.* **1982,** 104, 7549 ; (b) Fillaut, J.L. ; Astruc, D. ; *J. Chem. Soc. Chem Commun.* **1993,** 1320
15. (a) Matyajaszewski, K. ; Lin, C. ; Pugh, C. ; *Macromolecules* **1993,** 26, 2649. (b) Lin, C. ; Xiang, J. ; Matyjaszewski, K. ; *Macromolecules* **1993,** 26, 2785.
16. Pepper, D.C ; *J. Polym. Sci., Polym. Symp.* **1975,** 50, 51
17. Higashimura, T. ; Kishiro, O. ; *J. Polym. Sci., Polym. Chem. Ed.* **1974,** 12, 967

5

PHOTOINITIATION OF IONIC POLYMERIZATIONS

Wolfram Schnabel

Hahn-Meitner-Institut Berlin GmbH
Glienicker Str.100
D-14109 Berlin
Germany

I. INTRODUCTION

At present, there is growing interest in industrial applications of photopolymerizations which refers mainly to surface coating processes proceeding via free radical mechanisms [1]. However, acrylate- and methacrylate-based formulations which are largely employed for this purpose exhibit disadvantages due to shrinkage and incomplete monomer conversion. It is hoped to overcome these problems with the aid of formulations containing substances of different chemical nature such as compounds containing epoxide groups. However, the latter can be polymerized only cationically and, therefore, the photoinitiation of cationic polymerizations has recently been the subject of intense investigations at various places in the world [2,3].

Regarding the photo-induced initiation of cationic polymerizations several modes of action concerning the formation of cationic species capable of reacting with monomers can be distinguished, (a) generation by direct photolysis of the initiator, (b) generation by sensitized photolysis of the initiator and (c) oxidation of free radicals generated by photolysis or photoreactions of the initiator. Typical examples of the three modes of action will be presented in this chapter. Emphasis will be given to work performed in the author's laboratory. At the end of the chapter also the photoinitiation of anionic polymerizations which has been studied only scarcely will be dealt with briefly.

II. CATIONIC POLYMERIZATION

II.1. Generation of Cations by Direct Photolysis of Initiators

Table 1 shows a list of families of compounds which have been employed as photoinitiators operating in the UV wavelength range for the polymerization of vinyl ethers and oxiranes (see Table 2). Notably, the monomers listed in Table 2 cannot be polymerized by a free radical mechanism. Initiators which have been studied during recent years are

Macromolecular Engineering, Edited by M.K. Mishra et al.
Plenum Press, New York, 1995

67

Table 1. Photoinitiators for cationic polymerizations

Denotation	Name	Formula *
I	iodonium salts	$R_2I^+\ X^-$
II	sulfonium salts	$R_3S^+\ X^-$
III	selenium salts	$R_3Se^+\ X^-$
IV	pyrilium salts	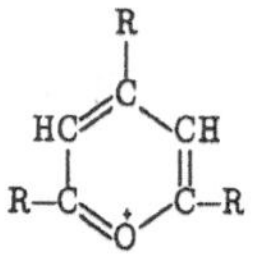
V	N-alkoxy pyridinium salts	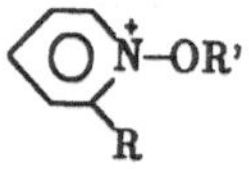
VI	N-alkoxy isoquinolinium salts	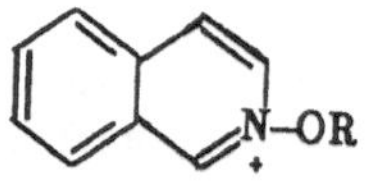
VII	phosphonium salts	$R_4P^+\ X^-$
VIII	arsonium salts	$R_4As^+\ X^-$
IX	iron cyclopentadiene arene complexes	$FeCp(\eta^6\text{-arene})^+\ X^-$

* X^-: BF_4^-, PF_6^-, SbF_6^-, AsF_6^-

compiled in Tables 3 to 7. Generally, radical cations $X^{+\bullet}$ are formed in the primary act of photolysis. The radical cations might be capable of initiating the polymerization:

$$X^{+\bullet} + M \longrightarrow \bullet X\text{-}M^+ \tag{1}$$

$$\bullet X\text{-}M^+ + nM \longrightarrow \bullet X\text{-}M_{n+1}^+ \tag{2}$$

Alternatively, radical cations can react with monomer or solvent molecules thus forming protons which can add to monomers and in this way form initiating cationic species:

$$X^{+\bullet} + RH \longrightarrow X + R\bullet + H^+ \tag{3}$$

Table 2. Monomers polymerizable by a cationic mechanism

Denotation	Name	Formula
X	cyclohexene oxide	
XI	3,4-epoxycyclohexenyl(methyl)- 3,4′-epoxycyclohexane carboxylate	
XII	4-vinyl cyclohexene dioxide	
XIII	1,4-butanediol diglycidyl ether	
XIV	glycidylphenyl ether	
XV *	epoxy silicone monomers	
XVI	n-butyl vinyl ether	$n\text{-}C_4H_9\text{-}O\text{-}CH=CH_2$

* for other silicon-containing epoxides see ref. [4]

$$H^+ + M \longrightarrow H\text{-}M^+ \qquad (4)$$

$$H\text{-}M^+ + nM \longrightarrow H\text{-}M_{n+1}^+ \qquad (5)$$

Typical reaction mechanisms referring to the photolysis of sulfonium salt IId and isoquinolinium salt VIa are presented in Schemes 1 and 2. Notably, in some cases evidence has been obtained for photolytical bond scission occurring both homolytically and heterolytically. Upon irradiation of sulfonium salt IId in acetonitrile solution both modes of bond scission were found to contribute almost equally to bond cleavage (see also Scheme 1) [21].

II.2. Generation of Cations by Direct Photolysis of Initiator Complexes

Certain ground state complexes of the electron acceptor-donor type form radical cations upon irradiation with UV light. For example, according to Hayashi et al. [22], radical cations capable of initiating the polymerization of X are formed upon the irradiation of

Table 3. Iodonium salt initiators

Denotation	Formula		Ref.
Ia		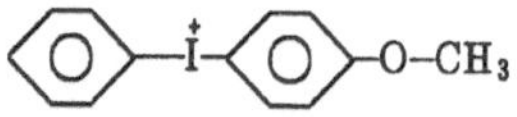 AsF_6^-	[5]
Ib		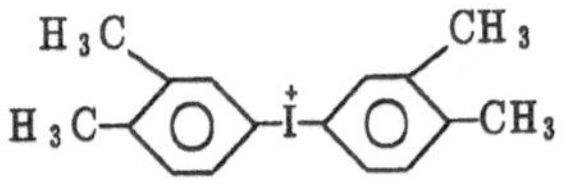 AsF_6^-; PF_6^-	[5]
Ic		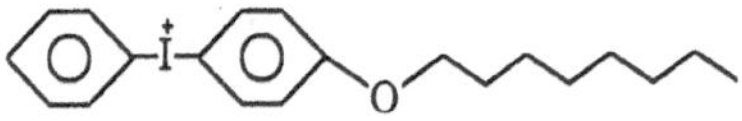 SbF_6^-	[3]
Id		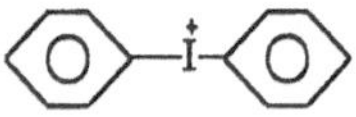 AsF_6^-; PF_6^-	[27, 34]

Table 4. Sulfonium salt initiators

Denotation	Formula		Ref.
IIa		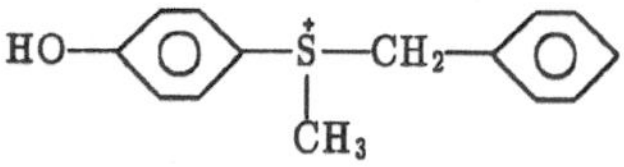 SbF_6^-	[6, 8, 9]
IIb			[7 - 9]
IIc			[39]
IId			[21]

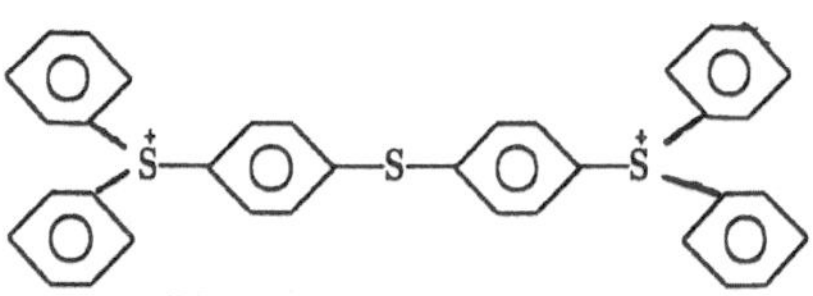

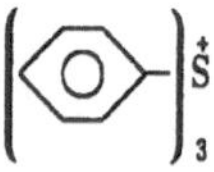

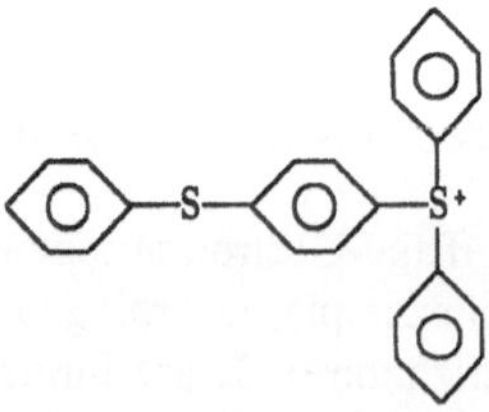

Table 5. Pyridinium and isoquinolinium salt initiators

Denotation	Formula	Ref.
Va	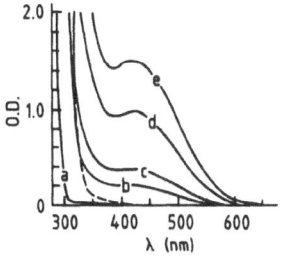 ⟨⟩Ṅ—OC$_2$H$_5$ PF$_6^-$ CH$_3$	[10 - 13]
Vb	NC—⟨⟩Ṅ—OC$_2$H$_5$ PF$_6^-$	[10 - 13]
Vc	⟨⟩—⟨⟩Ṅ—OC$_2$H$_5$ PF$_6^-$	[10 - 13]
Vd	—CH——CH$_2$——CH——CH$_2$— HC⫪CH HC⫪N—OC$_2$H$_5$ PF$_6^-$ HC⫪CH HC⫪CH H H	[10 - 13]
VIa	N—OC$_2$H$_5$ PF$_6^-$	[14]

complexes of 9,10-dicyanoanthracene and aromatic hydrocarbons (pyrene, phenanthrene, trans-stilbene, naphthalene and biphenyl) with light of $\lambda > 390$ nm. Also certain pyridinium ions are capable of forming ground state complexes with electron-rich donors such as methyl- and methoxy-substituted benzenes and naphthalenes [23]. Recently, CT complexes formed by mixing Va or VIa with hexamethylbenzene or 1,2,4-trimethoxybenzene were found to act as photoinitiators for the cationic polymerization of X and XII [24]. The complexes possess characteristic absorption bands at relatively long wavelengths where the components are transparent. This can be seen from Fig.1 which shows typical absorption spectra. Notably, these complexes are applicable for the photoinitiation of epoxide monomers but not for the photoinitiation of vinyl ethers and N-vinylcarbazol. The latter monomers are already polymerized in a dark reaction upon the addition of these complexes. Typical results demonstrating the photoinitiation of monomer X by complexes of Va and hexamethylbenzene are shown in Fig.2. Similarly, the bifunctional monomer XII, which like X, does

Figure 1. Optical absorption spectra of acetonitrile solutions containing VIIb (5×10^{-3} mol l^{-1}) and 1,2,4-trimethoxybenzene at different concentrations (denoted in mol l^{-1}): a (0), b (0.4), c (0.8), d (2.4) and e (4.0). Dotted curve: 1,2,4-trimethoxybenzene (0.5 mol l^{-1}) in the absence of VIIb.

Table 6. Phosphonium and arsonium salt initiators

Denotation	Formula *	Ref.
VIIa	$(Ph)_3\overset{+}{P}-CH_2-\overset{O}{\underset{\|}{C}}-CH_2-\overset{+}{P}(Ph)_3$ $\qquad$ 2 X⁻	[15 - 17]
VIIb	$\overset{O}{\underset{\|}{C}}-CH_2-\overset{+}{P}(Ph)_3$ $\quad$ X⁻	[15 - 17]
VIIc	$CH_3-\overset{O}{\underset{\|}{C}}-CH_2-\overset{+}{P}(Ph)_3$ $\quad$ X⁻	[15 - 17]
VIId	$R-\langle O \rangle-CH_2-\overset{+}{P}(Ph)_3$ $\quad$ X⁻ R: OCH₃, CH₃, NO₂, Cl, H	[18]
VIIe	$-CH_2-\overset{+}{P}(Ph)_3$ $\quad$ X⁻	[18]
VIIIa	$\overset{O}{\underset{\|}{C}}-CH_2-\overset{+}{As}(Ph)_3$ $\quad$ X⁻	[15 - 17]

* X⁻: PF₆⁻, SbF₆⁻, AsF₆⁻

not undergo a dark reaction with these CT complexes, readily forms an insoluble gel. A post-polymerization was not observed in these cases. Moreover, 2,6-di-tert-butylpyridine, which acts as a proton scavenger, did not significantly influence the polymerization. These findings suggest that radical cations TMB⁺ᵘ formed according to the mechanism depicted in Scheme 3 initiate the polymerization as shown by reaction (6):

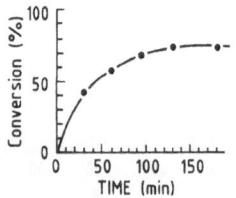

Figure 2. Photopolymerization of X in CH₂Cl₂ solution under argon at $\lambda_{inc} >$ 345 nm. Plot of monomer conversion vs. time of irradiation. [Va] = 5x10⁻³ mol l⁻¹, [hexamethylbenzene] = 0.5 mol l⁻¹.

Table 7. Iron cyclopentadienyl arene complex salt initiators

Denotation	Formula *	Ref.
IXa		[19 - 20]
IXb		[19 - 20]
IXc		[19- 20]

* X^-: PF_6^- or SbF_6^-

$$TMB^{+\cdot} + M \xrightarrow{} TMB\text{-}M^{\cdot\cdot} \xrightarrow{nM} TMB\text{-}M\text{-}(M)_{n-1}M^{\cdot\cdot} \qquad (6)$$

II.3. Sensitized Generation of Cations

The term "sensitized photopolymerization" generally refers to initiator systems consisting of two compounds that do not form ground state complexes. One of the compounds is transparent to the incident light. Commonly, sensitized photopolymerization is applied if light of relatively long wavelengths, i.e. near UV or visible light, is to be used. In the following sections three different cases will be discussed. Emphasis will be given to systems containing N-ethoxy pyridinium or isoquinolinium salts.

II.3.1. Sensitization by Anthracene, Perylene and Phenothiazine. As can be seen from Fig.3 anthracene, perylene and phenothiazine strongly absorb light at wavelengths up

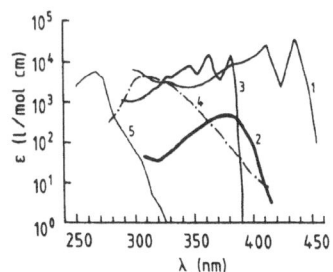

Figure 3. Optical absorption spectra of acetonitrile solutions containing Vb and XIV at different concentrations (denoted in mol l^{-1}): (a) 0, (b) 0.4, (c) 0.8, (d) 2.4 and (e) 4.0. Dotted curve: XIV (0.5 mol l^{-1}) in the absence of Vb.

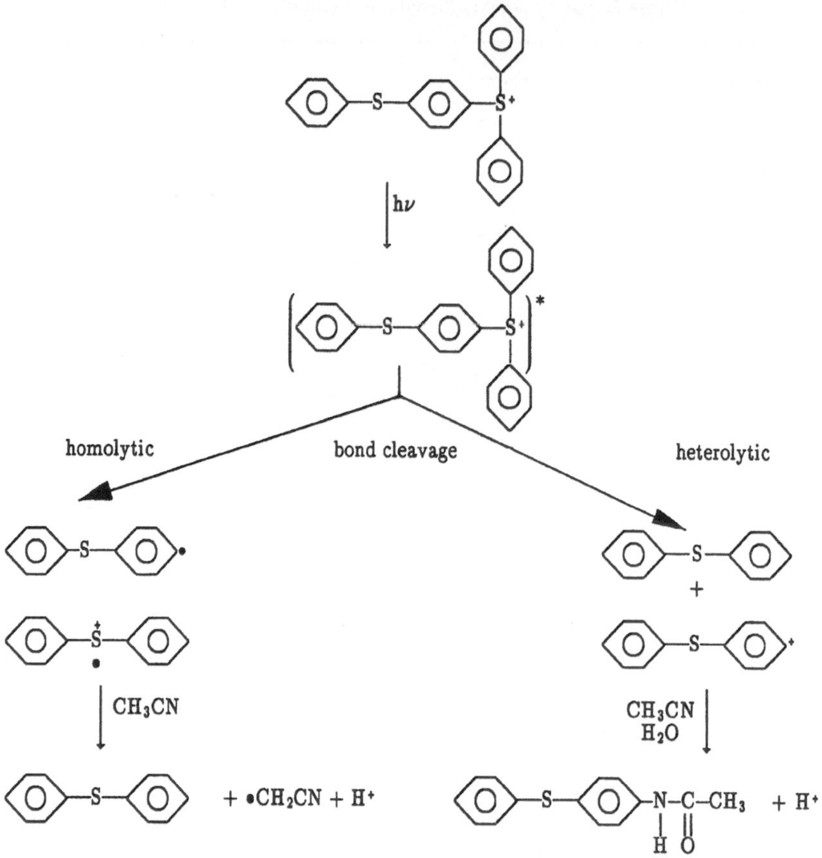

Scheme 1

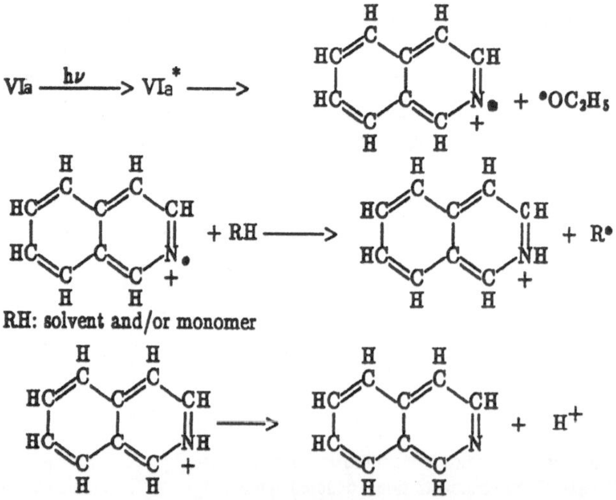

RH: solvent and/or monomer

Scheme 2

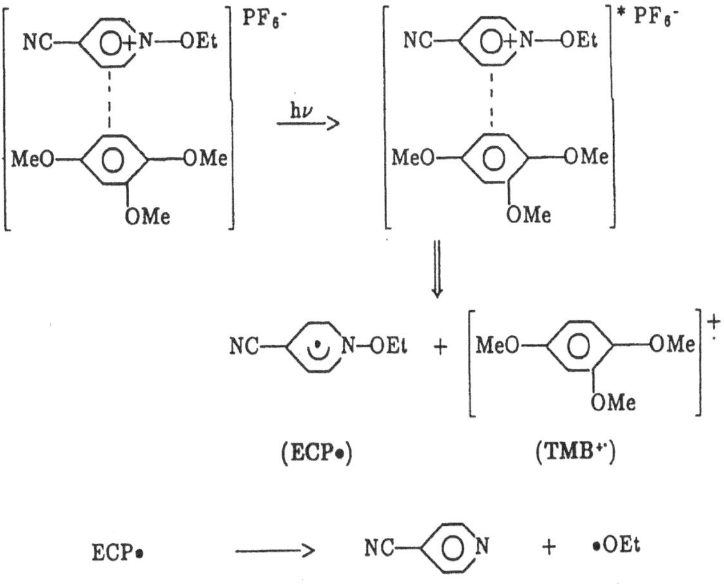

Scheme 3

to 450 nm. In conjunction with pyridinium salt Va the three compounds are capable of initiating the polymerization of cyclohexene oxide (X) as is demonstrated by typical results presented in Table 8 [24]. It seems that electronically excited states of the three photosensitizers (PS*) undergo electron transfer reactions with the onium ions of Va:

$$PS^* + On^+ \longrightarrow PS^{+\bullet} + On^\bullet \tag{7}$$

Table 8. Sensitized photopolymerization of X (5.8 mol l^{-1}) in CH$_2$Cl$_2$ solution containing Va (6.8x10^{-3} mol l^{-1}) at $\lambda_{inc} > 350$ nm [24]

Sensitizer		Conversion [a]	M$_w$ [b]
Denotation	Concentration (mol l^{-1})	(% w/w)	
Anthracene	5x10^{-3}	61.5	1.3x10^5
Perylene	5x10^{-3}	59.1	1.4x10^5
Phenothiazine	5x10^{-3}	29.1	1.2x10^5

[a] conversion after 30 min irradiation

[b] weight average molar mass determined by the light scattering method

Table 9. Triplet and singlet energies E(PS*) and half wave oxidation potentials E^{ox} (vs.SCE) of sensitizer

Sensitizer	E(PS*)	$E_{1/2}^{ox}$	ΔG [a]
	$(kJ\ mol^{-1})$	(V)	$(kJ\ mol^{-1})$
Benzophenone	290 (E_t)	2.7 [b]	+ 39.8
Acetophenone	308 (E_t)	2.9 [b]	+ 41.2
Thioxanthone	277 (E_t)	1.7 [c]	− 44.2
Anthracene	319 (E_s)	1.1 [d]	− 144.4
Perylene	277 (E_s)	0.9 [d]	− 121.8
Phenothiazine	239 (E_t)	0.6 [d]	− 112.9

[a] calculated with the aid of eq.(3) with $E_{1/2}^{red}$ = -0.7 V for cations of Va

[b] Miller, L.L., Nordbohm, D.G. and Mayeda, E.A., 1972, J.Org.Chem. 37, 916

[c] Kissinger, P.T., Holt, P.T. and Reilley, C.N., 1971, J.Electroanal.Chem. 33, 1

[d] Siegermann, H., 1975, in „Techniques of Electroorganic Synthesis", Weinberg.N.L., (ed.) Part II, Vol.V of „Techniques of Chemistry", Weissberger, A. (ed.), Wiley, New York

Actually, electron transfer from excited singlet or triplet states of the sensitizers to onium ions of Va is feasible because of the negative ΔG values (see Table 9) which are calculated on the basis of the Rehm-Weller equation:

$$\Delta G = f_c (E_{1/2}^{ox} - E_{1/2}^{red}) - E(PS*) \tag{8}$$

$E_{1/2}^{ox}$ and $E_{1/2}^{red}$: halfwave oxidation and reduction potentials of sensitizer and cation of Va, respectively, in V units; E(PS*): excitation energy of the sensitizer in $kJ\ mol^{-1}$ units; f_c: conversion factor $(97\ kJ\ mol^{-1}\ V^{-1})$. Flash photolysis studies [25] revealed that in the cases of anthracene and perylene excited singlet states, and in the case of phenothiazine excited triplet states effectively react with cations of Va. The absorption spectra of the relevant radical cations were observed in the three cases. According to the reaction mechanism depicted in Scheme 4 the reaction of the excited sensitizer with the pyridinium ions results in the formation of radical cations of the sensitizer and pyridinium radicals. The latter rapidly decompose into α-picoline and ethoxyl radicals. Actually, various routes can be discussed regarding the initiation step of the cationic polymerization: (a) the radical cations can directly react with the monomer:

$$PS^{+\bullet} + M \xrightarrow{\quad\quad} {}^{\bullet}PS\text{-}M^+ \xrightarrow{\ nM\ } {}^{\bullet}PS\text{-}M\text{-}(M)_{n-1}M^+ \tag{9}$$

(b) the radical cations can abstract hydrogens from monomer or solvent and disso-ciation of the resulting intermediate cation yields protons which can add to the monomer:

$$PS^{+\bullet} + RH \xrightarrow{\quad\quad} HPS^+ + R^{\bullet} \tag{10}$$

$$HPS^+ \xrightarrow{\quad\quad} PS + H^+ \tag{11}$$

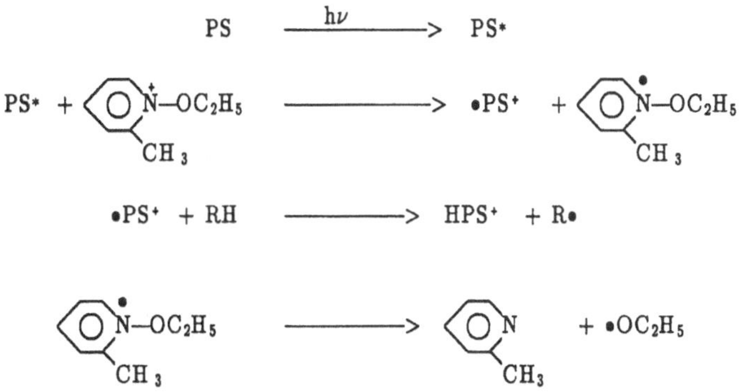

$$PS \xrightarrow{h\nu} PS\bullet$$

$$\bullet PS^+ + RH \longrightarrow HPS^+ + R\bullet$$

Scheme 4

$$H^+ + M \longrightarrow H\text{-}M^+ \tag{12}$$

(c) the radical cations can react with ethoxyl radicals thus forming cations which can add to monomer molecules:

$$PS^{+\bullet} + {}^\bullet OEt \longrightarrow EtO\text{-}PS^+ \tag{13}$$

$$EtO\text{-}PS^+ + M \longrightarrow EtO\text{-}PS\text{-}M^+ \tag{14}$$

Support for mechanism (c) was recently obtained [26] by reacting excited anthryl groups being the terminal groups of poly(tetrahydrofuran) with cations of Va in the presence of cyclohexene oxide (X). Size exclusion chromatograms recorded with the resulting product revealed the formation of a block copolymer which did not anymore possess the absorption bands characteristic of anthryl groups thus indicating the existence of the following structure:

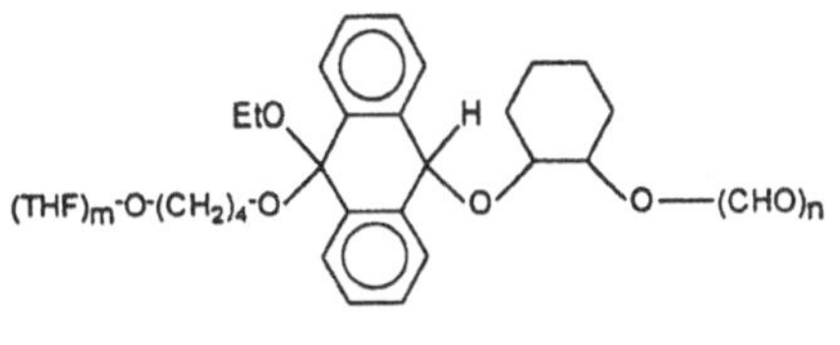

In recent studies concerning the photosensitization of the cationic polymerization of vinyl ethers by initiator systems consisting of anthracene and its derivatives and bis(4-do-decylphenyl)iodonium hexfluoroantimonate it was found that the reaction of the radical cation with monomer causes anthracene to lose its aromaticity at the 9 and 10 positions [27].

II.3.2. Sensitization by Thioxanthone. Upon irradiation with UV light thioxanthone (TX) and various of its derivatives form long-lived triplet states which can undergo hydrogen abstraction reactions with monomer molecules according to reaction (15):

$$\overset{|}{\underset{|}{C}}=O + \overset{|}{\underset{|}{H\text{-}C}} \xrightarrow{k_q} {}^\bullet\overset{|}{\underset{|}{C}}\text{-}OH + {}^\bullet\overset{|}{\underset{|}{C}}\text{-} \tag{15}$$

Thioxanthone triplets are only moderately reactive toward cyclohexene oxide (X). The rate constant $k_q({}^3TX^* + X) = 3.4\times10^4 \, l \, mol^{-1} \, s^{-1}$ [26] is not very high. Triplets of 2-chlorothiox-

$$\text{Scheme 5}$$

anthone (CTX) are more reactive: $k_q(^3TX^* + X) = 3\times10^6\,l\,mol^{-1}\,s^{-1}$. Onium ions react with TX triplets with the following rate constants: $k(^3TX^* + Va) = 4\times10^7\,l\,mol^{-1}\,s^{-1}$ [26] and $k_q(^3CTX^* + Id) = 2.4\times10^8\,l\,mol^{-1}\,s^{-1}$ [28]. In the latter case a transient absorption spectrum with a maximum around 430 nm was found which was assigned to the radical cation of CTX and this was taken as evidence for the occurrence of electron transfer in the reaction of onium ions with thioxanthones. Regarding the competition of the reactions of thioxanthone triplets with pyridinium ions of Va and with X it turns out that, inspite of the low k_q value, reaction (15) predominates if neat cyclohexene oxide containing TX and Va is irradiated. On the basis of the rate constants given above the probability for the occurrence of reaction (15) is 0.93 and 0.57 at $[Va] = 6.8\times10^{-4}$ and $6.8\times10^{-3}\,mol\,l^{-1}$, respectively. Fig.4 presents curves of the conversion of X into polymer vs. the time of irradiation obtained at these concentrations of Va. Presumably, under these conditions the initiation of the polymerization of X is due to protons rather than to radical cations of TX. The mechanism which is depicted in Scheme 5 suggests that ketyl radicals of TX which are formed by hydrogen abstraction reactions are oxidized by cations of Va and the resulting TXH^+ cations are converted into TX upon release of protons.

II.3.3. Sensitization by Dyes. The photoinitiation of polymerizations by visible light is of some practical interest regarding modern technologies based on the application of lasers emitting light at $\lambda > 400$ nm such as three-dimensional machining, photoimaging, holography or microlithography. In this respect a lot of work has been devoted to dye containing initiator systems that are applicable to free radical polymerizations [29-33]. Dye containing systems suitable for the photoiniation of cationic polymerizations have been studied already some time ago [34]. In that case the dye-sensitized photolysis of iodonium salts has been applied for the cationic polymerization of some oxiranes. Recent work performed in the

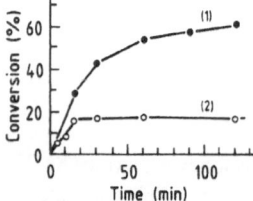

Figure 4. Photopolymerization of neat X at 30°C at $\lambda_{inc} > 345$ nm. Plot of monomer conversion vs. the time of irradiation. $[Va] = 6.8\times10^{-3}\,mol\,l^{-1}$ (1) and $6.8\times10^{-4}\,mol\,l^{-1}$ (2), $[TX] = 4.7\times10^{-4}\,mol\,l^{-1}$.

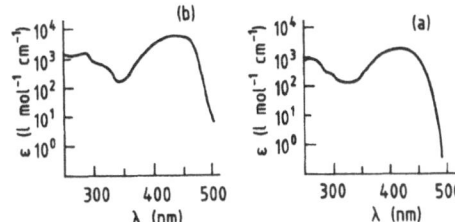

Figure 5. Optical absorption spectra of CHO solutions containing (a) coumarin 153 and (b) coumarin 7.

author's laboratory concerns the laser light-induced cationic polymerization of several oxiranes with the aid of coumarins [35]. Fig.5 shows the absorption spectra of coumarin 7 and coumarin 153, which have been examined to some extent.

Coumarin 7   **Coumarin 153**

Oxiranes such as X or XII, containing only coumarin 7 or 153, or only Id or VIa, did not polymerize upon irradiation at $\lambda_{inc} = 488$ nm. However, the oxiranes polymerized readily when both a coumarin and Id or VIa were present. Typical conversion-time plots referring to the polymerization of X and XII are shown in Fig.6. The polymerization is initiated essentially by protons that are formed with quantum yields $\Phi(H^+) = 1\text{-}2\times10^{-2}$. Notably, postpolymerization occurs, i.e. the polymerization continues after the end of the irradiation. Moreover, the polymerization of oxiranes can be initiated upon the addition of an irradiated coumarin solution.

II.4. Generation of Cations by Oxidation of Free Radicals

II.4.1. General Mechanism. The oxidation of carbon-centered free radicals by onium salts according to reaction (16):

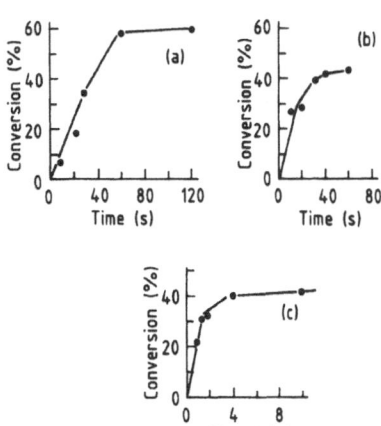

Figure 6. Laser light-induced bulk polymerization of CHO under argon ((a) and (b)) and of XII under air (c) at $\lambda_{inc} = 488$ nm and T = 23°C. Monomer conversion vs. the time of irradiation. Initiator system: (a) coumarin 153 (2.5x 10^{-3} mol l^{-1}) and VIa (0.11 mol l^{-1}), (b) coumarin 7 (2.5x10^{-3} mol l^{-1}) and VIa (0.11 mol l^{-1}), (c) coumarin 153 (2.5x 10^{-3} mol l^{-1}) and VIa (0.11 mol l^{-1}).

Table 10. The free energy ΔG of the reaction of 2-hydroxypropyl radicals with onium ions

Species	$E_{\frac{1}{2}}^{ox}$ (V)	$E_{\frac{1}{2}}^{red}$ (V)	ΔG [a] (V)
$(CH_3)_2COH$	-1.2		
cation of IIc		-1.1	-0.1
cation of Va		-0.7	-0.5
cation of VIa		-0.5	-0.7
cation of Vc		-0.5	-0.7
cation of Id		-0.2	-1.0

$$-\overset{|}{\underset{|}{C}}\cdot \ + \ On^+ \ \longrightarrow \ -\overset{|}{\underset{|}{C}}^+ \ + \ On\cdot \tag{16}$$

is an elegant method to generate reactive carbocations capable of initiating the cationic polymerization of appropriate monomers. The method which is frequently referred to as "free radical-promoted cationic polymerization" has been applied very successfully in conjunction with various free radical photoinitiators [36-38]. Provided the oxidation and reduction potentials of the free radical and the onium ion are known it can be estimated on the basis of the Rehm-Weller equation whether a certain radical can or cannot be oxidized. As a typical example, values of the free energy ΔG of the reaction of 2-hydroxypropyl radicals with various onium ions are compiled in Table 10. The Table shows that these radicals are prone to oxidation by the onium ions of Va and Id but they should be inert towards triphenyl sulfonium ions (IIc). In some cases radicals which are not readily oxidized can be converted into oxidizable radicals by reaction with vinyl compounds according to reaction (17):

$$R\cdot \ + \ CH_2{=}CH(R') \ \longrightarrow \ R\text{-}CH_2\text{-}CH(R') \tag{17}$$

It should be pointed out that radical promoted polymerizations based on photochemically generated free radicals have been studied for a long time in various laboratories. Therefore, in this article only two recent investigations will be dealt with briefly.

II.4.2. Oxidation Of Vinyl Radicals. Upon UV irradiation of substituted vinyl bromides (XVII) vinyl radicals are generated and bromine is released [39]:

$$R_2C{=}CR(Br) \ + \ h\nu \ \longrightarrow \ R_2C{=}CR\cdot \ + \ Br\cdot \tag{18}$$

Vinyl cations are formed when vinyl radicals undergo electron transfer reactions with appropriate onium ions, e.g. cations of Va or IIc:

$$R_2C{=}CR\cdot \ + \ On^+ \ \longrightarrow \ R_2C{=}CR^+ \ + \ On\cdot \tag{19}$$

Vinyl cations generated in this way can add to monomer molecules which results in the formation of an ionic species capable of initiating the propagation of the cationic polymerization of the monomer:

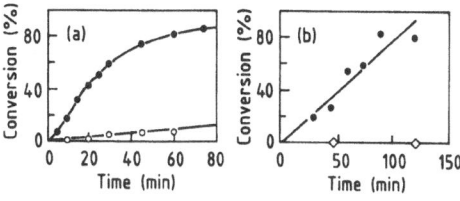

Figure 7. Photopolymerization of X (a), XI (b), XII (c) and IX (d) in CH$_2$Cl$_2$ solution at T = 35°C and λ_{inc} = 350 nm (nominal wavelength). I = 9.5x10^{-5} Einstein cm^{-2} s^{-1}. Plot of monomer conversion vs. time of irradiation. [1,2,2-triphenyl vinyl bromide] = 8.8x10^{-4} mol l^{-1}, [IIc] = 4 x10^{-3} mol l^{-1}. Open circles: monomer conversion in the absence of vinyl bromide.

$$R_2C=CR^+ \; + \; M \longrightarrow R_2C=CR(M)^+ \tag{20}$$

Alternatively, vinyl cations might be formed by the reaction of electronically excited vinyl bromide molecules with onium ions via electron transfer:

$$[R_2C=CR(Br)]^* \; + On^+ \longrightarrow [R_2C=CR(Br)]^{+•} \; + \; On• \tag{21}$$

$$[R_2C=CR(Br)]^{+•} \longrightarrow R_2C=CR^+ \; + \; Br• \tag{22}$$

Typical compounds examined in these studies are 1,2,2-triphenyl vinyl bromide (XVIIa) and 1,2,2-tri(4-methoxyphenyl) vinyl bromide (XVIIb) which absorb light appreciably up to wavelengths of about 360 -380 nm. Using XVIIa or XVIIb in conjunction with onium salt IIc the monomers X, XI, XII and XVI were photopolymerized to high yields. Typical results are presented in Fig.7.

II.4.3. Oxidation of A-*Amino Radicals.* In a system containing a xanthene dye, e.g. erythrosine, an aromatic amine, e.g. N,N-dimethylaniline, and an onium salt, e.g. diphenyliodonium hexafluoroantimonate, α-amino radicals are formed upon irradiation with visible light. According to the reaction mechanism depicted in Scheme 6 the α-amino radicals are oxidized by the onium ions and in this way converted into carbocations. The latter can initiate

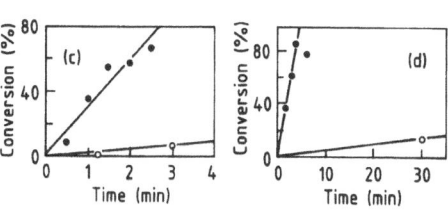

Scheme 6

the cationic polymerization of a suitable monomer such as X [40]. Since irradiations are performed with visible light, this method can be considered as another mode of dye-sensitized cationic photopolymerization. When X was polymerized in this way, dye molecules were incorporated into the polymer which was thought to be due to chain terminating processes.

III. ANIONIC POLYMERIZATIONS

Contrary to cationic polymerizations the photoinitiation of anionic polymerizations has been only occasionally attempted. Recently, the photopolymerization of α-cyano-ethyl acrylate initiated by isocyanate ions has been reported [41]. The isocyanate ions were generated by a light-induced ligand exchange reaction of Reineckate anions:

$$[Cr(NH_3)_2(NCS)_4]^- \, K^+ + h\nu + solv \longrightarrow$$

$$[Cr(NH_3)_2(NCS)_3solv] \; + \; NCS^- \; + \; K^+ \qquad (23)$$

In connection with anionic photopolymerization studies the anionic coordination polymerization of X catalyzed in a non-photochemical process by titanium tetraisopropoxide [Ti(O-i-prop)$_4$] in conjunction with a phenolic compound might also be mentioned [42]. In that work it was claimed that, while the ring-opening reaction is initiated by an isopropoxy group, the presence of a phenolic compound is necessary to control the electronic state of the titanium center for the coordination of the epoxide group. The initiation of the polymerization can be photochemically controlled if the phenolic compound is generated in the system by a photochemical reaction. In the reported case the photo-rearrangement of p-chlorophenyl o-nitrobenzyl ether led to the release of p-chlorophenol according to reaction (24):

$$(24)$$

REFERENCES

1. Pappas, SP (ed.), 1992, ,,Radiation Curing, Science and Technology", Plenum Press, New York.
2. Crivello, J.V., 1984, Adv.Polym. Sci.: 2, 3-8 .
3. Lohse, F. and Zweifel, H., 1986, Adv.Polym. Sci.: 78, 62-81.
4. Crivello,.J.V. and Lee, J.L., 1990, J.Polym.Sci., Part A : Polym.Chem.: 28, 479-503.
5. Fouassier, J.P., Burr, D., and Crivello, J.V., 1994, J.Macromol.Sci.Pure Appl.Chem. A 31, 677-701
6. Park, J., Kihara, N., Ikeda, T. and Endo, T. 1993, J.Polym.Sci., Part A : Polym.Chem.: 31, 1083-1085.
7. Decker, C. and Moussa, K., 1990, J.Polym.Sci., Part A : Polym.Chem.: 28, 3429-3443.
8. Manzhen, M.A., 1992, Intern.J.Polymeric Mater.: 18, 1-7.
9. Manzhen, M.A., 1992, Intern.J.Polymeric Mater.: 189-195.
10. Schnabel, W., Yagci, Y., Kornowski, A. and Massonne, K., 1992, EP 0498194 A1.
11. Yagci, Y., Kornowski, A. and Schnabel, W.,1992,. J.Polym.Sci., Part A: Polym.Chem.: 30, 1987-1991.
12. Yagci, Y., and Schnabel, W.,1994,. Macromol.Symp.: 85, 115-127.
13. Yagci, Y. and Schnabel, W., 1993, Macromol.Reports: A34 (Suppls.) 175-182
14. Karal, O., Önen, A. and Yagci, Y., 1994, Polymer: 35, 4694-4696
15. Neckers, D.C. and Abu-Abdoun, I.I., 1984, Macromolecules: 17, 2468-2473.
16. Abu-Abdoun,I.I. and Aale-Ali, 1992, Eur.Polym.J.: 28, 73-78.

17. Abu-Abdoun, I.I. and Aale-Ali, 1993, Eur.Polym.J.: 29, 1439-1443 and 1445-1450.

18. Takata, T., Takuma, K. and Endo, T., 1993, Makromol.Chem., Rapid Commun.: 14, 204-206.

19. Roloff, A., Meier, K., Riediker, M., 1986, Pure Appl. Chem.: 58, 1267

20. Meier, K. and Zweifel, H., 1986, I.Imaging Sci.: 30, 174-177

21. Wilpert, A., Doctoral Thesis, 1992, Alexander von Humboldt University, Berlin.

22. Ohtsuka, T., Yamamoto, Y. and Hayashi, K., 1988, J.Polym.Sci., Polym.Lett.Ed.: 26, 481-483

23. Wölfle, I., Lodaya, J., Sauerwein, B. and Schuster, G.B., 1992, J. Am.Chem.Soc.: 114, 9304-9309.

24. Hizal, G., Yagci, Y. and Schnabel, W.,1994, Polymer: 2428-2431

25. Yagci, Y., Lukac, I. and Schnabel, W., 1993, Polymer: 34, 1130-133.

26. Dossow, D., Yagci, Y. and Schnabel, W., publication in preparation

27. Nelson, E.W., Carter, T.P. and Scranton, A.B., 1994, Macromolecules: 27, 1013-1019.

28. Manivannan,G., Fouassier, J.P. and Crivello, J.V.,1992, J.Polym.Sci., Part A: Polym.Chem.:30, 1999-2001

29. Monroe B.M. and Weed. C.G., 1993, Chem.Rev.: 93, 435-448.

30. Neckers, D.C. and Valdes-Aguilera, O.M. 1993, Adv.Photochem.: 18, 315-394.

31. Gruber, 1992, Progr. Polym.Sci.: 12, 953-1044.

32. Dietlinker, K.H., 1991, in ,,Chemistry and Technology of UV and EB Formulations for coatings, Inks and Paints", Oldring, P.K.T. (ed.), Sita Techn.Ltd, London, Vol.3, Chapter 1

33. Timpe, H.-J. and Neuenfeld, S., 1990, Kontakte (Darmstadt): 2, 28.

34. Crivello, J.V. and Lam, J.H.W., 1978, J.Polym.Sci., Polym.Chem.Ed.: 16, 2441-2451

35. Zhu, Q.Q. and Schnabel, W., publication in preparation.

36. Abdul-Rasoul, F.A.M., Ledwith, A. and Yagci, Y., 1978, Polymer: 19, 1219-1222.

37. Ledwith, A., 1979, Makromol.Chem.Suppl.: 3, 348-358.

38. Ledwith, A., Al-Kass, S. and Hulme Lowe, A., 1984, ,,Cationic Polymerization and Related Processes", Goethals, E.J. (ed.), Academic Press, London, p.275 .

39. Johnen, N., Kobayashi, S., Yagci, Y. and Schnabel, W., 1993, Polym.Bull.: 30, 279-284

40. Bi, Y. and Neckers, D.C., 1994, Macromolecules: 27, 3683-3693.

41. Kutal, C., Grutsch, P.A. and Yang, D.B., 1991, Macromolecules: 24, 6872-6873.

42. Fukuchi,Y., Takahashi, T., Noguchi, H., Saburi, M. and Uchida, Y.,1987, Macromolecules: 20, 2316-2317.

6

SYNTHESIS AND PHOTOPOLYMERIZATION OF 1-PROPENYL ETHER MONOMERS

J.V. Crivello,* K.D. Jo, W.-G. Kim and S. Bratslavsky

Department of Chemistry
Rensselaer Polytechnic Institute
Troy, New York 12180

ABSTRACT

A variety of mono-, di-, and multifunctional 1-propenyl ethers were readily prepared in high yields by the condensation of alcohols with allyl halides followed by the base or transition metal catalyzed rearrangement of the resulting allyl ethers. These monomers in general display very high reactivity in cationic polymerizations. In our work, we have focused on photoinduced cationic polymerizations of these monomers using diaryliodonium and triarylsulfonium salt photoinitiators. To study these very fast photopolymerizations, extensive use of differential photoscanning calorimetry and real-time infrared spectroscopy were made. Employing these techniques, the effects of monomer and photoinitiator structure on the rates of polymerization were studied.

INTRODUCTION

Multifunctional monomers which can be rapidly and efficiently photopolymerized are of increasing technical importance. This is due to the widespread movement of industry away from the use of traditional solvent-based processes in compliance with the pressure applied by regulatory agencies to limit the amount of volatile organic compounds emitted into the atmosphere. Photopolymerizable multifunctional monomers, also called photocurable or UV curable monomers, do not require solvents and produce highly crosslinked polymers that can, in most cases, duplicate or exceed the properties available by traditional solvent-borne thermal curing processes. In addition, photocuring provides further benefits such as reduced energy usage, higher product throughput, lower capital investment and smaller floor space requirements. Until recently, most investigations have been focused on commercially available acrylate and methacrylate monomers which can be photopolymerized under free radical conditions.[1] The discovery of efficient and practical onium salt photoinitiators such as diaryliodonium and triarylsulfonium salts have made it possible to extend photopolymerizations to cationically polymerizable monomers.[2] Depicted in Scheme 1 is a simplified mechanism shown for triarylsulfonium salts whereby these onium

Macromolecular Engineering, Edited by M.K. Mishra et al.
Plenum Press, New York, 1995

salts behave as latent photochemical sources of strong Brønsted acids to initiate cationic polymerizations.

$$Ar_3S^+ \ X^- \xrightarrow{\ h\nu\ } [Ar_3S^+ \ X^-]^* \longrightarrow \begin{Bmatrix} Ar_2S^{+\cdot}X^- \ + \ Ar\cdot \\[2ex] Ar_2S \ + \ Ar^+ \ X^- \end{Bmatrix}$$

eq. 1

$$\begin{Bmatrix} Ar_2S^{+\cdot}X^- \ + \ Ar\cdot \\[2ex] Ar_2S \ + \ Ar^+ \ X^- \end{Bmatrix} \xrightarrow{\ RH\ } HX \ + \ \text{Products}$$

eq. 2

$$HX \ + \ n\,M \ \longrightarrow \ H(M)_{n-1}M^+ \ X^-$$

eq. 3

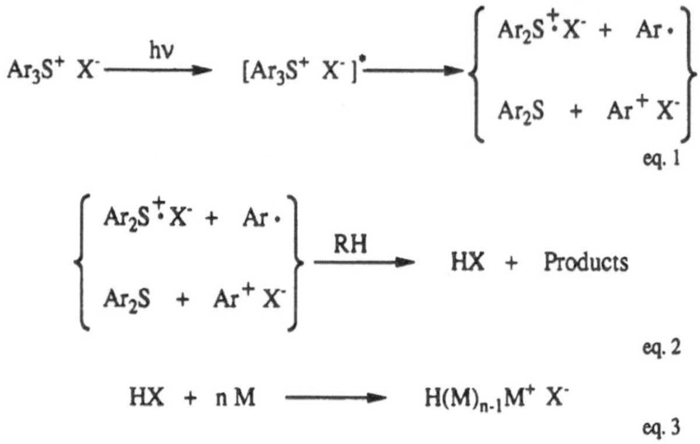

Scheme 1

In Scheme 1, monomer M is any cationically polymerizable monomer which encompasses both vinyl type monomers which undergo addition polymerization and heterocyclic monomers which polymerize by ring-opening mechanisms. In this laboratory our initial investigations centered about the use of photoinduced cationic polymerizations and employing commercially available difunctional epoxide[3] and vinyl ether monomers.[4] However, these commercially available monomers suffer from certain drawbacks. First, because they are not specifically designed or prepared for use in cationic polymerizations, they usually contain considerable amounts of inhibiting impurities and may undergo side reactions. Second, in the case of vinyl ether monomers, few multifunctional monomers are available and these are quite expensive.

For these reasons, we have been engaged in the design of novel classes of high reactivity, multifunctional, cationically photopolymerizable monomers and the development of general synthetic methods for their preparation.

RESULTS AND DISCUSSION

Synthesis of 1-Propenyl Ether Monomers

One class of monomers which meets the criteria of high reactivity and potential low cost are multifunctional 1-propenyl ethers. Higashimura and his group[5] prepared various monofunctional alkyl-substituted 1-propenyl ethers by the sequence of reactions given in Scheme 2.

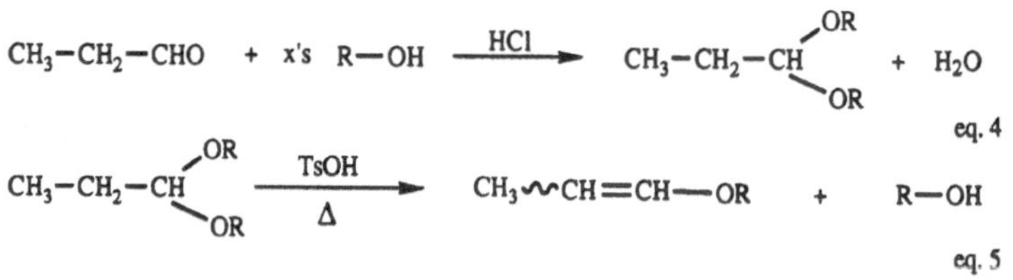

$$CH_3-CH_2-CHO \ + \ x's \ R-OH \xrightarrow{\ HCl\ } CH_3-CH_2-CH\begin{smallmatrix}OR\\[1ex]OR\end{smallmatrix} \ + \ H_2O$$

eq. 4

$$CH_3-CH_2-CH\begin{smallmatrix}OR\\[1ex]OR\end{smallmatrix} \xrightarrow[\Delta]{\ TsOH\ } CH_3\text{w}CH=CH-OR \ + \ R-OH$$

eq. 5

Scheme 2

In this scheme, propionaldehyde is condensed with an excess of the desired alcohol under acidic conditions to form the corresponding acetal. The acetal is then heated in the presence of p-toluene-sulfonic acid to generate the desired 1-propenyl ether with the elimination of one mole of the alcohol. Generally, the overall yields were in the range of 60-80% depending on the structure of the starting alcohol. When n-butanol is used a mixture composed of 75% of the E and 25% of the Z isomers of 1(1-propenoxy)butane (n-butyl 1-propenyl ether) which could be separated and purified by fractional distillation. While this method works well for the low molecular weight series of 1-propenyl ethers, complications due to side reactions make it less effective for the preparation of high molecular weight members of this series of monomers and particularly for multifunctional 1-propenyl ethers.

In Scheme 3 is shown the general synthetic methods which we have employed for the preparation of mono, di and multifunctional 1-propenyl ethers. First, the allyl ether precursors were prepared in very high yields by the phase transfer catalyzed condensation of alcohols with allyl chloride or bromide in the presence of base. Subsequently, the allyl ethers were quantitatively isomerized to the corresponding 1-propenyl ethers using two methods: either strong base in the presence of dimethylsulfoxide[6,7] or by treatment with a catalytic amount of tris(triphenylphosphine)ruthenium (II) dichloride.[8,9]

$$CH_2 = CH-CH_2-Br \ + \ R-OH \xrightarrow[\text{(n-C}_4\text{H}_9)_4\text{N}^+ \text{ Br}^-]{\text{KOH,}} CH_2 = CH-CH_2-O-R \qquad \text{eq. 6}$$

$$CH_2 = CH-CH_2-O-R \xrightarrow[\text{DMSO}]{\text{t-BuOK,}} \ \overset{CH_3}{\underset{H}{\diagdown}} C = C \overset{O-R}{\underset{H}{\diagup}} \qquad \text{eq. 7}$$
$$Z$$

$$CH_2 = CH-CH_2-O-R \xrightarrow[\text{120-130°C}]{(Ph_3P)_3RuCl_2} \ \overset{CH_3}{\underset{H}{\diagdown}} C = C \overset{O-R}{\underset{H}{\diagup}} \ + \ \overset{CH_3}{\underset{H}{\diagdown}} C = C \overset{H}{\underset{O-R}{\diagup}} \qquad \text{eq. 8}$$
$$Z \qquad \qquad E$$

Base catalyzed isomerization results in the exclusive formation of the Z isomer while the corresponding ruthenium catalyzed reaction gives a mixture of Z and E isomers. The course of the isomerization reactions can be conveniently followed by ^{1}H-NMR. As shown in Figure 1, the rearrangement of 1,2-diallyloxyethane to 1,2-di(1-propenoxy)ethane (IV) in the presence of potassium t-butoxide in DMSO gives rise to a new methyl quartet at 1.6 ppm and resonances at 4.4 and 5.98 ppm assigned to the olefinic protons of the 1-propenyl ether double bond.

Listed in Table 1 are a few of the many 1-propenyl ether monomers which were prepared in high yields (90-100%) during the course of this investigation.[10] All the monomers (mixtures of E and Z isomers) were isolated as colorless, low viscosity liquid monomers and purified by distillation or by flash column chromatography. Certain substrates, such as furfuryl allyl ether (I), could be isomerized only using base catalysts, while at the same time, d-sorbitol hexaallyl ether underwent extensive decomposition under the same strongly basic conditions. In contrast, all six allyl groups of this latter compound were smoothly isomerized to give sorbitol hexa-(1-propenyl) ether (XIV) by tris(triphenylphosphine)-ruthenium (II)

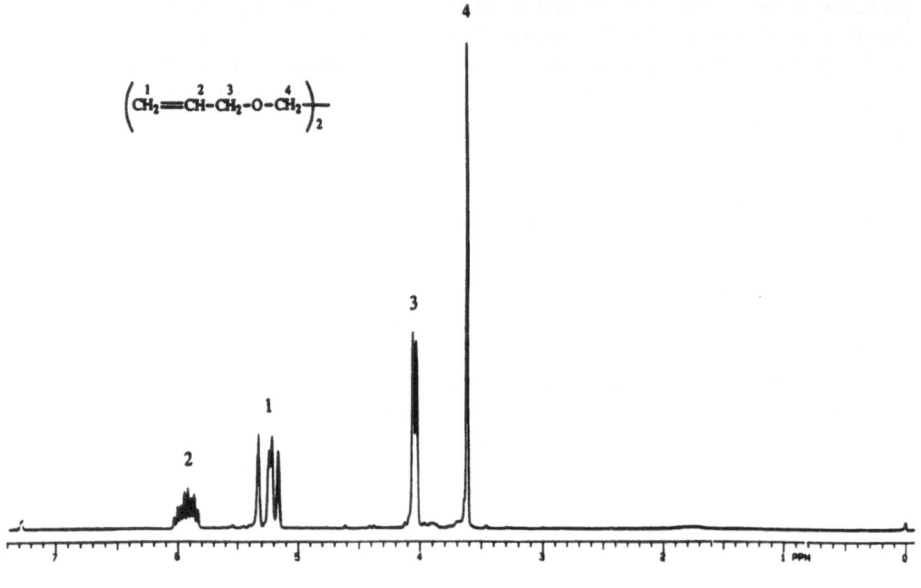

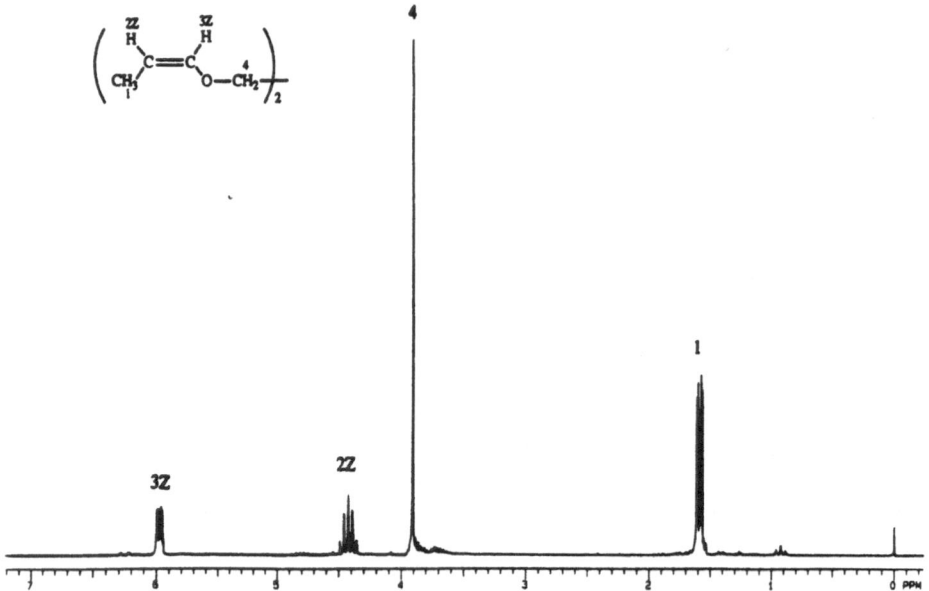

Figure 1. ^{1}H-NMR Spectra in CDCl$_3$ of 1,2-diallyloxyethane and Z,Z-1,2-di(1-propenoxy)ethane.

dichloride at 120-130°C over the course of 6 h. Thus, in contrast to their vinyl ether analogues, mono-, di-, and multifunctional 1-propenyl ethers can be easily prepared using these isomerization techniques.

The above synthetic methods are very general and a wide variety of functional groups may be tolerated. This includes esters, urethanes, amides, amines and epoxides.

Table 1. Structure and elemental analyses of multifunctional 1-propenyl ethers

Notation	1-Propenyl Ether	Elemental Analysis	
		%C	%H
I	Furfuryl (1-propenyl) ether	68.69 (69.55)	7.23 (7.30)
II	1(1-Propenoxy)decane	78.51 (78.72)	13.33 (13.21)
III	1(1-Propenoxy)dodecane	79.37 (79.58)	13.54 (13.36)
IV	1,2-Di(1-propenoxy)ethane	67.17 (67.57)	9.94 (9.92)
V	1,4-Di(1-propenoxy)butane	69.66 (70.55)	10.66 (10.66)
VI	1,6-Di(1-propenoxy)hexane	72.32 (72.68)	11.28 (11.18)
VII	1,10-Di(1-propenoxy)decane	75.40 (75.54)	12.03 (11.89)
VIII	Neopentylglycol di(1-propenyl) ether	71.25 (71.70)	10.91 (10.94)
IX	Diethylene glycol di(1-propenyl) ether	64.27 (64.49)	9.57 (9.74)
X	Triethylene glycol di(1-propenyl) ether	61.82 (62.59)	9.73 (9.63)
XI	Tetraethylene glycol di(1-propenyl) ether	60.40 (61.29)	9.63 (9.55)
XII	1,2,3-Tri(1-propenoxy)propane (Glycerol tri(1-propenyl) ether)	67.12 (67.89)	9.42 (9.50)
XIII	Pentaerythritol tetra(1-propenyl) ether	68.15 (68.89)	9.54 (9.52)
XIV	Sorbitol hexa(1-propenyl) ether	67.62 (68.22)	9.07 (9.06)

For example, allyl glycidyl ether is isomerized (equation 9) to 1-propenyl glycidyl ether (**XV**) in 95% yield in the presence of tris(triphenylphosphine)ruthenium (II) dichloride in 5 h at 120°C.[11]

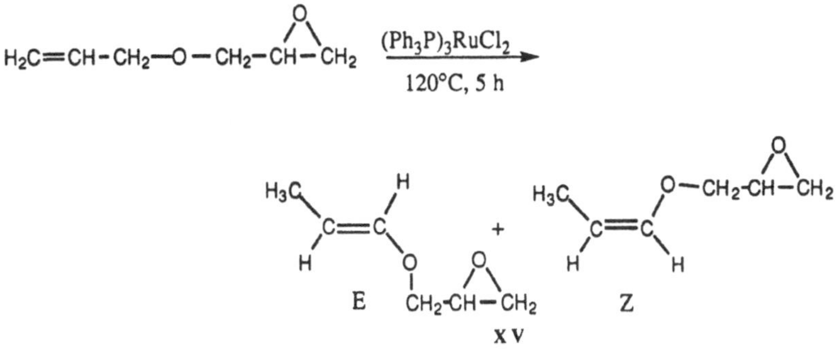

eq. 9

Allyl glycidyl ether or 1-propenyl glycidyl ether can be used as starting materials for the preparation of other reactive monomers containing the 1-propenyl ether group.[12] Two examples are shown in Schemes 4 and 5.

In both schemes, allyl glycidyl ether is condensed with a mono- or difunctional phenol to give an allyl ether bearing pendant hydroxyl group(s). Subsequent phase transfer catalyzed allylation followed by isomerization gives either the di (**XVI**) or tetrafunctional (**XVIIa-d**) 1-propenyl ethers. Employing similar schemes, a wide variety of additional 1-propenyl ether functional monomers and oligomers can be prepared.

Cationic Photopolymerization of 1-Propenyl Ethers

Only a few studies of the cationic polymerization of 1-propenyl ether monomers have been reported in the literature and the majority of these were carried out using $BF_3 \cdot Et_2O$ as the initiator.[5,13-19] From this prior work it may be concluded that 1-propenyl ethers are at least as reactive as the analogous vinyl ethers due to the higher electron density of the double bond. Further, it has been shown that the Z 1-propenyl ether isomers are more reactive than

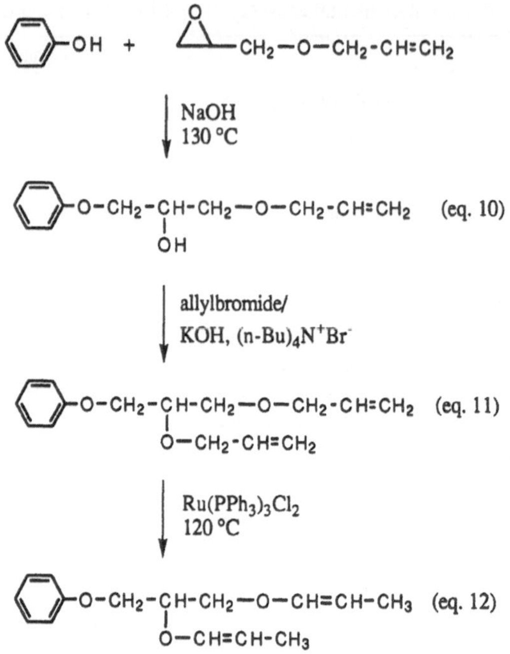

Scheme 4

the corresponding E isomers and that the E isomers undergo predominant syn opening of the double bond while the Z isomers undergo a mixture of syn and anti double bond openings.

We have been investigating the photoinduced polymerization of 1-propenyl ether monomers and oligomers using the different diaryliodonium and triarylsulfonium salt photoinitiators shown in Table 2. Since many of these monomers are rather non-polar, the

Table 2.

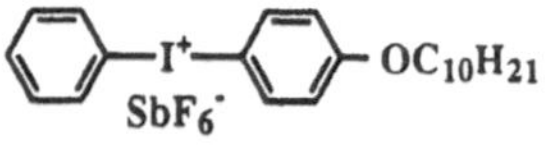

IOC

(λ_{max}: 247 nm)

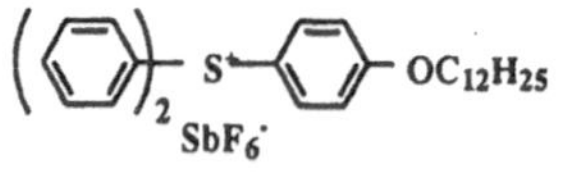

SOC

(λ_{max}: 262 nm)

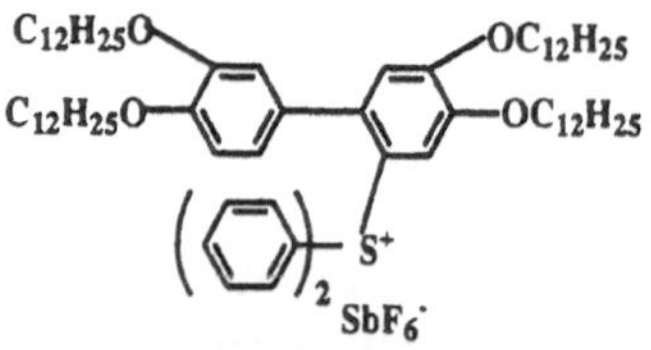

SOOC

(λ_{max}: 282 nm)

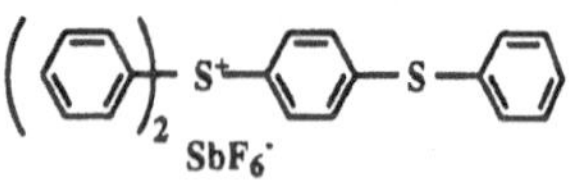

SS

(λ_{max}: 246 and 305 nm)

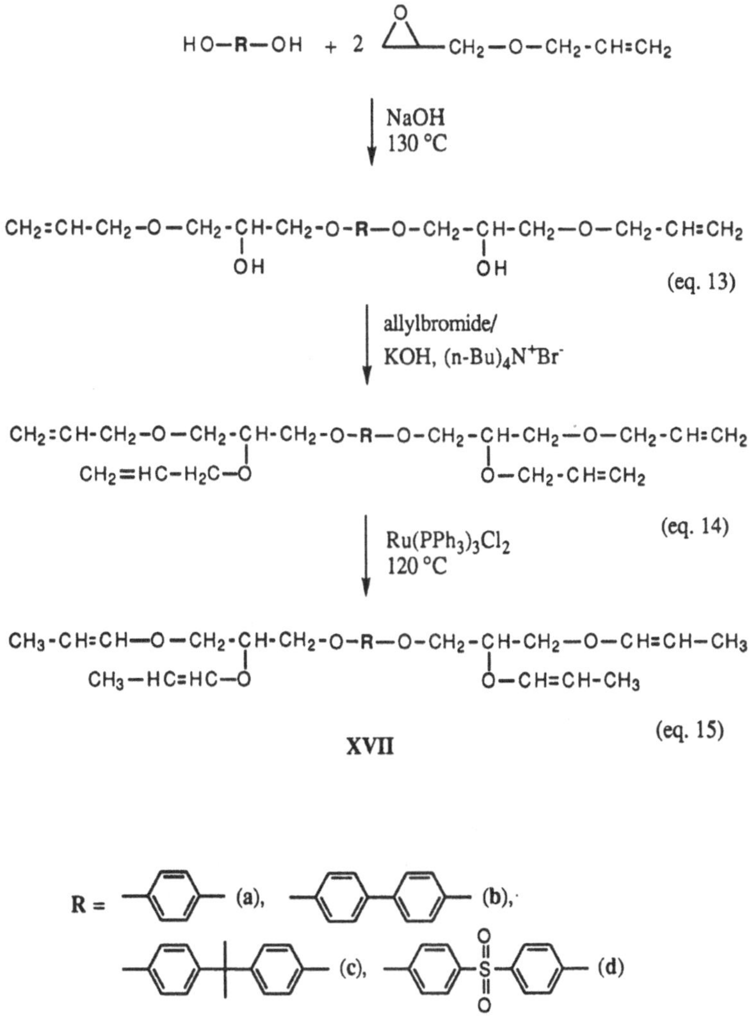

HO—R—OH + 2 △—CH₂—O—CH₂-CH≡CH₂

$$\downarrow \text{NaOH} \atop 130\,°C$$

CH₂≡CH-CH₂–O—CH₂-CH-CH₂-O-R—O-CH₂-CH-CH₂—O—CH₂-CH≡CH₂
                              | |
                             OH   OH

(eq. 13)

$$\downarrow \text{allylbromide/} \atop \text{KOH, (n-Bu)}_4\text{N}^+\text{Br}^-$$

CH₂≡CH-CH₂-O—CH₂-CH-CH₂-O-R—O-CH₂-CH-CH₂—O—CH₂-CH≡CH₂
                 | |
        CH₂=HC-H₂C—O   O—CH₂-CH≡CH₂

(eq. 14)

$$\downarrow \text{Ru(PPh}_3)_3\text{Cl}_2 \atop 120\,°C$$

CH₃-CH≡CH—O—CH₂-CH-CH₂-O-R—O-CH₂-CH-CH₂—O—CH≡CH—CH₃
                 | |
       CH₃—HC≡HC—O   O—CH≡CH-CH₃

XVII

(eq. 15)

R = ◯—(a),   ◯—◯—(b),

◯—†—◯—(c),   ◯—S(=O)(=O)—◯—(d)

Scheme 5

solubility of the photoinitiator in the monomer is problematic. However, this difficulty was overcome by employing modified photoinitiators bearing long alkoxy groups such as IOC, SOC and SOOC.

All the monomers shown in Table 1 and in Schemes 4 and 5 undergo facile cationic polymerization in the presence of the photoinitiators given in Table 2. The choice of the specific photoinitiator depended predominantly on its solubility in a given monomer. One mil (25 μm) films of the di and multifunctional 1-propenyl ether monomers spread onto glass or steel substrates underwent facile crosslinking polymerization in under one second when exposed to UV light using a 300 W Fusion Systems microwave actuated mercury arc lamp. In general, the films derived from multifunctional monomers were transparent, colorless and hard after photopolymerization.

While it was possible to demonstrate the fact that various 1-propenyl ether monomers undergo facile photoinitiated cationic polymerization, more quantitative data was desired.

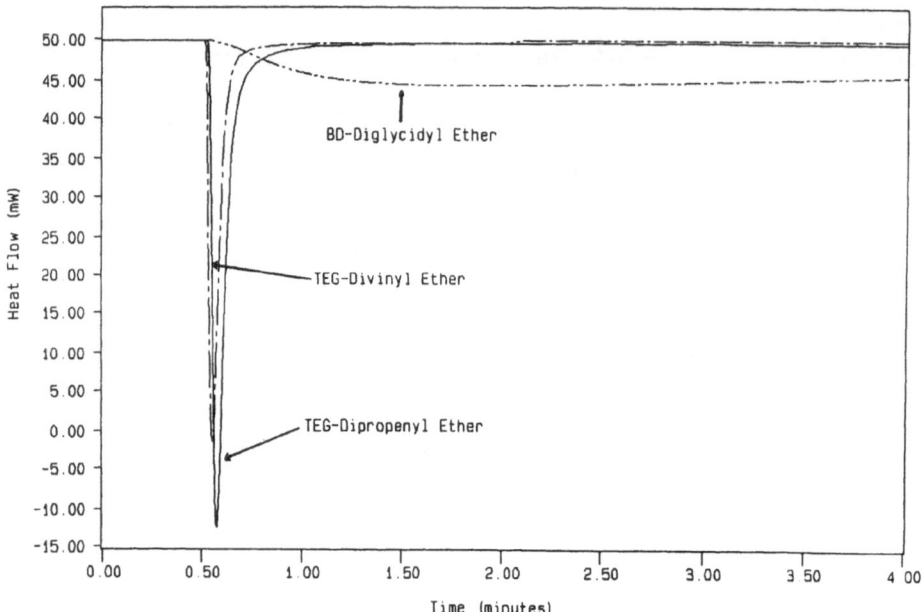

Figure 2. Isothermal (30°C) DSP Study of the photopolymerization of triethyleneglycol di(1-propenyl) ether, triethyleneglycol divinyl ether and 1,4-butanediol diglycidyl ether using 5 mole % IOC as the photoinitiator.

Previously, we have employed differential scanning photocalorimetry (DSP) as a technique for directly monitoring the rate and extent of the cationic photoinitiated polymerizations of epoxide monomers.[20] This method was employed as shown in Figure 2 to determine the general order of 1-propenyl ether reactivity with respect to vinyl ethers and epoxides of similar structure using identical molar amounts of the same photoinitiator. The reactivity of a monomer can be judged by the sharpness of the exothermic peak when the shutter of the instrument is opened and by the rapidity with which the peak returns to the baseline. A comparison of the three curves in Figure 1 shows that triethylene glycol di(1-propenyl) ether (**X**) is very much more reactive than epoxides and as reactive as the analogous vinyl ether. These same conclusions can be drawn from similar DSP studies of other 1-propenyl ether monomers. While it has been found possible to determine the relative reactivity of 1-propenyl ethers, vinyl ethers and epoxides in photoinitiated cationic polymerization using the DSP, this technique is not sufficiently sensitive enough to allow it to be used for distinguishing between the reactivity of a series of related 1-propenyl ether monomers.

Real time infrared spectroscopy (RTIR) is an excellent method for determining both the rate and extent of very rapid free radical and cationic photopolymerizations because of its rapid response time. This method involves monitoring the decrease or increase of a distinctive infrared band with time during simultaneous UV irradiation as a monomer is converted to polymer during a photopolymerization. In these studies, the decrease of the characteristic band at 1660-1680 cm^{-1} due to the carbon-carbon double bond of the 1-propenyl ether group was monitored for the 1-propenyl ether monomers. A thin, 25 μm layer of the liquid monomer containing the photoinitiator was placed between two 25 μm films of polyethylene which were then mounted in 5 cm X 5 cm slide frames. The apparatus used for these measurements consisted of a Buck Scientific Model 500 Infrared Spectrometer which was equipped with a UVEXS Co. Model SCU 110 UV lamp fitted with a fiber optic cable. The probe of the fiber optic cable was positioned so that the UV irradiation was directed onto the sample window of the spectrometer. The intensity of the irradiation could

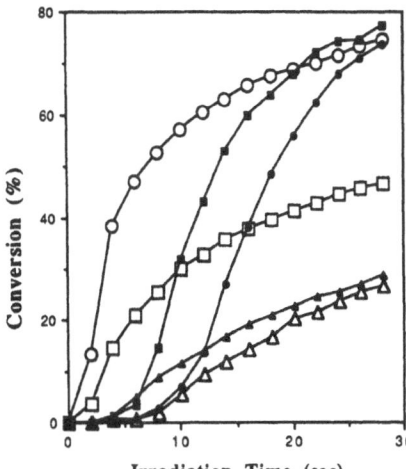

Figure 3. RTIR Study of the rates of photopolymerization of monofunctional 1-propenyl ether monomers using various photoinitiators. □ Z-furfuryl 1-propenyl ether, IOC; O Z-furfuryl 1-propenyl ether, SS; D Z-furfuryl 1-propenyl ether, SOC; ▲ Z-furfuryl 1-propenyl ether, SOOC; ■ 1(1-propenoxy) decane, SOOC; ● 1(1-propenoxy) dodecane, SOOC.

be varied by fixing the probe at various distances from the sample. Typical irradiation intensities used were 13-20 mJ/cm². A fast strip chart recorder was used to monitor the change in the selected IR band. Thereafter, the curves were digitized and transferred to a computer using an optical scanner and then replotted as conversion versus time curves.

In Figures 3-6 are shown the conversion to polymer versus time plots for various 1-propenyl ether monomers obtained in these studies. The data presented in these figures were confined to the early stages (first 30 seconds) of the polymerization. A direct impression of the reactivity and extent of polymerization for these novel monomers can be obtained by a simple inspection of the curves shown in Figures 3-6. It should also be noted, that many of the 1-propenyl ether monomers are poor solvents even for the modified diaryliodonium and triarylsulfonium salt photoinitiators. Among these photoinitiators, SOOC is the most soluble in 1-propenyl ether monomers, while IOC and SOC exhibit considerably poorer solubility characteristics and SS is completely soluble at the 0.5 mole % level only in relatively polar monomers such as 1,2-di(1-propenoxy)ethane (**IV**) and diethylene glycol di(1-propenyl) ether (**IX**). Except for SOOC, all the photoinitiators are insoluble in sorbitol hexa(1-propenyl) ether.

A comparison of the RTIR curves for the photopolymerization of the three monomers, Z-furfuryl 1-propenyl ether (**I**), 1(1-propenoxy)decane (**II**) and 1(1-propenoxy)dodecane (**III**) is given in Figure 3 using SOOC as the photoinitiator. While **II** and **III** are very reactive monomers, **I** is considerably less reactive than the other two monomers. On the basis of infrared and ¹H-NMR evidence, it has been shown that this monomer is very reactive but undergoes an oxy-Cope rearrangement instead of a conventional cationic polymerization reaction.

In Figure 4 is shown a study of the reactivity by RTIR of five difunctional 1-propenyl ether monomers using the diaryliodonium salt, IOC. This photoinitiator has good solubility in all of the monomers except for neopentylglycol dipropenyl ether (1,3-dipropenoxy-2,2-dimethylpropane) (**VIII**). All of the monomers with the exception of **VIII** have nearly the same order of reactivity. Inspection of the curves in Figure 4 shows that all of these latter monomers polymerize to high conversions.

As already noted, previous investigators[18,19] have shown that the Z 1-propenyl ether isomers are more reactive than the isomers having the E configuration. Figure 5 compares the conversion versus time curves for the photopolymerization of the E and Z double bonds of 1,2,3-tris(1-propenoxy)propane (**XII**) using SOC as the photoinitiator.[21] From the RTIR

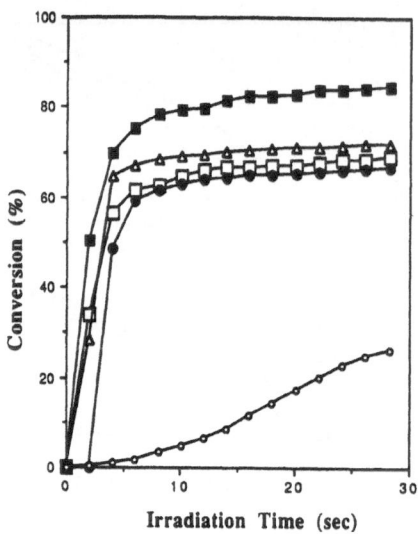

Irradiation Time (sec)

Figure 4. RTIR Study of the rates of photopolymerization of difunctional 1-propenyl ether monomers in the presence of IOC. □ 1,2-di(1-propenoxy)ethane; ■ 1,4-di(1-propenoxy)butane; D 1,6-di(1-propenoxy)hexane; ● 1,10-di(1-propenoxy)decane; O neopentylglycol di(1-propenyl) ether.

curves shown in Figure 5, it can be seen that the *Z* isomer is more reactive than the *E* isomer. Calculations based on the slopes of the RTIR curves indicate that the *Z* isomer is 2.2 times more reactive than the *E* isomer.

A series of four related di(1-propenyl) ethers derived from ethylene, diethylene, triethylene and tetraethylene glycols were prepared. The monomers are all free flowing colorless liquids which are excellent solvents for a wide variety of onium salt photoinitiators. The reactivities of these monomers are displayed in Figure 6 using SS as the photoinitiator. This this series of monomers is particularly reactive in photoinitiated cationic polymerization. Despite the uniform high reactivity of all of these monomers, some general trends can be discerned. It appears that the reactivity of the monomers increases as the number of ethylene glycol units are increased. This would appear to be due to the presence of the increasing number of aliphatic ether units present in the monomer.

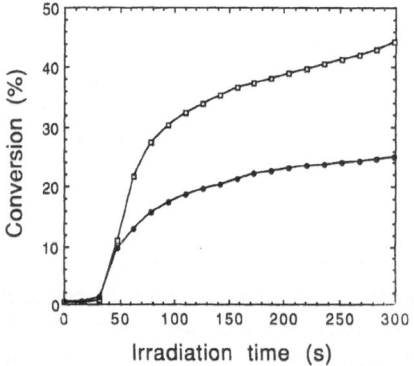

Irradiation time (s)

Figure 5. Comparison of the RTIR curves for the cationic photopolymerization of 1,2,3-tris(1-propenoxy)propane in the presence of 0.5 mol% SOC: □ *Z* 1-propenylether double bonds, ● *E* 1-propenyl ether double bonds.

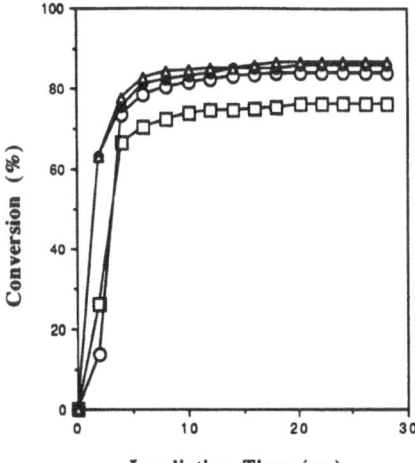

Figure 6. Photopolymerization of polyethyleneglycol di(1-propenyl) ethers in the presence of photoinitiator SS □ 1,2-di(1-propenoxy)ethane; O diethylene glycol di(1-propenyl) ether; ● triethylene glycol di(1-propenyl) ether; D tetraethylene glycol di(1-propenyl) ether.

An interesting study shown in Figure 7 is the comparison of the photopolymerization of monomer **IV** in the presence of four different photoinitiators. Photopolymerizations conducted in the presence of the triphenylsulfonium salts, SOC, and SOOC, show a definite and reproducible induction period while those polymerizations carried out using the alkoxy substituted diaryliodonium salt, IOC and triarylsulfonium salt, SS, do not. RTIR studies employing other diaryliodonium salts with various lengths of the alkoxy side chains confirmed that these diaryliodonium salt photoinitiators similarly do not display an induction period in the polymerization of multifunctional 1-propenyl ethers. The presence of an induction period which is due to the character of the photoinitiator has not previously been reported in the cationic photopolymerization of epoxides, 1-propenyl ethers or vinyl ethers.

As the RTIR study depicted in Figure 8 shows, the addition of either t-butylhydroquinone or nitrobenzene, which are typical free radical inhibitors and retarders, to the IOC photoinitiated polymerization of triethylene glycol di(1-propenyl ether) (**X**) results in a considerable lengthening of the induction period. In both these studies, nitrobenzene was found to be a much better inhibitor than t-butylhydroquinone.

The above results suggest the involvement of free radicals in these photoinduced cationic polymerizations. A mechanism, consistent with the results has been proposed and is shown in Scheme 6. The photolysis of the onium salt photoinitiator results in the decomposition of the onium salt photoinitiator to give many different species as shown in

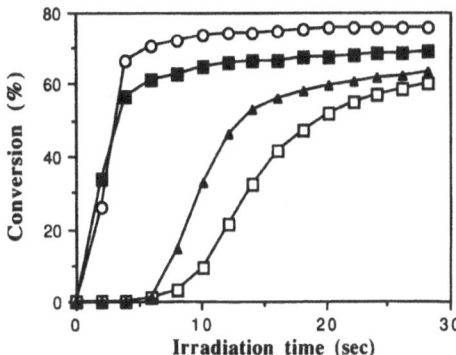

Figure 7. RTIR Study of the rates of photopolymerization of 1,2-di(1-propenoxy)ethane in the presence of different photoinitiators. ■ IOC; O SS; □ SOC; ▲ SOOC.

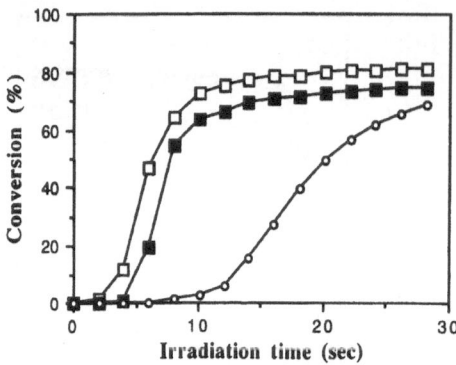

Figure 8. RTIR Study of the effect of inhibitors and retarders on the photopolymerization of triethylene glycol di(1-propenyl) ether initiated by IOC. $\square$, IOC; $\blacksquare$, IOC with t-butylhydroquinone; $\bigcirc$, IOC with nitrobenzene.

Scheme 1. Among these species are cation-radicals (equation 16) which in a subsequent series of reactions can generate protonic acids (equation 17) responsible for initiating cationic polymerization (equation 18). In addition to the onium cation-radical, aryl free radicals are also generated in the photolysis of onium salt photoinitiators. It is proposed that these aryl radicals can abstract the easily removed α-ether protons of the 1-propenyl ether monomers as shown in equation 19 to give the corresponding α-ether radicals. Subsequently, these stabilized α-ether radicals can interact with the onium salt photoinitiator to cause its reduction (equation 20). The reduction of the onium salt results in the production a new aryl radical which can cycle back into equation 19. The overall conclusion which can be drawn from the proposed mechanism shown in Scheme 6 is that the onium salt photoinitiator may undergo a non-photochemical i.e. dark free radical-induced chain decomposition reaction in addition to its direct photolysis.

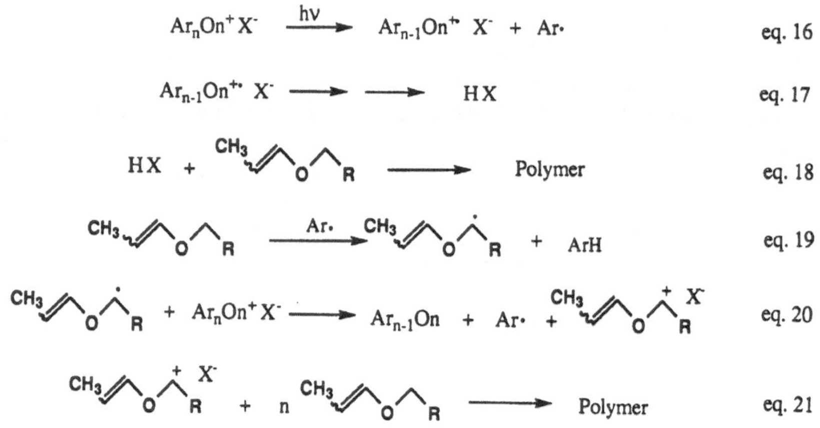

$$Ar_nOn^+X^- \xrightarrow{\ h\nu\ } Ar_{n-1}On^{+\bullet}\ X^-\ +\ Ar\bullet \qquad\qquad eq.\ 16$$

$$Ar_{n-1}On^{+\bullet}\ X^- \longrightarrow\ \longrightarrow\ HX \qquad\qquad eq.\ 17$$

$$HX\ +\ \overset{CH_3}{\diagup\!\!\diagup}\!\!\diagdown_O\!\!\diagup\!\!\diagdown_R \longrightarrow\ Polymer \qquad\qquad eq.\ 18$$

$$\overset{CH_3}{\diagup\!\!\diagup}\!\!\diagdown_O\!\!\diagup\!\!\diagdown_R \xrightarrow{\ Ar\bullet\ } \overset{CH_3}{\diagup\!\!\diagup}\!\!\diagdown_O\!\!\diagup\!\!\diagdown_R\ +\ ArH \qquad\qquad eq.\ 19$$

$$\overset{CH_3}{\diagup\!\!\diagup}\!\!\diagdown_O\!\!\diagup\!\!\diagdown_R\ +\ Ar_nOn^+X^- \longrightarrow\ Ar_{n-1}On\ +\ Ar\bullet\ +\ \overset{CH_3}{\diagup\!\!\diagup}\!\!\diagdown_O\!\!\overset{+}{\diagup}\!\!\diagdown_R\ X^- \qquad eq.\ 20$$

$$\overset{CH_3}{\diagup\!\!\diagup}\!\!\diagdown_O\!\!\overset{+\ X^-}{\diagup}\!\!\diagdown_R\ +\ n\ \overset{CH_3}{\diagup\!\!\diagup}\!\!\diagdown_O\!\!\diagup\!\!\diagdown_R \longrightarrow\ Polymer \qquad\qquad eq.\ 21$$

Scheme 6

A key step in the above mechanism involves the oxidation of the α-ether free radical shown in equation 20 by the onium salt. This reaction is governed by the free energy of the reaction, which depends on the difference in the oxidation potential of the free radical, E_r^{ox} and the reduction potential of the onium salt, E_o^{red}.

$$\Delta G = E_r^{ox} - E_o^{red}$$

eq.22

The E_o^{red} values for diphenyliodonium salts such as IOC and triphenylsulfonium salts such as SOC and SOOC have been found to be -14 kJmol^{-1}, and -112 kJmol^{-1} respectively.[22] At the same time, the reduction potential for diphenyl(4-thiophenoxyphenyl)sulfonium salts have been estimated from the half-wave potentials to be ~-89 kJmol^{-1} or at least 23 kJmol^{-1} lower than that of the corresponding triphenylsulfonium salts.[23] For the reduction to take place, the value of ΔG must be negative (i.e., exothermic) by at least 40 kJmol^{-1} or more.

The oxidation potentials of α-ether radicals such as those derived from diethyl ether, benzyl methyl ether and tetrahydrofuran are of the order of -78 kJmol^{-1}.[24,25] Thus, one would predict that while an α-ether radical is capable of reducing a diaryliodonium salt, it should not reduce a triphenylsulfonium salt such as SOC or SOOC. At the same time, a reliable prediction cannot be made for such borderline cases as the sulfonium salt SS since there exists considerable error in the methods for the determination of the E_o^{red} values and in addition, the real values would be expected to have a considerable dependance on the actual solvent (monomer) in which the photolysis is carried out. Further, in at least one case, indications of the existence of a free-radical decomposition of SS has been postulated. Ledwith[26] has reported a quantum yield of 3 of the photolysis of SS in cyclohexene oxide. This suggests that the proposed catalytic cycle shown in equations 19 and 20 may in this case be operative for sulfonium salt SS.

It is well known that alkyl ethers containing α-protons undergo spontaneous free radical autoxidation to form hydroperoxides. It has been observed in this laboratory that 1-propenyl ether monomers containing diaryliodonium salts such as IOC, will undergo slow polymerization even in the dark. The instability is most pronounced when these mixtures contain 1-propenyl ether monomers derived from poly(ethylene glycols). This can now be attributed to a slow autoxidation of the monomers which results in the formation of α-ether radicals that subsequently initiate cationic polymerization by a radical-induced chain decomposition of the photoinitiator. This same effect explains the order of reactivity which was observed earlier in Figure 6 with polyethylene glycol di(1-propenyl) ethers.

CONCLUSIONS

Mono, di and multifunctional 1-propenyl ethers can be readily prepared by the straightforward catalytic rearrangement of the corresponding allyl ethers. This new synthetic method makes these monomers readily available in good yields. The 1-propenyl ethers may be cationically photopolymerized using special diaryliodonium and triarylsulfonium salt photoinitiators bearing alkoxy groups which enhance their solubility in these monomers. Several interesting effects were observed during the study of these cationic photopolymerizations. First, comparative studies with analogous vinyl ether monomers showed that the 1-propenyl ethers were of comparable reactivity under identical conditions of irradiation. Second, the Z 1-propenyl ether isomers were more reactive than the corresponding E isomers. Third, photopolymerizations using diaryliodonium salt photoinitiators exhibit essentially no induction period while the same polymerizations carried out with triarylsulfonium photoinitiators showed considerable induction periods. This effect was attributed to a free radical induced chain decomposition of the photoinitiator in the case of the diaryliodonium salts which results in a higher effective quantum yield for decomposition as compared to the triarylsulfonium salts which do not participate in this reaction. Confirmation of this mechanism was afforded by additional studies which showed that when typical radical inhibitors and retarders were included in diaryliodonium salt photoinitiated polymerizations of 1-propenyl ether monomers they also exhibited induction periods. Because of their excellent reactivity in photoinduced cationic polymerizations, as well as the ease of preparation from readily available sources, 1-propenyl ether monomers should find abundant uses in many application areas.

REFERENCES

1. Pappas, S.P. and McGinniss, V.J., 1978, *UV Curing Science and Technology.* Technology Marketing, Stamford, CT., p. 1-21.
2. Crivello, J.V., 1993, Latest Developments in the Chemistry of Onium Salts, *Radiation Curing in Polymer Science and Technology Vol. II*, J.P. Fouassier and J.F. Rabek, editors, Elsevier Appl. Sci., New York, p. 435-471.
3. Crivello, J.V. and Lam, J.H.W., 1979, The Photoinitiated Cationic Polymerization of Epoxy Resins, *Epoxy Resin Chemistry, ACS Symp. Ser.*, 114, R.S. Bauer, Editor, Am. Chem. Soc., Washington, D.C., p.1.
4. Crivello, J.V., and Conlon, D.A., 1983, Photoinitiated Cationic Polymerization With Multifunctional Vinyl Ether Monomers, *J. Radiation Curing*, 1:6-12.
5. Mizote, A., Kusudo, S., Higashimura, T. and Okamura, T., 1967, Cationic Polymerization of a,b-Disubstituted Olefins. Part II. Cationic Polymerization of Propenyl n-Butyl Ether, *J. Polym. Sci.*, 5:1727-1738.
6. Prosser, T.J., 1961, The Rearrangement of Allyl Ethers to Propenyl Ethers, *J. Am. Chem. Soc.*, 83:1701-1704.
7. Price, C.C. and Snyder, W.H., 1960, Solvent Effects in the Base-Catalyzed Isomerization of Allyl to Propenyl Ethers, *J. Am. Chem. Soc.*, 83:1773.
8. Corey, E.J. and Suggs, J.W., 1973, Selective Cleavage of Allyl Ethers Under Mild Conditions by Transition Metal Reagents, *J. Org. Chem.*, 83:1961.
9. Zoran, A., Sasson, Y. and Blum, 1980, Catalytic Double Bond Isomerization by Polystyrene-Anchored $RuCl_2(PPh_3)_3$, *J. Org. Chem.*, 46(2):255-260.
10. Crivello, J.V. and Jo, K.D., 1993, Propenyl Ethers I. The Synthesis of Propenyl Ether Monomers, *J. Polym. Sci., Polym. Chem. Ed.*, 31:1473-1482.
11. Crivello, J.V. and Kim, W.-G., 1994, Synthesis and Photopolymerization of 1-Propenyl Glycidyl Ether, *J. Polym. Sci., Polym. Chem. Ed.*, 32:1639-1648.
12. Crivello, J.V. and Kim, W.-G., 1994, Synthesis and Photopolymerization of Multifunctional Propenyl Ether Monomers, *J. Macromol. Sci., Pure and Appl. Chem.*, A31(9):1105-1119.
13. Okuyama, T., Fueno, T. and Furukawa, J., 1968, Cationic Polymerization of α,β-Disubstituted Olefins. II. Reactivity of Propenyl Alkyl Ethers in Cationic Polymerization, *J. Polym. Sci., Polym. Chem. Ed.*, 6:993-1000.
14. T. Okuyama, T. Fueno, J. Furukawa and K. Uyeo, 1968, Structure and Reactivity of α,β-Unsaturated Ethers. III. Cationic Copolymerizations of Alkenyl Alkyl Ethers, *J. Polym. Sci., Polym. Chem. Ed.*, 6:1001-1007.
15. Fueno, T and Okuyama, T., 1969. Structure and Reactivity of α,β-Unsaturated Ethers. VII. Cationic Copolymerizations of Alkenyl Alkyl Ethers, *J. Polym. Sci., Polym. Chem. Ed.*, 7:3219-3228.
16. Okuyama, T. and Fueno, T., 1971, Structure and Reactivity of α,β-Unsaturated Ethers. XIII. Cationic Copolymerizations of Alkenyl Alkyl Ethers, *J. Polym. Sci., Polym. Chem. Ed.*, 9:629 638.
17. Higashimura, T., Kusudo, S., Ohsumi, Y. and Okamura, T., 1968, Cationic Polymerization of α,β-Disubstituted Olefins. V. Reactivity of Propenyl Alkyl Ethers in Cationic Polymerization, *J. Polym. Sci., Polym. Chem. Ed.*, 6:2523-2531.
18. Yamamoto, K. and Higashimura, T., 1974, Cationic Copolymerization of Phenyl Propenyl Ethers, *J. Polym. Sci., Polym. Chem. Ed.*, 12:613-626.
19. Kennedy, J.P. and Maréchal, E., 1982, *Carbocationic Polymerization*, Wiley-Interscience, New York, pp. 108-82, 346.
20. Crivello, J.V. and Lee, J.L., 1990, The Synthesis, Characterization and Photoinitiated Cationic Polymerization of Silicon-Containing Epoxy Resins, *J. Polym. Sci., Polym. Chem. Ed.*, 28:479-503.
21. Crivello, J.V. and Bratslavsky, S., 1994, Synthesis and Photoinitiated Cationic Polymerization of 1,2,3-Tris(1-propenoxy)propane, *J. Polym. Sci., Polym. Chem. Ed.*, 32:2929-2930.
22. Sundell, P.-E., Jönsson, S and Hult, A, 1990, Effect of Chemical Structure on Electron Beam-Induced Decomposition of Onium Salts, *Polymer Preprints*, 31(2):373-374.
23. Sundell, P.-E., 1990, Cationic Polymerization of Vinyl Ethers Using Iodonium and Sulfonium Salts, Thesis, Dept. Polym. Tech. Royal Inst. of Techn., Stockholm, Sweden.
24. Arai, S. and Sauer, C. Jr.,1966, Absorption Spectra of the Solvated Electron in Polar Liquids: Dependence on Temperature and Composition of Mixtures, *J. Chem. Phys.*, 44, 2297-2305.

25. Sundell, P.-E., Jönsson, J and Hult, A, 1991, Photo-Redox Induced Cationic Polymerization of Divinyl Ethers, *J. Polym. Sci., Polym Chem. Ed.,* 29:1525-1533.
26. Ledwith, A., 1982, Photochemically Initiated Epoxide Polymerizations, *Polymer Preprints*, 23(1):323-324.

7

DESIGN OF MACROMOLECULAR PRODRUG FORMS OF ANTITUMOR AGENTS

Tatsuro Ouchi

Department of Applied Chemistry
Faculty of Engineering
Kansai University
Suita, Osaka 564
Japan

1. INTRODUCTION

In recent years, many drugs having high potential were synthesized or purified. Therefore, drug delivery systems (DDS) which allow such new drugs to perform up to their potentials, became more and more important. DDS are the systems ideally devised to disseminate a drug when and where it is needed, and at minimum dose levels.

Drug delivery systems involve two major aspects as follow:

1. Controlled release

2. Targeting

The technique of macromolecular prodrug is one of DDS techniques. The terms "polymer drugs" or "polymeric drugs" have been widely used for pharmaceutically active macromolecules.[1] The term "polymer drugs" means the polymer itself shows its own pharmaceutical activity, although the corresponding monomeric species are biologically inactive. On the other hand, in the polymeric drug carrier, polymer acts as the drug carrier. Such polymers are defined as "polymeric drugs". In other words, they can also be termed "macromolecular prodrugs" or "polymeric prodrugs".[2] Macromolecular prodrugs are designed to protect against rapid elimination or metabolization by adding a protective polymer to the therapeutic material. Therapeutic activity is usually lost with this attachment and reinstated with the removal of the protective group. The protective group is designed to be easily removed, usually by hydrolysis. Temporary attachment of a pharmacon is necessary if the drug is active only in the free form. A drug which is active only after being cleaved form the polymer chain is called a prodrug. Temporary attachment usually involves a hydrolyzable bond such as an anhydride, ester, acetal, or orthoester. Permanent attachment of the drug moiety is generally used when the drug exhibits activity in the attached form. The pharmacon is usually attached away from the main polymer chain and other pendent groups by means of a spacer moiety that allows for more efficient hydrolysis. After a model

Macromolecular Engineering, Edited by M.K. Mishra et al.
Plenum Press, New York, 1995

101

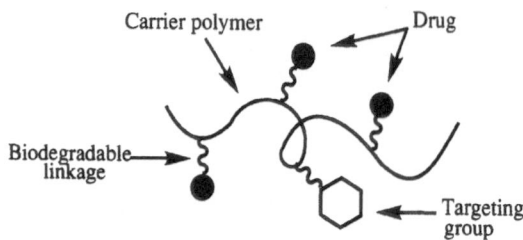

Figure 1. Schematic presentation of macromolecular prodrug.

for pharmacological active macromolecular prodrug was elucidated by Ringsdorf (Fig. 1),[3] many investigations have been devoted to this field.[4-8]

The design and development of therapeutically active macromolecular prodrug conjugate require considerably precise knowledge on the mechanism of interaction of polymers with various body organs and cells. Although the science of the pharmacokinetics or molecular pharmacology of polymers is still in its infancy, the design of macromolecular conjugate has already resulted in some observations which appear to suggest that the concept can be used to improve the therapeutic action and minimize the toxicity problem. Such featues as the appropriate selection of the backbone polymer, the method of drug attachment, and the introduction of a targeting moiety have been evaluated.

As mentioned above, a general scheme showing principal configurations of a polymer/drug conjugate system is portrayed in Fig. 1. In this schematic representation, the polymer backbone, which is mostly biodegradable but sometimes biostable, is modified by at least three functional moieties. One area of polymer is used to fix the therapeutic agents via a covalent linkage. Covalent bonds may be either permanent or temporary attachments depending on whether the therapeutic agent can exert its pharmacological effect while still bound to the carrier. A spacer arm may be necessary to provide physical separation between drug and polymer which may improve conjugation efficiency, drug-receptor interaction efficiency, or improve the cleavability of polymer-drug linkage. Targeting moieties can be incorporated in the conjugates as a device for guiding or "homing" the therapeutic component to the target site. Another part of the conjugate can be utilized for controlling the physicochemical properties of the totoal conjugate molecule. Molecular weight, solubility, hydrophilic/lipophylic balance, and electric charge are properties which can be controlled by molecular design.

Arrival routes of parent drug in macromolecular prodrug to the active patch can be divided into two routes (Fig. 2); 1) the diffusion route as a free drug released into the cells and 2) endocytosis route as a polymer-drug conjugate. The more ideal route for macromolecular prodrug is the endocytosis route.

The technique of macromolecular prodrug conjugate of low-molecular weight drug is one method of DDS, which is aimed at 1) the improvement of its movement in the body by changing solubility and molecular size, 2) retaining the appropriate concentration of drug by means of slow release from carrier, 3) the site-specific transport of drug by using recognition moiety to target cells, 4) the promotion of drug incorporation into cells *via* endocytosis, and 5) co-effect by means of hybrization of two kinds of drug or drug with bioactive polymer carrier. In order to exhibit effectively such aims, the drug is not only attached to carrier polymer but also the fine molecular design of conjugate have to be made by considering the physical property of conjugate, the biochemical property of carrier polymer and the appropriate attachment method of drug. Thus, macromolecular carriers, because of their size, electric charges, hydro- or lipophilicity, and specific transmembrane

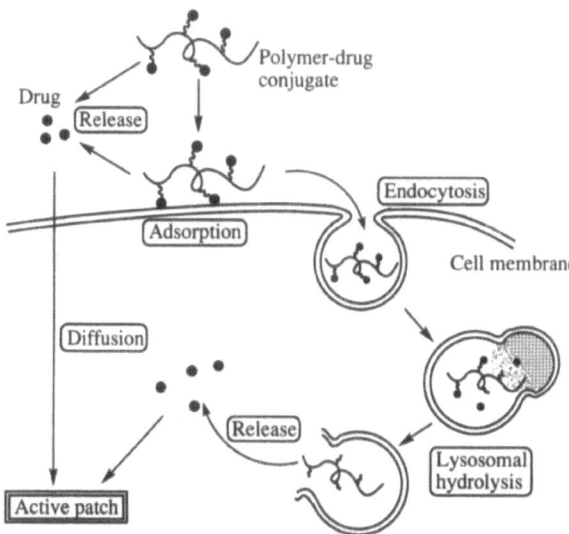

Figure 2. Arrival routes of parent drug in macromolecular prodrug.

transport capability,may well alter pharmacological and immunological activities of the agent they deliver. Consequently, the design of macromolecule/drug conjugates or macromolecular prodrugs is a rather sophisticated problem.

The present article concerns with the results of studies on especially the macromolecular prodrug of antitumor agent from standpoints of the release of drug from macromolecular prodrug, especially the lysosomally cleaved macromolecular prodrug and the macromolecular prodrug with targeting ability.

2. RELEASE OF DRUG FROM MACROMOLECULAR PRODRUG

Chitin and chitosan, which are also biodegradable and biocompatible, have attracted much interest as a component polysaccharide in the macromolecular prodrug system.[9] In particular, it has been reported that partially N-acetylated chitosan is selectively accumulated in certain tumor cells to inhibit their growth.[10] Therefore, chitin and chitosan are of interest in the application for polymer drug. On the other hand, α-1,4-polygalactosamine (PGA) is a novel basic polysaccharide purified from the culture fluid of *Paecilomyces sp.I-1* and devoid of acute toxicity.[11] PGA and N-acetyl-α-1,4-polygalactosamine (NAPGA) are nontoxic, non-immunogenic and biodegradable polymers. And water-soluble chito-oligosaccharide[12] and N-acetyl-galactosamino-oligosaccharide[13] have also been reported to act as immunomodulaters. So, these oligosaccharides might exhibit immunological activities or synergistic effects when they are used as carrier polymers in macromolecular prodrug.

5-Fluorouracil (5FU) has a strong antitumor activity,[14] which is accompanied, however, by undesirable side-effects.[15] In order to provide a macromolecular antitumor prodrug with reduced side-effects and strong antitumor activity, we synthesized chitin/5FU, chitosan/5FU, chitosamino-oligosaccharide/5FU conjugates, NAPGA/5FU, PGA/5FU and galactosamino-oligosaccharide/5FU conjugates (Figs. 3 and 4), and investigated their antitumor activities.

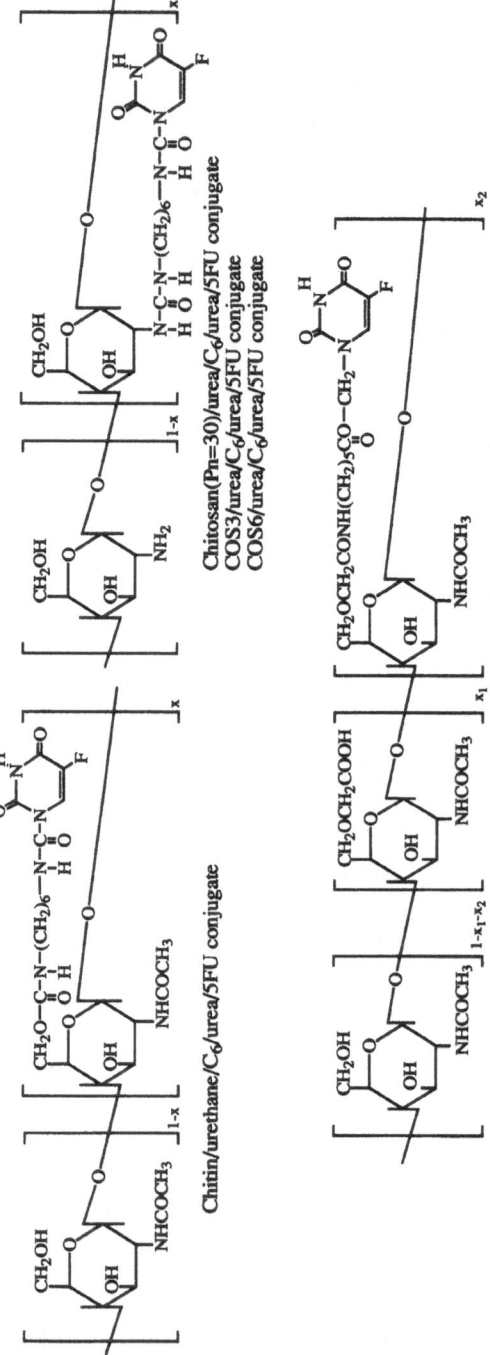

Figure 3. Structure of Chitin/urethane/C6/urea/5FU[16], Chitosan/urea/C6/ urea/5FU[16] and CM-chitin/amide/C5/ester/C1/5FU[18] conjugates.

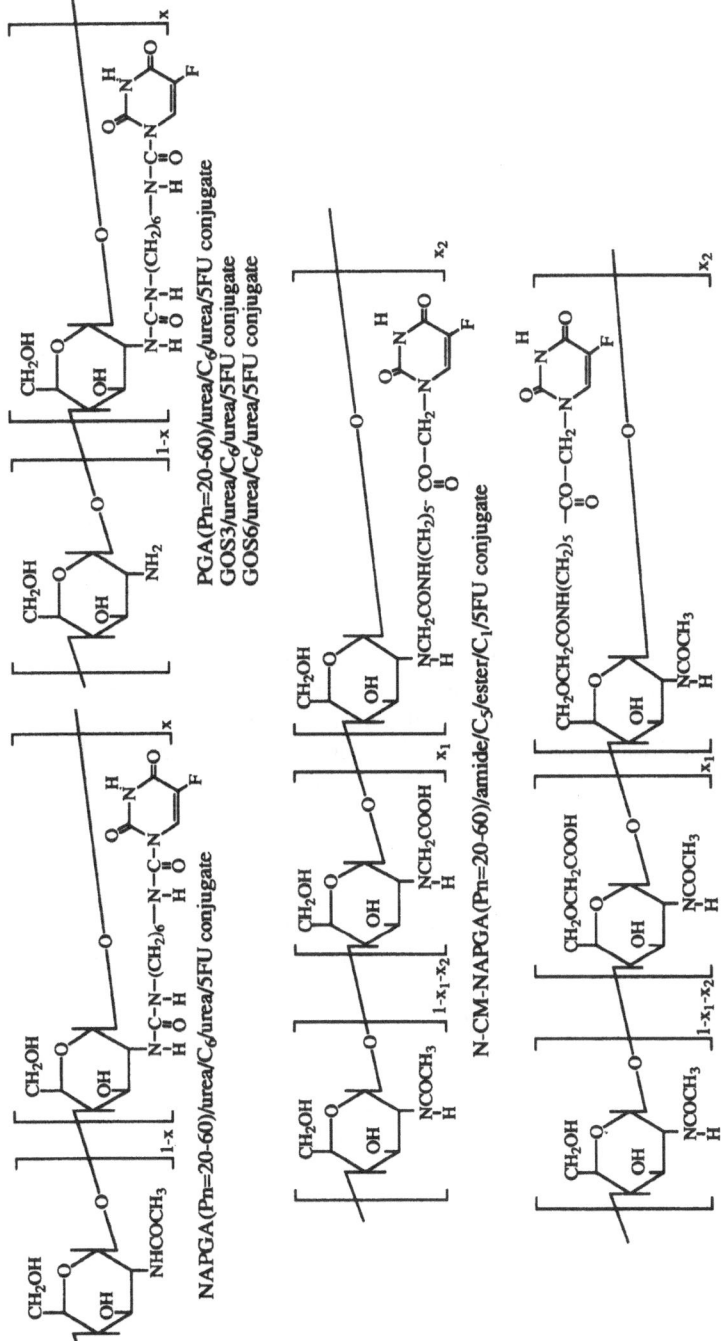

Figure 4. Structure of NAPGA/urea/C6/urea/5FU,[17] PGA/urea/C6/urea/ 5FU[17] and CM-NAPGA/amide/C5/ester/C1/5FU[19] conjugates.

Firstly, the antitumor activities of the conjugates of 5FU attached to chitin, chitosan or chitosamino-oligosaccharide through hexamethylene spacers *via* urethane/urea or urea/urea bonds were evaluated *in vivo*. The conjugate of 5FU attached to chitosan or chitosamino-oligosaccharide at 2-positions through hexamethylene spacer groups *via* urea/urea bonds and the conjugate of 5FU attached to chitin at 6-positions through hexamethylene spacer groups *via* urethane/urea bonds were found to exhibit significant survival effects against p388 *lymphocytic leukemia* in female CDF_1 mice *i.p./i.p.* [16] Secondly, the antitumor activities of NAPGA with 5FUs attached *via* urea bonds and PGA with 5FUs attached *via* urea/urea bonds were evaluated *in vivo*. The NAPGA/urea/C_6/urea/ 5FU conjugate, and PGA/urea/C_6/urea/5FU conjugate showed also remarkable survival effects against p388 *lymphocytic leukemia* in mice *i.p./i.p.* [17]

Moreover, the chitosan/urea/C_6/urea/5FU conjugate and chitosamino-oligosaccharide/urea/C_6/urea/5FU conjugates exhibited stronger growth-inhibitory effects on Meth-A *fibrosarcoma* or MH-134Y *hepatoma* in SPF-C3H/He mice by subcutaneous (*s.c.*) implantation/intravenous (*i.v.*) injection than free 5FU, chitosan, chitosamino-oligosaccharide and their blends (Fig. 5).[16] The NAPGA/ urea/C_6/urea/5FU conjugate and galactosamino-oligosaccharide/urea/ C_6/urea/5FU conjugates exhibited strong growth-inhibitory effects against Meth-A *fibrosarcoma* in male Balb/c SPF mice *s.c./i.v.* These results suggested to these chitin or chitosan/5FU and NAPGA or PGA/5FU conjugates acted as hybrid type of macromolecular prodrugs. However, these conjugates were water-insoluble which restricted the capacity of doses and the way of their injections.

In order to obtain a water-soluble macromolecular prodrug of 5FU, we used CM-chitin, N-carboxymethyl-N-acetyl-α-1,4-polygalactosamine (N-CM-NAPGA) and O-carboxymethyl-N-acetyl-α-1,4-polygalactosamine (O-CM-NAPGA) as drug carriers. The results of the survival effect for CM-chitin/amide/C_5/ester/C_1/ 5FU conjugate and 1-[(amino-*n*-pentyl)-ester]-methylene-5FU(monomeric 5FU derivative) against p388 *lymphocytic leukemia* in female CDF_1 mice *i.p./i.p.*, as well as that of free 5FU for comparison, are shown in Fig. 6, compared with those for 5FU.[18] Although the survival effect for 5FU derivative was much lower than that of 5FU, the survival effects for CM-chitin/amide/C_5/ester/C_1/ 5FU conjugate was higher than that of 5FU. The N-CM-NAPGA/amide/C_5/ester/C_1/5FU conjugate exhibited a significant survival effect against p388 *lymphocytic leukemia* in mice *i.p./i.p.*, compared with that for monomeric 5FU derivative.[19] The survival time for these CM-chitin/5FU and N-CM-NAPGA/5FU conjugates tended to increase with increase in dose. The conjugates did not cause rapid decrease of body weight of the treated mice even in the high dose range tested; they did not display an acute toxicity in such high dose ranges. The conjugation 5FU in CM-chitin/5FU and N-CM-NAPGA/5FU appears to depress the side-effects of 5FU. Thus, such water-soluble CM-chitin/5FU conjugate and CM-NAPGA/5FU conjugate can be expected to be used as tumor therapeutical agent by intravenous injection.

From the standpoint of therapeutic mechanism, the release behavior of the drug from the carrier polymer is important. Generally, water-soluble macromolecules are expected to be uptaken into cells by endocytosis or phagocytosis. Tumor cells show a higher degree of uptake efficiency than normal tissue cells. After uptake of the macromolecules, a phagosome is formed to fuse with a lysosome, and subsequently the macromolecules are exposed to a mixture of at least forty kinds of digestive enzymes at an acidic pH (4.0-5.5). So, it is one of the promising approach to design the macromolecular prodrug system which can release drug in lysosomal condition only after cellular uptake.

Kopecek and Duncan synthesized models of macromolecular prodrug, HPMA copolymers fixing p-nitroanilin (NAp) residues as model drugs through various peptidyl spacer groups, and investigated the enzymatic release behavior of NAp from HPMA copolymers by lysosomal proteases, *cathepsin B*, H and L *in vitro*, and then they demonstrated that the

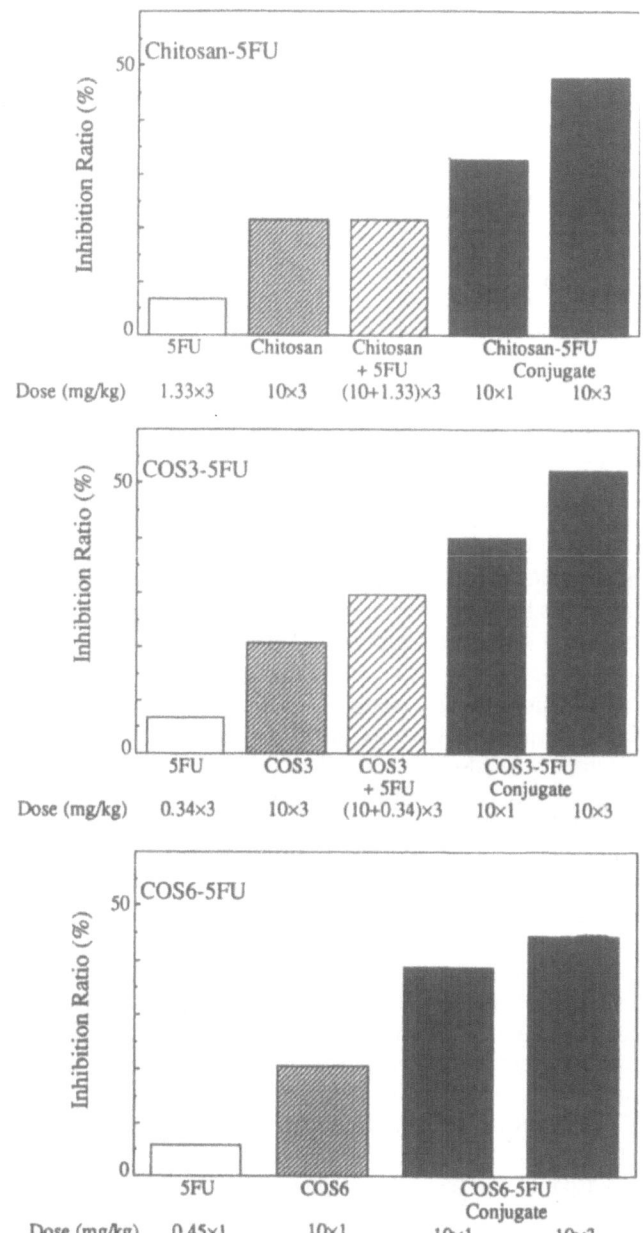

Figure 5. Growth-inhibitory effect by Chitosan(Pn=30)/urea/C6/urea/5FU (D5FU=27mol%), COS3/urea/C6/urea/5FU(D5FU=34mol%), COS6/urea/C6/urea/5FU(D5FU=44mol%) conjugates against MH134Y *hepatoma* or Meth-A *fibrosarcoma* in mice by *s.c./i.v.*[16]

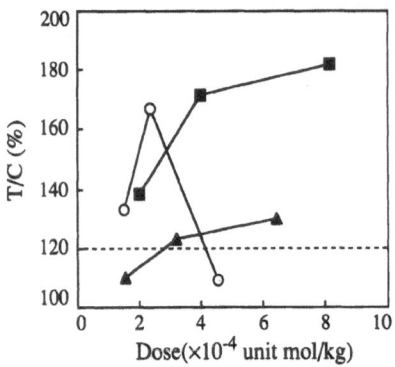

Figure 6. Survival effect for CM-chitin/amide/C5/ester/C1/5FU conjugate and monomeric 5FU derivative against p388 *lymphocytic leukemia* in mice *i.p./i.p.*[18] ■ CM-chitin (Mw = 10000)/amide/C5/ester/C1/5FU conjugate (D5FU[a]=35.2mol%,DCM[b]=34.8mol%), ▲ 1-(amino-n-pentyl-ester)-methylene-5FU, ○ 5FU. [a]Degree of substitution of 5FU per saccharide unit. [b]Degree of free carboxymethylation per saccharide unit.

tetrapeptide spacer, Gly-Phe-Leu-Gly, can facilitate specific drug release in lysosomal condition.[20] They also reported that the HPMA copolymers fixing antitumor agents through the tetrapeptide spacer groups showed good therapeutic effects against mice bearing tumor cells *in vivo*.[21] However, the polymers they employed were non-biodegradable vinyl polymers.

In order to provide a biodegradable macromolecular prodrug of doxorubicin (DXR; adriamycin) reducing the side-effects and exhibiting the high antitumor activity, we designed CM-chitin fixing DXRs through covalent bonds. DXR is one of the most prominent clinical antitumor agents. However, its very strong side-effects have also been cited. DXR has low stability and low water-solubility. By conjugation of DXR to the water-soluble polysaccharide derivative, CM-chitin, the low stability and water-solubility of DXR should be improved. Especially, we employed the lysosomally digestible tetrapeptide, Gly-Phe-Leu-Gly, group as a spacer between CM-chitin and DXR. This spacer is expected to be stable in the blood stream, to be cleaved specifically by lysosomal proteases after the uptake into tumor cells and to release drug efficiently. The present paper concens with the synthesis of CM-chitin/Gly-Phe-Leu-Gly/DXR conjugate having lysosomally digestible tetrapeptide spacers and CM-chitin/C$_5$/DXR conjugate having lysosomally non-digestible pentamethylene spacers. It was reported that the amount of *cathepsin B* was larger in the tumor cells than that in normal tissue cells.[22,23] So, it must be a promising approach for cancer chemotherapy to design the macromolecular prodrug system which can release drug in response to *cathepsin B*. Therefore, we also studied on the release behavior of free DXR from the conjugates fixing DXRs through lysosomally digestible peptide spacer or lysosomally non-digestible spacer in the presence or absence of lysosomal enzyme, *cathepsin B*, at 37°C *in vitro*.[24] Moreover, the antitumor activity of CM-chitin/Gly-Phe-Leu-Gly/DXR conjugate and CM-chitin/C$_5$/DXR conjugate against tumor cells were investigated *in vitro*.

CM-chitin/Gly-Phe-Leu-Gly/DXR conjugate and CM-chitin/C$_5$/DXR conjugate were synthesized according to the reaction steps shown in Schemes 1 and 2, respectively. The values of the degree of introduction of DXR (DDXR) for conjugates were determined by the absorption at 495nm in water.

The release behavior of free DXR from the conjugates was investigated by reversed-phase HPLC system under the following conditions at 37°C *in vitro*. (1) 1/15M KH$_2$PO$_4$-Na$_2$HPO$_4$ buffer solution (pH=7.4), (2) 0.2M Na$_2$HPO$_4$-0.1M citric acid buffer solution (pH=5.5), (3) in the presence of *cathepsin B* (lysosomal thiol-protease) (pH=5.5), (4) in the presence of leupeptin as inhibitor for *cathepsin B*. The time course of the release behavior of free DXR from the conjugates is shown in Fig. 7. The release rates of free DXR from both conjugates were very low at pH7.4 and slightly enhanced at pH5.5. On the other hand, a drastic increase of the release rate of free DXR from CM-chitin/Gly-Phe-Leu-Gly/DXR

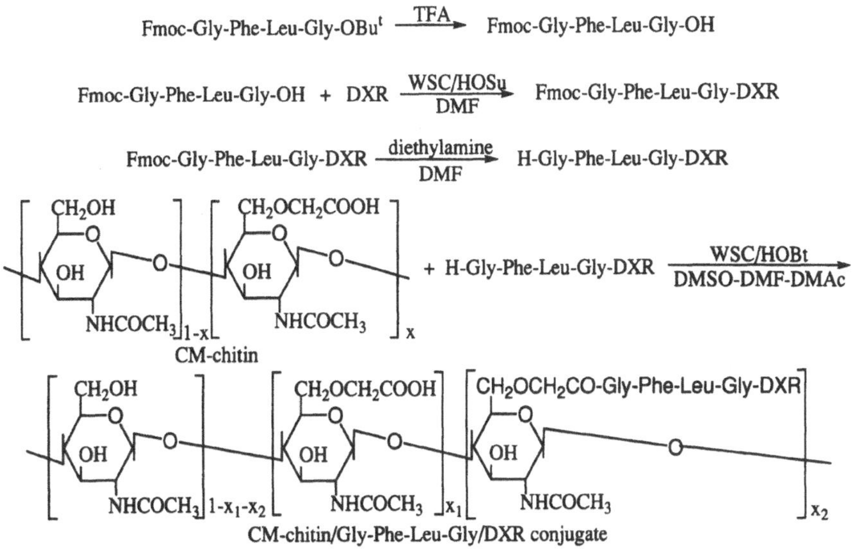
Fmoc-Gly-Phe-Leu-Gly-OBu^t
TFA
Fmoc-Gly-Phe-Leu-Gly-OH
Fmoc-Gly-Phe-Leu-Gly-OH + DXR
WSC/HOSu
DMF
Fmoc-Gly-Phe-Leu-Gly-DXR
Fmoc-Gly-Phe-Leu-Gly-DXR
diethylamine
DMF
H-Gly-Phe-Leu-Gly-DXR
CH2OH
CH2OCH2COOH
OH
NHCOCH3
1-x
x
+ H-Gly-Phe-Leu-Gly-DXR
WSC/HOBt
DMSO-DMF-DMAc
CM-chitin
CH2OCH2CO-Gly-Phe-Leu-Gly-DXR
1-x1-x2
x1
x2
CM-chitin/Gly-Phe-Leu-Gly/DXR conjugate

Scheme 1

conjugate was observed in the presence of *cathepsin B* (pH=5.5). However, the release rate of free DXR from CM-chitin/C_5/DXR conjugate in the presence of *cathepsin B* (pH=5.5) was not different from that in the absence of *cathepsin B* (pH=5.5). Moreover, by the addition of leupeptin ($2.1\infty10^{-1}$mol/l), the inhibition effect against *cathepsin B* was observed, the release rate of free DXR from the CM-chitin/Gly-Phe-Leu-Gly/DXR conjugate was decreased to the same level as non-enzymatic condition (pH=5.5). Therefore, the CM-chitin/Gly-Phe-Leu-Gly/DXR conjugate is expected not to release in blood stream (pH=7.4)

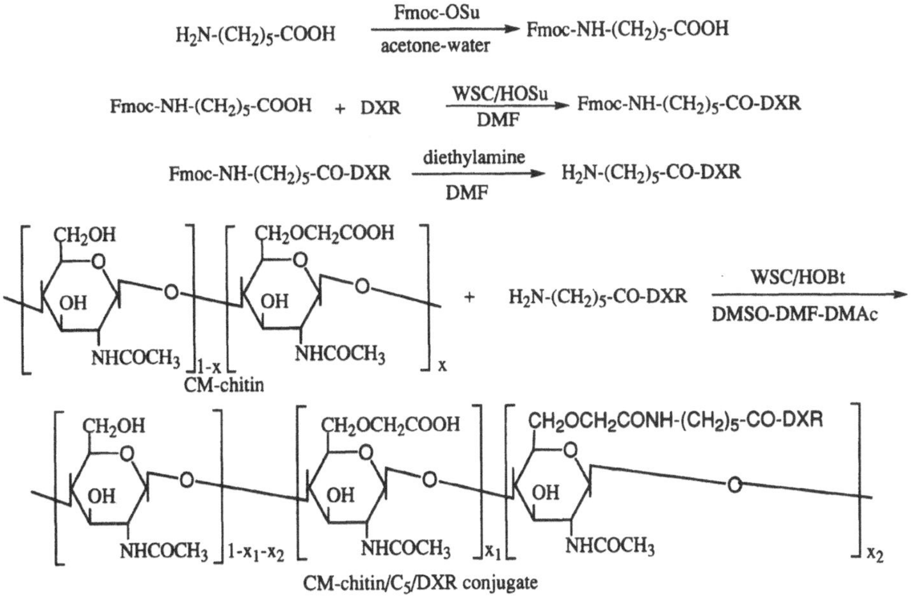
H2N-(CH2)5-COOH
Fmoc-OSu
acetone-water
Fmoc-NH-(CH2)5-COOH
Fmoc-NH-(CH2)5-COOH + DXR
WSC/HOSu
DMF
Fmoc-NH-(CH2)5-CO-DXR
Fmoc-NH-(CH2)5-CO-DXR
diethylamine
DMF
H2N-(CH2)5-CO-DXR
CH2OH
CH2OCH2COOH
OH
NHCOCH3
1-x
x
+ H2N-(CH2)5-CO-DXR
WSC/HOBt
DMSO-DMF-DMAc
CM-chitin
CH2OCH2CONH-(CH2)5-CO-DXR
1-x1-x2
x1
x2
CM-chitin/C5/DXR conjugate

Scheme 2

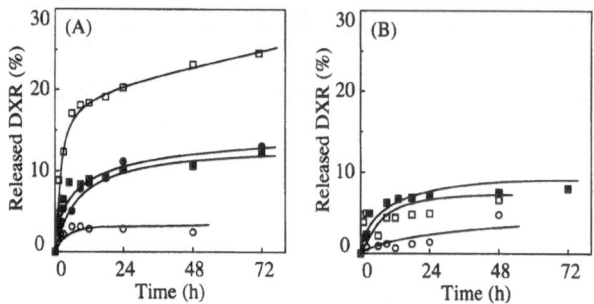

Figure 7. The release behavior of free DXR from (A) CM-chitin/Gly-Phe-Leu-Gly/DXR conjugate and (B) CM-chitin/C_5/DXR conjugate in PBS at 37°C.[24] O: pH=7.4, ●: pH=5.5, □: pH=5.5, [*cathepsin B*]=1.9×10⁻⁷mol/l, [GSH]=2.5×10⁻²mol/l, [EDTA]=1.0×10⁻⁶mol/l, ■: pH=5.5, [*cathepsin B*]=1.9×10⁻⁷ mol/l, [GSH]=2.5×10⁻²mol/l, [EDTA]=1.0×10⁻⁶mol/l, [leupeptin]=2.1×10⁻¹mol/l.

and to release effectively free DXR after uptake into tumor cells through hydrolysis of oligopeptide spacer by lysosomal enzyme *in vitro*.

In order to evaluate the stability of DXR residue, UV absorbance at 495 nm for CM-chitin/Gly-Phe-Leu-Gly/DXR conjugate and CM-chitin/C_5/DXR conjugate, the mixture of free DXR and CM-chitin, and free DXR was monitored in PBS (pH=7.4) (Fig. 8). In the cases of free DXR and the mixture of free DXR and CM-chitin, the decrease of absorbance was observed. These decrease in absorbance were attributable to the formation of precipitate of degradation product of DXR by the attack of hydroxyl ion. However, the absorbance for CM-chitin/Gly-Phe-Leu-Gly/DXR conjugate and CM-chitin/C_5/DXR conjugate did not decrease drastically. These results suggested that the stability of DXR residue was improved by conjugation to CM-chitin because of shielding effect by polymer chain.

The results of cytotoxic activity of the conjugates measured against p388D$_1$ *lymphocytic leukemia*, L1210 *leukemia*, HLE human *hepatoma* and Hela *utrocervical carcinama* cells *in vitro* are shown in Table 1.[24] The cytotoxic activities of conjugates were lower than that of free DXR, while the CM-chitin/Gly-Phe-Leu-Gly/DXR having oligopeptide, Gly-Phe-Leu-Gly, spacer groups tended to exhibit slightly higher cytotoxic activity than the CM-chitin/C_5/DXR conjugate having pentamethylene spacer groups. Table 2 summarizes the results of the survival effect in the CM-chitin/Gly-Phe-Leu-Gly/DXR conjugate, CM-chitin/C_5/DXR conjugate and free DXR against mice bearing p388D$_1$ *lymphocytic leukemia* by i.p./i.v.[25] The CM-chitin/Gly-Phe-Leu-Gly/DXR conjugate having oligopeptide, Gly-

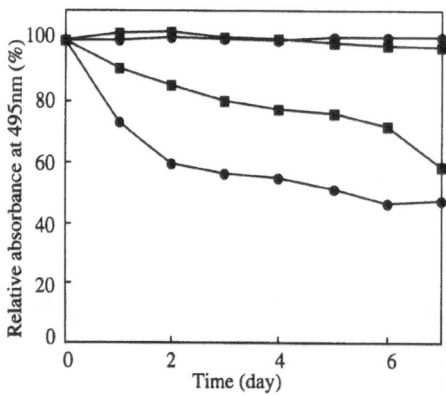

Figure 8. Stability of CM-chitin/Gly-Phe-Leu-Gly/DXR conjugate and CM-chitin/C5/DXR conjugate in PBS.[24] ●: CM-chitin/Gly-Phe-Leu-Gly/DXR conjugate, □: CM-chitin/C5/DXR conjugate, ■: mixture of CM-chitin and DXR, O: DXR.

Table 1. IC_{50} value of CM-chitin/Gly-Phe-Leu-Gly/DXR conjugate, CM-chitin/C_5/DXR conjugate and free DXR against varius tumor cells for 48h in vitro[25]

	IC_{50}(mol/l)[a]			
DXR	p388D$_1$	L1210	HLE	Hela
	1.9×10^{-7}	2.0×10^{-7}	7.0×10^{-7}	3.0×10^{-7}
CM-chitin/Gly-Phe-Leu-Gly/DXR conjugate	1.4×10^{-7}	1.2×10^{-7}	1.5×10^{-7}	1.0×10^{-7}
CM-chitin/C5/DXR conjugate	3.0×10^{-7}	3.8×10^{-7}	2.3×10^{-7}	6.0×10^{-7}

a)IC_{50}: concentration at which the cytotoxic activity reaches 50%

Phe-Leu-Gly, spacer groups exhibited a higher survival effect than the CM-chitin/C_5/DXR conjugate having pentamethylene spacer groups against mice bearing p388D$_1$ *lymphocytic leukemia i.p./i.v.* The conjugates did not exhibit rapid decrease of body weight of treated mice, even in the high dose range; the CM-chitin/Gly-Phe-Leu-Gly/DXR did not display an acute toxicity in high dose ranges.

These results mean that the CM-chitin/Gly-Phe-Leu-Gly/DXR having Gly-Phe-Leu-Gly tetrapeptide spacer groups can be presumed not to release in blood stream of pH=7.4 and to release effectively free DXR after the uptake into tumor cells through hydrolysis of tetrapeptide spacer by lysosomal enzymes.

3. MACROMOLECULAR PRODRUG WITH TARGETING ABILITY

Macromolecular carriers fall into two broad groups: (1) carriers that do not have any specific affinity to targeted malignant cells and (2) carriers that manifest highly specific binding to cell-surface receptors.

As to the aspect of targeting, the discovery of monoclonal antibody put forth the concept of "missile drug". Drugs can be targeted to specific organs or cells by controlling the size and physicochemical properties of polymeric carrier system (passive targeting) or by the use of homing or targeting devise which has specific affinity to the target cells or organs (active targeting). Many studies of drug targeting using polymer-drug conjugates having monoclonal antibody, lectin, hormone or saccharide as homing devices have been carried out.

Table 2. Comparison of prolonged life-span effects of CM-chitin/Gly-Phe-Leu-Gly/DXR conjugate, CM-chitin/C5/DXR conjugate and DXR against mice bearing p388*leukemia i.p./i.v.*[a)25]

Compounds	Dose[b] (mg/kg)	No. of mice	Body wt. loss (g)	Survival time (days±SE)	ILS (%)
Vehicle	0×5	8	0	7.4±0.2	0
CM-chitin/Gly-Phe-Leu-Gly/DXR conjugate	6.4×5	6	0	8.2±0.2**	11
	12.8×5	6	0	9.2±0.5*	24
	25.6×5	6	0.8	10.3±0.2***	39
CM-chitin/C5/DXR conjugate	6.4×5	6	0.5	8.2±0.2	11
	12.8×5	6	0.9	8.2±0.2	11
	25.6×5	6	2.0	8.2±0.3	11

ILS(%)=Increased life-span; T/C-100.
a): i.v. ¥5(day 1-5). b): The values showed the dosage of DXR.
*, **, ***: $p < 0.05$, 0.01, 0.001 when compared with vehicle gorup by Student's *t*-test.

Table 3. Saccharide residues recognized by cell surface receptor

Saccharide residues	Cells
β-galactose, N-acetyl-b-galactosamine	Liver parenchimal cells
α-mannose	Liver kupffer's cells
α-mannose	Alveolus macrophages
mannose-6-phosphate	Fibroblasts
α-galactose	3LL Lewis lung carcinoma
α-fucose	L1210 Leukemia

Certain saccharides were found to play important roles in biological recognition processes.[26] Recently, receptors specific to some saccharides were detected on cell membranes. For example, it was recognized that liver parenchymal cells have receptors specific to galactose, phagocytic cells have receptors specific to mannose, L1210 *leukemia* cells have receptors to fucose, and fibroblasts have receptors specific to mannose-6-phosphate (Table 3). Therefore, it is expected that saccharide residues act as targeting groups specific to certain organs or cells in macromolecular prodrug systems.

One of the most successful polymer-drug-saccharide conjugate system has been reported by Kopecek, Duncan and their co-workers.[27] The carrier polymer used in their system was poly[N-(2-hydroxypropyl)methacrylamide] (PHPMA), a biocompatible polymer that had been used as a plasma expander. Doxorubicin (DXR) is one of the most prominent clinical antitumor agents; however, its undesirable side-effects have also been cited. They have evaluated a number of polymer/DXR conjugates for cancer chemotherapy with excellent therapeutic results. DXRs are attached to the polymer backbone through a peptidyl spacer arm. These peptide links are found to be stable in plasma but to be readily hydrolyzed intracellularly by lysosomal enzymes. PHPMA copolymers bearing fucosylamine or galactosamine in addition to DXR were synthesized to promote the cell-specific targeting. L1210 *leukemia* cell was reported to display a cell-surface receptor which recognizes fucose[28]; similarly, the hepatocyte membrane contains a receptor which recognizes galactose.[29] Galactose has been used as a targeting moiety for drug conjugates e. g. protein/DXR conjugate for treatment of primary hepatocellular *carcinoma*. Kopecek and Duncan *et al.* prepared the water-soluble PHPMA/DXR/saccharide conjugates in which both DXR and saccharide were attached through oligopeptide links as lysosomally hydrolyzable spacer groups.[27] The survival effect against mice bearing L1210 *leukemia i.p./i.v.* was found to be promoted by using fucosyl group. Moreover, the conjugate having galactosamine groups was reported to be collected into the liver selectively.[30] However, they used the non-biodegradable vinyl polymer as a backbone polymer.

In order to achieve the active targeting of bioadsorbable macromolecule/drug conjugate, we have employed the biodegradable poly(malic acid) as a carrier and the saccharide residue as a targeting moiety. Poly(malic acid) is of interest as a biodegradable and bioadsorbable poly(lactide) type drug carrier. As poly(malic acid) has reactive carboxyl group, it can covalently attach both drug and targeting moiety. As a carrier, we used poly(α-malic acid)[31] synthesized by our original method of ring-opening polymerization of malide dibenzyl ester. We investigated the effectiveness of poly(α-malic acid)/5FU/saccharide[32] and poly(α-malic acid)/DXR/saccharide conjugates (Fig. 9) were investigated.[33] Monosaccharides such as galactosamine, glucosamine and mannosamine were used as targeting moieties. The survival effect against mice bearing tumor cells and the growth inhibition against various tumor cells *in vitro* were tested. The conjugates of poly(α-malic acid)/5FU/saccharide exhibited significant survival effects against p388 *lymphocytic leuke-*

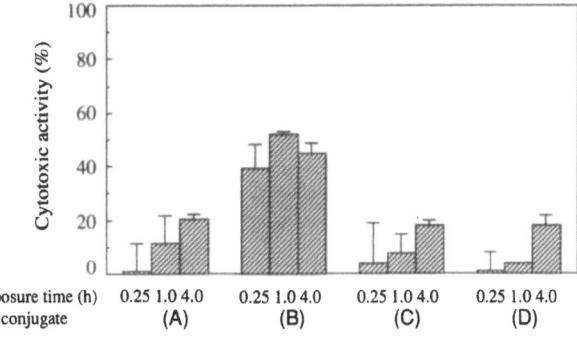

Figure 9. Structures of poly(α-malic acid)/amide/ DXR/saccharide conjugates, (A) (degree of substitution of saccharide = 0 : y=0), (B) (S : galactosamine), (C) (S : glucosamine), (D) (S : mannosamine).[33]

mia in mice *i.p./i.p.* These conjugates did not display an acute toxicity even in high dose ranges.

Poly(α-malic acid)/DXR/galactosamine conjugate showed a stronger cytotoxic activity than poly(α-malic acid)/DXR against *human hepatoma* cells *in vitro*. The results of cytotoxic activity by conjugates against HLE *human hepatoma* cells *in vitro* through short exposure time at 4°C are shown in Fig. 10. The cytotoxic activity by poly(α-malic acid)/DXR/galactosamine conjugate having galactosamine residues was higher than that of the other conjugates having the other kind of saccharide residues or no saccharide residue. The results mean that the number of poly(α-malic acid)/DXR/galactosamine conjugate binding on the surface of HLE *human hepatoma* cells was larger than that of the other conjugates. The inhibitory effect of poly(α-malic acid)/galactosamine conjugate having no drug group on the cytotoxic activity of poly(α-malic acid)/DXR/galactosamine conjugate or free DXR against HLE *human hepatoma* cells is shown in Fig. 11. The cytotoxic activity of free DXR was not affected by the addition of poly(α-malic acid)/galactosamine conjugate, however, the cytotoxic activity of the poly(α-malic acid)/DXR/galactosamine conjugate was inhibited by the addition of poly(α-malic acid)/ galactosamine conjugate. These results suggested that the high cytotoxic activity of poly(α-malic acid)/DXR/galactosamine conjugate resulted from the receptor-mediated uptake of poly(α-malic acid)/DXR/galactosamine conjugate into HLE *human hepatoma* cells. Therefore, the galactosamine residue was confirmed to act as a targeting moiety to *hepatoma* cells. Thus, poly(α-malic acid)/Gly-Phe-Leu-Gly//DXR/galactosamine conjugate can be expected to be used as the biodegradable and water-soluble macromolecular prodrug of DXR having targetability to *hepatoma* cells, releasing free DXR by lysosomal enzymes after uptake into *hepatoma* cells. Figure 12 shows schematically the transoprt of drug uptake into the cells through lysosomal digestion.

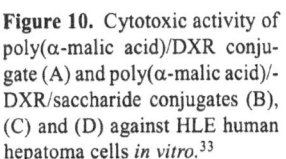

Figure 10. Cytotoxic activity of poly(α-malic acid)/DXR conjugate (A) and poly(α-malic acid)/-DXR/saccharide conjugates (B), (C) and (D) against HLE human hepatoma cells *in vitro*.[33]

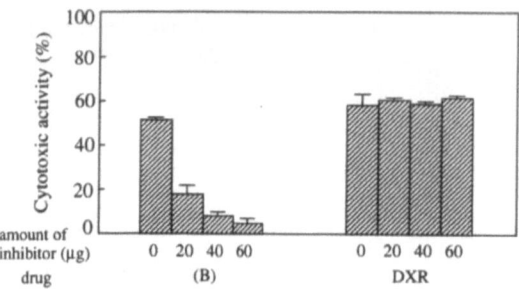

Figure 11. Effect of addition of poly(α-malic acid)/galactosamine conjugate on the cytotoxic activity of poly(α-malic acid)/DXR/galactosamine conjugate (B) and free DXR against HLE human hepatoma cells *in vitro*.[33]

4. CONCLUSIONS

The specificity attributable to macromolecular drugs results mostly from their high molecular weight and site-specific binding which hinder their passage across various barriers in the body. For example, limited renal excretion of high molecular weight polymers appears to be a serious problem. Although many chemists have been encouraged to synthesize polymer-bound drugs because they show persistent retention in the body, the limitation of the usage of nonbiodegradable and the high molecular weight of polymers present real problems in their potential clinical usage.

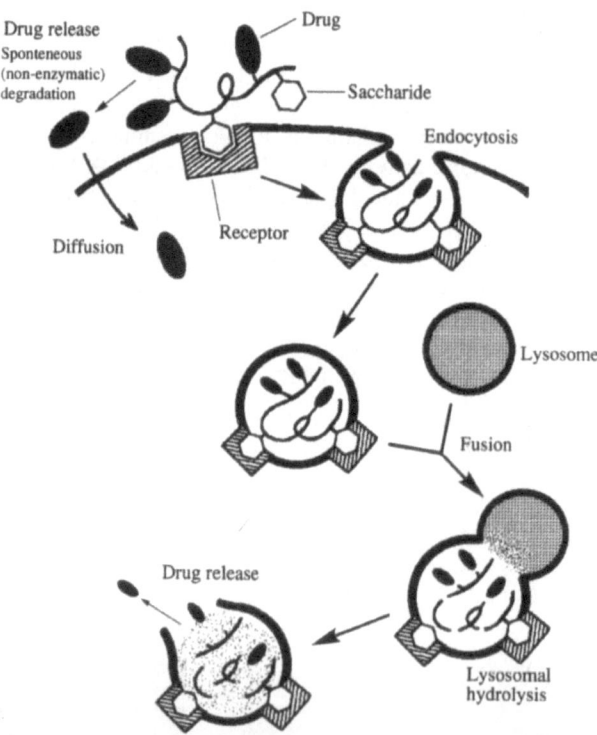

Figure 12. Transport of drug uptaken into the cells through lysosomal digestion.

There are some lectins or lectin-like receptors binding specific oligo- or monosaccharides in various organs or cells of various organisms. This ability of specific recognition of carbohydrates has been applied in various ways, such as in bio-assays similar to the antigen-antibody recognition. The macromolecular prodrug using saccharides as targeting moieties will be successfully employed in clinical situations in the near future.

REFERENCES

1. Donaruma, L. G. and Vogl, O., 1978, Polymeric Drugs, Academic Press.
2. Akashi, M., 1993, Polymers for Pharmaceutical and Biomolecular Engineering, *Biomedical Applications of Polymer Matetials* (Tsuruta, T., Hayashi, T., Kataoka, K., Ishihara, K., and Kimura, Y. eds.), pp. 371-406, CRC Press, Boca Raton.
3. Ringsdorf, H., 1975, Structure and Properties of Pharmacologically Active Polymers, *J. Polym. Sci. Polym. Symp.* 51:135-153.
4. Dumitriu, S., Popa, M., and Dumitriu, M., 1988, Polymeric biomaterials as enzyme and drug carriers Part I: Immobilization of enzymes, *J. Bioact. Compat. Polym.* 3:243-312.
5. Dumitriu, S., Popa, M., and Dumitriu, M., 1988, Polymeric biomaterials as enzyme and drug carriers Part II: Immobilization of coenzymes and other biological components, *J. Bioact. Compat. Polym.* 3:403-437.
6. Dumitriu, S., Popa, M., and Dumitriu, M., 1989, Polymeric biomaterials as enzyme and drug carriers Part III: Polymeric drug and drug delivery systems. *J. Bioact. Compat. Polym.* 4:57-73.
7. Dumitriu, S., Popa, M., and Dumitriu, M., 1989, Polymeric biomaterials as enzyme and drug carriers Part IV: Polymeric drug carrier systems. *J. Bioact. Compat. Polym.* 4:151-197.
8. Dunn, L. and Ottenbrite, R. M. eds., 1991, *Polymeric Drug and Drug Delivery Systems; ACS Symp. Ser. 469.*
9. Muzzarelli, R. A. A., Weckx, M., Filippini, O., and Lough, C., 1989, Characteristic properties of N-carboxybutyl chitosan. *Carbohydr. Polym.* 11:307-320.
10. Sirica A. E. and Woodman, R. J., 1971, Selective aggregation of L1210 leukemia cells by the polycation chitosan. *J. Natl. Cancer Inst.* 47:377-388.
11. Takagi, H. and Kadowaki, K., 1985, Purification and chemical properties of a flocculant produced by Paecilomyces sp. I-1. *Agric. Biol. Chem.* 49:3159-3164.
12. Ishitani, K., Suzuki, S., and Suzuki, M., 1988, Antitumor activity of polygalactosamine isolated from Paecilomyces sp. I-1 stain. *J. Pharmacobio. Dyn.* 11:58-65.
13. Suzuki, K., Tokoro, A., Okawa, Y., Suzuki S., and Suzuki, M., 1986, Effect of N-acetylchito-oligosaccharides on activation of phagocytes. *Microbiol. Immunol.* 30:777-787.
14. Waxman, S. and Scanlon, K. J., 1982, *Clinical Interpretation & Practice of Cancer Chemotherapy* (Greenspan, E. M. eds.), p. 39, Raven Press, New York.
15. Bounous, G., Pageau, R., and Regoli, P., 1978, The role of diet on 5-fluorouracil toxicity, *Int. J. Clin. Pharmacol. Biopharm.* 16:519-521.
16. Ouchi, T., Banba, T., Matsumoto, T., Suzuki S., and Suzuki, M., 1989, Antitumor activity of chitosan and chitin immobilized 5-fluorouracils through hexamethylene spacers via carbamoyl bonds. *J. Bioact. Compat. Polym.* 4:362-371.
17. Ohya, Y., Huang, T. Z., Ouchi, T., Hasegawa, K., Tamura, J., Kadowaki, K., Matsumoto, T., Suzuki, S., and Suzuki, M., 1991, Synthesis and antitumor activity of α-1,4-polygalactosamine and N-acetyl-α-1,4-polygalactosamine immobilized 5-fluorouracils through hexamethylene spacer groups via urea, urea bonds. *J. Control. Rel.* 17:259-266.
18. Ohya, Y., Inosaka, K. and Ouchi, T., 1992, Synthesis and antitumor activity of 6-O-carboxymethyl chitin fixing 5-fluorouracils through pentamethylene, monomethylene spacer groups via amide, ester bonds. *Chem. Pharm. Bull.* 40:559-561.
19. Ouchi, T., Inosaka, K., and Ohya, Y., 1994, Use of α-1,4-Plygalactosamine as a Carrier of Macromolecular Prodrug of 5-fluorouracil, *Polymeric Drugs and Drug Administration; ACS Symp. Ser.* (Ottenbrite, R. M. ed.), pp.204-213.
20. Subr, V., Kopecek, J., and Pohl, J., 1988, Cleavage of Oligopeptide Side-Chains in N-(2-hydroxypropyl)methacrylamide, *J. Control. Rel.*, 8:133-140.
21. Duncan, R., Kopeckova, P., Strohalm, J., Hume, I. C., Lloyd, J. B., and Kopecek, J., *Br. J. Cancer*, 57:147-156.
22. Sloane, B. F., Dunn, J. R., and Honn, K. V., 1981, Lysosomal Cathepsin B : Correlation with Metastatic Potential, *Science*, 212:1151-1153.

23. Keppler, D., Fondaneche, M. C., Dalet-Fumeron, V., Pagano, M., and Burtin, P., 1988, Immunohisto-chemical and biochemical study of a cathepsin B-like proteinase in humancolonic cancers, *Cancer Res.* 48:6855-6862.

24. Ohya, Y., Nonomura, K., Hirai, K., and Ouchi, T., 1994, Synthesis of 6-O-carboxymethylchitin immobi-lizing doxorubicins through tetrapeptide spacer groups and its enzymatic release behavior of doxorubicin in vitro, *Macromol. Chem. Phys.* 195:2839-2853.

25. Ohya, Y., Nonomura, K., and Ouchi, T., In Vivo and in Vitro Antitumor Activity of CM-chitin Immobi-lizing Doxorubicins through Lysosomally Digestible Tetrapeptide Spacer Groups, *J. Bioact. Compat. Polym.* in press.

26. Ouchi, T., Ohya, Y., 1994, Drug Delivery Systems Utilizing Carbohydrate Recognition, Neoglycoconju-gates : *Preparation and Application,* Lee, Y. C. and Lee, R. T. eds., Academic Press, pp.465-498.

27. Duncan, R., Hume, I. C., Kopeckova, P., Ulbrich, K., Strihalm, J., and Kopecek, J., 1989, Anticancer agents coupled N-(2-hydroxypropyl)methacrylamide copolymers. 3. Evaluation of adriamycin conju-gates against mouse leukemia L1210 in vivo. *J. Control. Rel.* 10:51-63.

28. Monsigny, M., Roche, A.-C., and Midoux, P., 1984, Uptake of neoglycoproteins via membrane lectin(s) of L1210 cells evidenced by quantitative flow cytofluorometry and drug targeting. *Biologie Cellulaire* 51:187-196.

29. Hudgin, R. L. and Ashwell, G., 1974, Studies on the role of glycosyltransferases in the hepatic binding of asialoglycoproteins. *J. Biol. Chem.* 249:7369-7372.

30. Seymour, L. W., Duncan, R., Chytry, V., Strohalm, J., Ulbrich, K., and Kopecek, J., 1991, Intraperitoneal and subcutaneous retention of a soluble polymeric drug carrier bearing galactose. *J. Control. Rel.* 16:255-262.

31. Ouchi, T., Fujino, A., Tanaka, K., and Banba, T., 1990, Synthesis and antitumor activity of conjugates of poly(α-malic acid) and 5-fluorouracils bound via ester, amide or carbamoyl bonds. *J. Control. Rel.* 12:143-153.

32. Ohya, Y., Kobayashi, H., and Ouchi, T., 1991, Design of poly(α-malic acid)-5FU-saccharide conjugate exhibiting antitumor activity. *React. Polym.* 15:153-163.

33. Ouchi, T.,Kobayashi, H., Hirai, K., and Ohya, Y., 1993, Design of poly(α-malic acid)-antitumor drug-saccharide conjugate exhibiting cell specific antitumor activity, *Polymeric Delivery Systems, Properties and Applications; ACS Symp. Ser. 520,* (EI-Nokaly, M. A., Piatt, D. M., and Charpentier, B. A. eds.), pp. 382-394.

8

TRANSPARENT MULTIPHASIC OXYGEN PERMEABLE HYDROGELS BASED ON SILOXANIC STATISTICAL COPOLYMERS

C. Robert,[1] C. Bunel,[1] M.A. Dourges,[1] J.P. Vairon,[1] and F. Boué[2]

[1] Laboratoire de Chimie Macromoléculaire
CNRS-URA 24
Université P. et M. Curie
4, Place Jussieu
75005 Paris, France
[2] Laboratoire Léon Brillouin
CNRS-UMR 12
C.E.N., Saclay, France

ABSTRACT

Highly oxygen permeable siloxanic hydrogels with Dk up to $170 \cdot 10^{-11}$ cm^3(O$_2$) cm cm$^{-2} \cdot$ s$^{-1} \cdot$ mmHg^{-1} and 20-30 wt% hydration can be obtained by copolymerization of acrylic acid (AA) with tris(trimethyl siloxy) γ methacryloxy propylsilane) (TRIS) or dimethacryloxybutyl polydimethylsiloxane (DMPDMS). Optical properties of these materials depend directly on size of hydrophilic demixed domains as it had been shown by small angle neutron scattering experiments.

INTRODUCTION

Oxygen permeability is of particular importance for biomedical polymers which are used in contact with living tissues. For some definite applications, like in contactology, these materials generally have also to be hydrophilic, colourless, transparent, and it is not simple to fulfill these four conditions.

The first idea is to play with the intrinsic oxygen permeability of water in homogeneous hydrogels, and thus to increase the hydration level of the material. According to Yasuda and Lamaze [1], the diffusion of small size solutes in water-swollen membranes can be related to the free volume fraction in the polymer, which, for hydrogels, is proportional to the degree of hydration. For solutes of smaller size than the pore size of the membrane, the permeability is exponentially depending on $[(1/H)-1]$ where H is the degree of hydration defined as the ratio of the mass of uptaken water to that of the swollen material $(0 < H < 1)$. The validity

Macromolecular Engineering, Edited by M.K. Mishra et al.
Plenum Press, New York, 1995

117

of this theory was confirmed for highly swollen networks (%H$_2$O > 30%)[2], and experiments on a series of hydrogels ranging from 30 to 80% water content led to the following relationship between hydration H and oxygen permeability defined as the product of diffusion coefficient (D) by oxygen solubility (k) [3,4]:

$$Dk = 1{,}39 \ 10^{-11} \ exp(4.42 \ H) \qquad at \ 22°C$$

Dk is expressed in cm^3(O$_2$) cm cm^{-2}s^{-1} (mmHg)$^{-1}$.

These results are in good agreement with the estimation of Refojo [5] who showed that an increase in temperature from 25 to 34°C would increase the permeability coefficient by 10%. Thus, oxygen permeability coefficients of hydrogels depend on water content and temperature, but they are obviously limited to the permeability of pure water (Dk ≈ 80 10^{-11} cm^3(O$_2$) cm cm^{-2}s^{-1}(mmHg)$^{-1}$ at r.t.). The commercial hydrogel soft lenses -the classical or the new disposable ones- are not adapted to the prolonged wear, as a permeability of about 120-150.10^{-11} is needed for safety of the cornea.

A second possibility consists in choosing chemical structures which give to the polymer network itself a high oxygen permeability. Some polymers are reported in the literature as highly permeable. They are mainly i) fluorinated polymers[6], ii) poly (alkyl or aryl fumarates)[7], iii) porphyrin derivatives[8], iiii) polysiloxanes. These polymers are inert compounds diversely used as biomaterials and most generally hydrophobic.

The very flexible polysiloxanes are the most oxygen permeable, particularly the polydimethylsiloxane (Dk ≥ 500.10^{-11} cm^3(O$_2$) · cm · cm^{-2}s^{-}1 · mmHg^{-1})[9]. But their lack of hydrophilicity, leading to adhesion to cornea, is the major hurdle for their use as material for soft contact lenses. Thus the idea merged to conceive water absorbing polysiloxane derivatives in which the siloxanic network constitutes the continuum needed for oxygen permeation. It has been proposed either to copolymerize hydrophilic monomers with siloxanic monomers[10,11], with difunctional polysiloxane macromers eventually bearing hydrophilic side-chains[12,13,14], or even to blend silicone rubber with a slightly crosslinked hydrogel precursor[15]. This gives heterophasic materials when swollen at equilibrium in water, the physical properties of which are depending on composition. A majority of siloxanic moieties will lead to an oxygen permeable continuous phase with hydrated demixed domains, the size and refractive index of which governing the transparency. The observed permeabilities (Dk) mostly range from a few to 90.10^{-11} even if some values are reported to exceed 100, and the hydrations are very variable, the lowest giving the highest permeability.

Following the above concept we prepared series of heterophasic hydrogels composed of a continuous siloxanic matrix containing demixed hydrophilic domains. The materials were obtained through statistical copolymerization, block copolymerization and interpenetrated networks elaboration[16,17].

We shall limit our report here to the syntheses, morphologies and properties of some hydrophilic polysiloxanes derivatives obtained by statistical copolymerization of acrylic acid (AA) with a siloxanic side-chain methacrylate(TRIS), and with a polysiloxanic dimethacrylic macromonomer (DMPDMS).

EXPERIMENTAL

1. Materials

- The tris(trimethylsiloxy) γ-methacryloxypropylsilane (TRIS) (**1**) was supplied from Roth Sochiel Co. and was washed before use, first with NaOH 10% and then with H$_2$O.

$$CH_2{=}\underset{\underset{O}{\parallel}}{\overset{\overset{CH_3}{|}}{C}}{-}C{-}O{-}(CH_2)_3{-}\underset{\underset{\underset{|}{CH_3}}{\overset{\overset{H_3C-\underset{|}{Si}-CH_3}{|}}{O}}}{\overset{\overset{CH_3}{\overset{|}{H_3C-\overset{|}{Si}-CH_3}}}{Si}}{-}O{-}\underset{\overset{|}{CH_3}}{\overset{\overset{CH_3}{|}}{Si}}{-}CH_3 \qquad (1)$$

- The dimethacryloxybutyl polydimethylsiloxane (DMPDMS) **(2)** was synthesized from 1,3bis(4-hydroxybutyl)tetramethyl disiloxane, supplied from Roth Sochiel Co. and purified by distillation. Methacryloyl chloride, supplied from Fluka, was distilled under reduced pressure (80 mmHg) in the presence of copper turnings. Octamethylcyclotetrasiloxane (D4), supplied from Aldrich was distilled under 15 mmHg, and pyridine was purified by distillation in the presence of KOH (115°C).

$$H_2C{=}\underset{\underset{O}{\parallel}}{\overset{\overset{CH_3}{|}}{C}}{-}C{-}O{-}(CH_2)_4{-}\underset{\overset{|}{CH_3}}{\overset{\overset{CH_3}{|}}{Si}}{-}O{-}{\left[\underset{\overset{|}{CH_3}}{\overset{\overset{CH_3}{|}}{Si}}{-}O\right]}_{n\text{-}2}{-}\underset{\overset{|}{CH_3}}{\overset{\overset{CH_3}{|}}{Si}}{-}(CH_2)_4{-}O{-}\underset{\underset{O}{\parallel}}{\overset{\overset{CH_3}{|}}{C}}{-}C{=}CH_2 \qquad (2)$$

- Acrylic acid, supplied from Aldrich, was purified by distillation in the presence of copper turnings.

2. Preparation of DMPDMS

This difunctional macromonomer was prepared as previously reported in literature[13].

2.1. Synthesis of the Precursor 1,3bis(4-Methacryloxybutyl)tetramethyl Disiloxane (DMDS). 10g (0.036 mole) of 1,3bis(4-hydroxybutyl)tetramethyldisiloxane, 12g (0.15 mole) of dry pyridine and 36 ml of hexane were charged into a reaction flask equipped with a magnetic stirrer. The mixture was chilled to 0°C and then 15 g (0.15 mole) of methacryloyl chloride was added dropwise, under inert atmosphere. The mixture was then stirred continuously overnight. After reaction the solution was extracted consecutively with 10% aqueous solutions of HCl and NH₃ in order to remove excess reagents and pyridine hydrochloride. The resulting solution of DMDS in hexane was dried over anhydrous MgSO₄. The drying salt was then filtered off, and the solvent was removed under reduced pressure.

2.2. Synthesis of Difunctional Macromonomer (DMPDMS). 5g (0.012 mole) of DMDS and 45 g (0.15 mole) of octamethylcyclotetrasiloxane (D4) were mixed under nitrogen. Trifluoromethane sulfonic acid (73μl) (Aldrich Chemical Cy) was added into the polymerization flask. The reaction mixture was stirred for about 15 hours under nitrogen. The acidic polymerization mixture was then neutralized by addition of "Terre de Clarcel" (Rhône Poulenc) dispersed in 100 ml of chloroform. The solution was filtered and the excess solvent was removed under reduced pressure. Low molecular weight unreacted cyclic siloxanes were removed by heating the monomer to 110°C under 0.2 mmHg. The resulting macromonomer still contains about 25 wt% of macrocycles formed during the chain extension. These macrocycles were characterized by SEC and NMR (absence of protons from butyl end groups).

3. Copolymerizations

Acrylic acid (AA) was copolymerized with TRIS or DMPDMS. In the first case, ethylene glycol dimethacrylate (EGDMA, Aldrich Chemical Cy) was used as crosslinker. With the difunctional DMPDMS crosslinking resulted from the siloxanic monomer itself.

The mixture containing monomers and 1% by weight of photoinitiator (Darocur 1173, Merck-Ciba) was poured on a polypropylene mold (disc shape, 15 mm diam., 0.1-0.2 mm thickn.) and covered with a thin glass plate. The material was then irradiated with UV source (Hg, 100W) for 30 to 60 minutes. After photopolymeri-zation, the resulting discs were extracted with chloroform for 4 hours, dried and then extracted for 15 hours with distilled water. In the case of DMPDMS-stat-AA copolymers, chloroform extraction elimi-nates the noncopolymerizable macrocycles present in the macromonomer.

4. Oxygen Permeability

Dry samples were swollen to equilibrium in distilled water. Oxygen permeability coefficients were determined by FATT's polarographic method [18]. Oxygen transmissibil-ity was measured with a permeometer 201 T (Rehler Development) at 35°C, in a water saturated box to avoid samples dehydration. Oxygen permeability coefficient (Dk) is computed from :

$$Dk = T \cdot e$$

where T and e are respectively the oxygen transmissibility and thickness of the sample. Transmissibility and thickness measurements were performed at the Research Center of Essilor International (Creteil, France).

5. Small Angle Neutron Scattering (S.A.N.S.)

The AA-stat-TRIS and AA-stat-DMPDMS copolymers were swollen to equilibrium in heavy water (Aldrich, 99,8 D%). The swollen specimens of 0.2 to 0.3 mm thickness were placed between quartz discs, sealed by compression against a glass gasket. Neutron scatter-ing experiments were performed at Leon Brillouin Laboratory, Orphee Reactor, CEN Saclay, France. Each scattering spectrum was corrected from transmission, sample thickness and detector efficiency using 1 mm water cell. The incoherent background was then substracted.

RESULTS AND DISCUSSION

1. Characterization of the Difunctional Macromonomer (DMPDMS)

The number average molecular weight of the DMPDMS was determined by ^{1}H NMR. The monomer comprises an average amount of about 40 repeating ($-Me_2Si-O-$) units which corresponds to $Mn \approx 3400$, and NMR agrees with the expected structure:

2. Copolymerization of Acrylic Acid with Siloxanic Methacrylates

2.1. Copolymerization with TRIS The different copolymers (samples 1-5) were for-mulated as shown in Table 1. To the mixture of monomers were added 1-5 percents by weight of EGDMA. The hydration (%) was calculated from the overall weight of the material swollen at equilibrium.

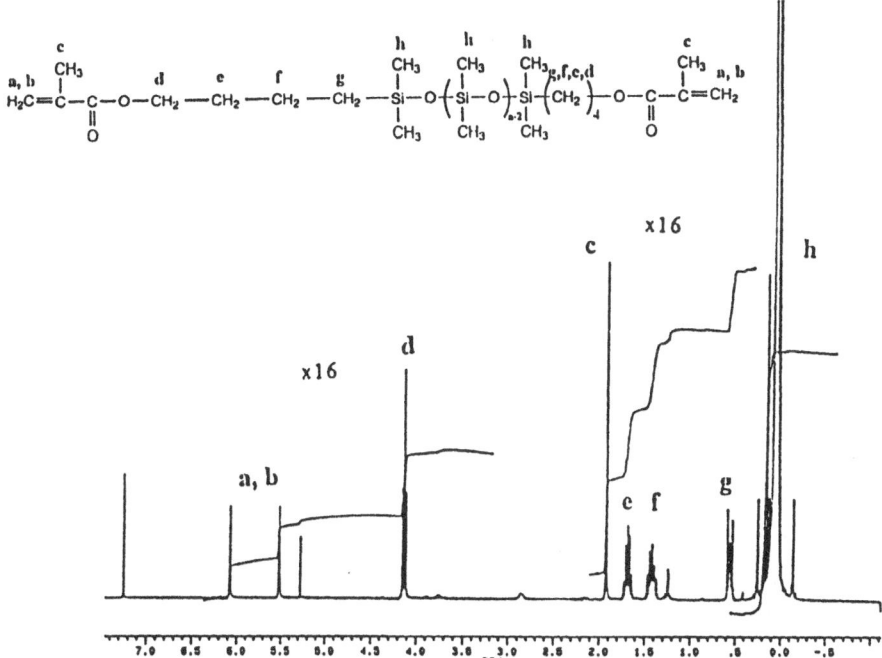

Figure 1. ^{1}H NMR spectrum (300 MHz) of DMPDMS in CDCl$_3$.

The Alfrey-Price copolymerization parameters were computed for TRIS using the experimental reactivity ratios we measured at 60°C for the pair TRIS/N-vinylpyrrolidone ($r_{TRIS} = 3.5$, $r_{NVP} = 0.08^{(16)}$) :

$$Q_{TRIS} = 0.49 \qquad e_{TRIS} = -0.024$$

This allowed, using the Q-e values for AA [19], to compute the reactivity ratios for the pair TRIS/AA :

$$r_{TRIS} = 0.41, r_{AA} = 1.27.$$

These values show that the chains are progressively enriched in siloxanic units as copolymerization proceeds, and average polyAA sequences are calculated to be composed

Table 1. Composition (wt%), hydration at equilibrium (wt%) and oxygen permeability at 35°C of TRIS-stat-AA copolymers

Sple No.	% TRIS	% AA	% EGDMA	% H$_2$O	Dk . 10^{11} (*)
1	75	25	5	10	63
2	75	25	2.5	11	70
3	75	25	1	14	80
4	60	40	1	25	68
5	50	50	5	32	50

(*) in cm^3(O$_2$) cm cm^{-2} s^{-1} mmHg^{-1}, accuracy ± 2

Table 2. Composition (wt%), hydration (wt%) and oxygen permeability at 35°C of copolymers DMPDMS-stat-AA

Sple No.	% DMPDMS	% AA	% H_2O	$Dk \cdot 10^{11}$ (*)
6	80	20	15	170
7	75	25	20	150
8	70	30	25	150
9	65	35	32	150

(*) $cm^3(O_2)$ cm cm^{-2} s^{-1} $mmHg^{-1}$, accuracy ± 30

of about 4 to 8 units, depending on the copolymer compositions considered in Table 1. Thus, in the hydrated state, the demixed polyAA domains should remain small and nonscattering. It appears experimentally that most of hydrated samples are clear (§ 4) but all of them are rigid whatever the water content. This agrees with the nature of the network which in the continuum should be comparable to that of the rigid homo(polyTRIS)[20]. For a constant monomers to cross-linker ratio (samples 1 and 5 ; 3 and 4), the hydration increases as expected with AA content. For a constant TRIS to AA ratio (samples 1, 2 and 3), the hydration remains roughly the same even if it is slightly influenced by the concentration of EGDMA.

2.2. Copolymerization with DMPDMS

Composition of copolymers is given in Table 2. All materials are soft either in the dry state or when hydrated at equilibrium, which agrees with a continuum mainly made up from flexible polysiloxane sub-chains. The water uptake roughly follows the AA content, which corresponds well with an increase of hydrophilic component and simultaneous decrease in crosslinking density.

3. Oxygen Permeability

For TRIS-stat-AA materials oxygen permeability coefficients are lower than that of pure water (Table 1). Their variations with initial TRIS/AA/EGDMA composition remain limited but are nevertheless coherent and significant. For a constant monomers/EGDMA ratio (samples 1 and 5; 3 and 4) the Dk increases when the percentage of TRIS increases even though the water content decreases. This is a non-classical behavior for hydrogels for which the Dk increases exponentially with their equilibrium water content as was considered before. This is easily explained in the present case as now the oxygen permeability is mainly depending on siloxanic monomer content and no more on hydration. For a constant TRIS/AA ratio (samples 1, 2 and 3), the Dk values increase when the EGDMA content is lowered, which also agrees with the expected increase of permeability when decreasing the cross-linking density of the network.

The copolymerization of DMPDMS with AA leads to discs which are much more oxygen permeable than the previous copolymers as the observed Dk are twice that of pure water (Table 2). The different behaviors for the two systems can be explained by the chemical natures of the respective networks. In this last case the continuum surrounding the hydrophilic demixed domains is mainly constituted of flexible polysiloxanic sub-chains ($DP_n \approx$ 40), the free volume of which is considerably enhanced with respect to that of the methacrylic backbone of TRIS copolymers. The DMPDMS matrix obviously resembles more than that of a classical silicone rubber, even if polyAA sub-chains are participating to the mesh and if their partial demixing introduces a supplementary physical crosslinking.

Table 3. Optical properties of water swollen copolymers TRIS-stat-AA (samples 1 to 5) and DMPDMS-stat-AA (samples 6 to 9)

Sple No.	% H_2O	Optical properties
1	10	Clear
2	11	Clear
3	14	Clear
4	25	Clear
5	32	Hazy
6	15	Clear
7	20	Clear
8	25	Hazy
9	32	Hazy

4. Optical Properties and Morphologies of the Hydrated Materials

Optical properties of the two types of copolymers are obviously depending on the polyAA content and consequently on hydration level (Table 3), but it is remarkable that transparency is kept for water content as high as ≈ 30 wt%. This could mean that up to this limit, the demixed domains, hydrated at equilibrium, are numerous with an average size small enough to prevent light scattering. Such a morphology agrees with the statistical nature of the copolymers and the rather short lengths of the hydrophilic and hydrophobic sequences, but it is not clear how the water can attain small, isolated polyAA domains dispersed in a siloxanic matrix.

A second and different explanation for transparency is related to the respective refractive indexes of the separate phases. Depending on hydration, the index of the polyacrylic acid phase decreases from $n_D = 1.51$ (dry state) towards that of water ($n_D = 1.33$). It might become close enough from index of the polysiloxanic continuum ($n_D = 1.40$) and thus limit scattering. In this case transparency should be no longer related to the size of demixed domains and an eventual percolation could explain the easy and rather high hydration of the materials.

In order to clarify this point we undertook a study of the morphologies of the different copolymers with respect to their composition by small angle neutron scattering technique. After hydration at equilibrium in heavy water the samples Nos. 4, 7, and 8 have been studied over a wide range of the scattering vector q ($q = (4\pi/\lambda) \sin(\theta/2) \approx 2\pi/\lambda D$). We used a sample to detector distance $D \approx 3.2$ m and the three wavelengths 5, 12.8 and 16 Å, leading to a satisfactory connection between the three spectra. The scattering is large enough for good counting statistics, but low enough to rule out any multiple scattering. The scattering is central, i.e. shows no maximum as a function of q and decreases monotonically. On ln(I) vs ln(q) plots (Figures 2 and 3), it is characterized for the 3 samples by a plateau at low q with a crossover towards a slope relatively close to -4 at large q.

Such a behavior is close to the one predicted by the DEBYE and BUECHE's relation[21]:

$$I(q) = \frac{8\pi \, \xi^3 K_f^2 \Phi(1-\Phi)}{\left(1+q^2\xi^2\right)^2} = \frac{C}{\left(1+q^2\xi^2\right)^2} \qquad [1]$$

for a random biphasic material with compact inclusion of one species inside the other. K_f is the maximum contrast that would correspond to a complete separation of (PAA+D_2O) in one

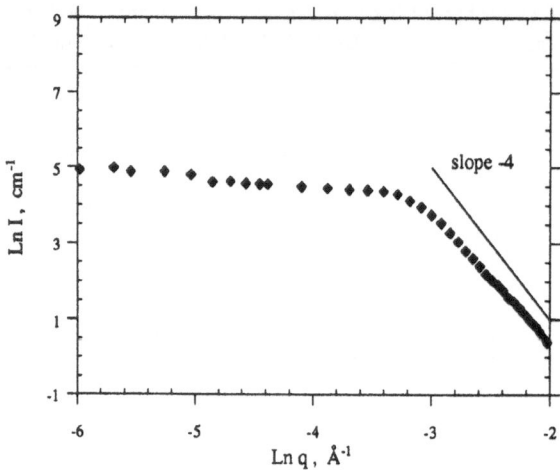

Figure 2. Ln I(q) versus Ln q for TRIS-stat-AA 60/40 (sample 4).

phase and TRIS or DMPDMS in another one, Φ is the volume fraction of the (PAA+D_2O), and ξ is the correlation length giving the size of the demixed domain.

The variation at low q can be reanalysed using a plot $1/\sqrt{I}$ vs q^2, yielding the ξ value obtained for each sample (Table 4).

The large q behavior ($q > \xi^{-1}$) is the same as for an isolated particle. We note the crossing of the curves for the two samples 7 and 8, which yields correlation lengths ξ_7 and ξ_8, with $\xi_7 < \xi_8$ (Figure 3, Table 4).

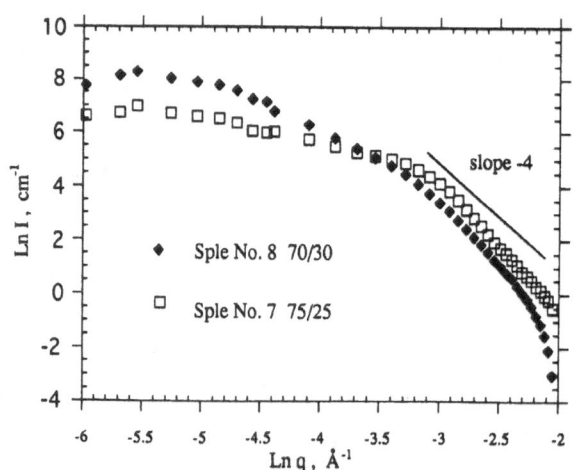

Figure 3. Ln I(q) versus Ln q of DMPDMS-stat-AA samples 7 and 8.

Table 4. Small angle neutron scattering experiments. Correlation lengths and optical properties for samples hydrated in heavy water

Sple No.	Composition	% D_2O	Optical properties	ξ (Å)
4	TRIS-AA= 60-40	25	Clear	30
7	DMPDMS-AA= 75-25	20	Clear	25
8	DMPDMS-AA= 70-30	25	Hazy	>150

This is interpreted in the following way:

- at low q (q → 0), the scattering becomes proportional to ξ^3. So, if the contrast remains reasonably constant i.e. if the degree of demixion is constant, $\xi_8^3 > \xi_7^3$ implies $I_8 > I_7$. The value of ξ^3 corresponds to the average volume of the demixed domain.
- at large q, the scattering intensity becomes proportional to $1/\xi q^4$ (see eq. 1) and thus $1/\xi_8 q^4 < 1/\xi_7 q^4$, so $I_8 < I_7$. The $1/q^4$ variation can be rewritten as $2\pi K^2 \phi(1-\phi)(S/V)q^{-4}$ where S/V is the specific area of the demixed domain. It can be seen on the two figures that there is a small deviation from the slope -4, indicating that the interface is not perfectly sharp. We find however that this specific area, proportional to ξ^{-1}, is lower for large objects.

To summarize, the scattering of the three considered samples is close to the one of a well-defined biphasic medium, even if the separating interface appears not completely sharp. Figure 3 and Table 4 show that the less crosslinked sample (No. 8) presents quite a larger low angle scattering intensity and a larger characteristic size. The crosslinking has an obvious effect of governing the size of the demixed domains. For samples 4 and 7 the hydrated domains are small enough to preserve the transparency. For sample 8 we increased the AA content and correlatively decreased the crosslinking density by lowering the DMPDMS content. The domains are expected to be bigger, which is observed (ξ > 100 Å), and they don't allow to preserve the transparency of the material. Thus, the optical properties are in good agreement with the calculated correlation lengths.

CONCLUSION

Clear, hydrophilic and oxygen permeable materials can be obtained through statistical copolymerization of AA with siloxanic monomer or macromonomer like TRIS or DMPDMS. The compositions, for which the balance between hydration, oxygen permeability and optical properties is the best, are TRIS/AA = 60-40 and DMPDMS/AA = 75-25 (wt %). In all cases, the materials are clear if the size of hydrated demixed domains is small enough, as it has been shown by small angle neutron scattering experiments. An increase in crosslinking density leads to a decrease of hydrated domains size and to better optical properties but, on the opposite, it induces a decrease of water content and oxygen permeability.

The highest permeability coefficients are obtained when AA is copolymerized with a polysiloxanic network precursor like DMPDMS (Dk $\approx$ 150·10^{-11}), and the maximal water content which preserves transparency is 20-25 % by weight. These materials are very flexible and might be considered for soft, prolonged wear contact lenses.

ACKNOWLEDGMENT

The supports of Ministry of Research (MESR) and of Essilor International are gratefully acknowledged.

REFERENCES

1. Yasuda H., Lamaze C.E., 1971, J. Macromol. Sci. Phys., 5: 111
2. Yang W.H., Smolen W.F., Peppas N.A., 1981, J. Membr. Sci., 9: 53
3. Sarver M.D.,Baggett D., Harris M.G., Louie K., 1981, Am. J. Optom. Physiol. Opt., 58: 386
4. Caston J., Fatt I., 1981, J. Am. Optom. Ass., 52: 237
5. Holly F.J., Refojo M.F., 1972, J. Am. Optom. Ass., 43: 1173
6. Riess J.G., 1987, J. Chim. Phys., 84(9): 1119
7. Murata Y., Hirano J., 1985, Chem. Econ. Stud. Eng. Rev. 17(10): 18
8. Nishide H., Kawakami K., Suzuki T., Azechi Y, Tsuchida, E., 1990, Macromolecules, 23(15): 3714 (1990)
9. Laroche J., Lalanne F., Michot G., Colloque CNRS, Membranes à perméabilité sélective, Ed. CNRS, Paris 1969.
10. Chang S.H., 1980, U.S. Pat. No. 4,182,822; 1988, WO 88 05060
11. Bambury R.E., Seelye D.E., 1991, U.S. Pat. No. 5,070,215
12. Mueller K.F. et al., 1979, U.S. Pat. No. 4,136,250
13. Chromecek R.C., Deichert W.G., Falcetta J.J., VanBuren M.F.,1981, U.S. Pat. No.4,276,402
14. Keogh P.L., Kunzler J.F., Niu G.C.C., 1981, U.S. Pat. No. 4,260,725
15. Cifkova I., Lopour P., Vondracek P., Jelinek F. 1990, Biomaterials, 11: 393
16. Robert C., 1993, Thesis, University P. et M. Curie, Paris
17. Robert C., Bunel C., Vairon J.P., 1993, Fr. Pat. No. 9310763
18. Fatt I., St Helen R., 1971, Am. J. Optom., 48: 545
19. Young L.J., 1975, Polymer Handbook, Sec. Ed., II-394, Wiley
20. Ali M. I., 1992, WO 92 07013
21. Debye P., Anderson H.R., Bumberger H., 1957, J. Appl. Phys., 28: 679

9

PREPARATION OF TUBULAR POLYMERS FROM CYCLODEXTRINS

Akira Harada,[*] Jun Li, and Mikiharu Kamachi

Department of Macromolecular Science
Faculty of Science
Osaka University
Toyonaka, Osaka 560
Japan

1. INTRODUCTION

In recent years, polymers having unique structures, such as comb like polymers[1], double helical polymers[2], dendrimers[3], and polymers with bilayer membranes structures[4], have been prepared and attracted much attention because of their unique properties and functions. Much effort has been devoted to obtain other polymers with unique structures. In spite of strenuous efforts, tubular polymers have not been obtained until recently. Tubular polymers are ubiquitous in nature, especially in the living systems. Microtubules, flagella, and ion channels have tubular structures of various sizes. Tubular polymers in the living systems have various functions, not only as channels for membranes, but also as devices for mechanical movements. However, there have been no synthetic tubules made of organic materials of nanometer size.

Carbon nanotubes have been prepared by an arc-discharge method similar to that of fullerene synthesis[5]. These tubes range from about 1 to 30 nanometers in diameter. In order to design and construct smaller tubes, subnano-tube, direct chemical synthesis may be a more convenient approach. One of the most promising methods to construct subnanotubes chemically is to pile up ring molecules. Among the ring molecules, cyclodextrins(CDs), which are cyclic oligomers of glucose linked through α-1,4-glycosidic linkages, are one of the most suitable candidates for this purpose, because cyclodextrins have cavities with diameters of 0.45 nm for α-CD, 0.7 nm for β-CD, and 0.85 nm for γ-CD, respectively[6]. Moreover, glucose rings of CD are almost perpendicular to the cavity plane and the cavities of CDs have some depth (0.7 nm). CDs have many hydroxy groups on the both sides of the rings, which might enable CDs to link together to form a channel.

Previously, we prepared inclusion complexes of CDs with polymers[7] and polyrotaxanes[8], in which many α-CDs are threaded on a polymer chain. Now we have succeeded in

---

[*] Address correpondence to Dr. Harada or Dr Kamachi.

Macromolecular Engineering, Edited by M.K. Mishra et al.
Plenum Press, New York, 1995

Table I. Comparison of hydrophilic polymers with various chain cross-sectional areas in formation of crystalline complexes with cyclodextrins

Polymer	Structure	Yield (%)		
		α-CD[a]	β-CD[b]	γ-CD[c]
PEG (MW 1000)	$+CH_2CH_2O+_n$	92	0	trace
PPG (MW 1000)	$+CH_2CHO+_n$ $\quad\ \ CH_3$	0	96	80
PMVE (MW 20000)	$+CH_2CH+_n$ $\quad\quad O$ $\quad\quad CH_3$	0	0	67

[a] α-CD saturated aqueous solution, 1.5 mL.; polymers, 15 mg. [b] β-CD saturated aqueous solution, 7 mL; polymers, 15 mg. [c] γ-CD saturated aqueous solution, 2 mL; polymers, 15 mg.

preparing a subnano-tube by crosslinking neighboring CDs of polyrotaxane and removing stoppers and the polymer chain. We call these subnano-tubes a "molecular tube"[9].

2. COMPLEX FORMATION BETWEEN CYCLODEXTRINS AND POLYMERS

We conducted some experiments to see whether cyclodextrins would form complexes with some water-soluble nonionic polymers by the same procedure as that for low molecular weight compounds. **Table I** shows the results of the formation of the complexes of cyclodextrins with some nonionic polymers. We found that cyclodextrins did not form complexes with some nonionic, water-soluble polymers, such as poly(vinyl alcohol) (PVA) and polyacrylamide (PAAm). However, we found that α-CD forms complexes with poly(ethylene glycol) (PEG) in high yields in a crystalline state[10]. β-CD did not form complexes with PEG of any molecular weight . β-CD formed complexes with poly(propylene glycol) (PPG) to give crystalline compounds in high yields[11], although α-CD did not form complexes of PPG of any molecular weight. Moreover, we found that γ-CD formed complexes with poly(methyl vinyl ether) (PMVE)[12], although α- and β-CDs did not form complexes with PMVE. There is a good correlation between the cross-sectional areas of polymers and sizes of CDs.

Recently, we found that cyclodextrins form complexes with not only hydrophilic polymers but also hydrophobic polymers, such as oligoethylene[13] and polyisobutylene[14]. **Table II** shows the yields of the complexes between cyclodextrins and some hydrophobic polymers. α–CD forms complexes with oligoethylenes, although β-, and γ-CD did not form complexes oligoethylenes under the same conditions. On the other hand, β-, and γ-CD formed complexes with polyisobutylene (PIB), although α-CD did not form complexes with PIB. There is a good correlation between the sectional areas of polymers and sizes of CDs, again.

Carbohydrate polymers, such as dextran and pullulan, did not form insoluble complexes with PEG. Amylose and dextrin did not form insoluble complexes with PEG. Glucose, methyl glycoside, maltose, and maltotriose did not form complexes with PEG. Cyclodextrin derivatives, such as glucosyl-α-CD, maltosyl-α-CD, and soluble polymers of α-CD, did not form insoluble complexes with PEG.

Table II. Formation of solid-state complexes between cyclodextrins and hydrophobic polymers/oligomers with various chain sectional areas

polymer/oligomer	structure	molecular weight	Yield (%)		
			α-CD	β-CD	γ-CD
OE(20)	—CH$_2$CH$_2$—	563	63	0	0
squalane	—CH$_2$CHCH$_2$CH$_2$— 　　　| 　　　CH$_3$	423	0	62	24
PIB	CH$_3$ 　　　| —CH$_2$C— 　　　| 　　CH$_3$	~800	0	8	90

2.1. Effects of the Molecular Weight of Polymers on the Yields of the Complexes

When aqueous solutions of PEG were added to a saturated aqueous solution of α-CD at room temperature, the solution became turbid and the complexes were formed as precipitates. The complexes were isolated by filtration or centrifugation. Figure 1 shows the yields of the complexes of α-CD with PEG of various molecular weights. The yields are calculated on the basis of 2:1 (ethylene glycol unit : α-CD) stoichiometry which will be mentioned in the following section. α-CD did not form complexes with the low molecular weight analogs. α-CD formed complexes with PEG of molecular weight higher than 200. The yields increase with an increase in the molecular weight. The complexes were obtained almost quantitatively with PEG of molecular weight over 1000. β-CD did not form complexes with PEG of any molecular weight. Although PEG of molecular weight over 1000 formed complexes with α-CD slowly, they gave high yields after several hours.

PPG formed complexes with β-CD when the molecular weight is higher than 400. The yields increased with increase in the molecular weight, reached the maximum at the molecular weight 1000, and then decreased with the molecular weight. This is in contrast to

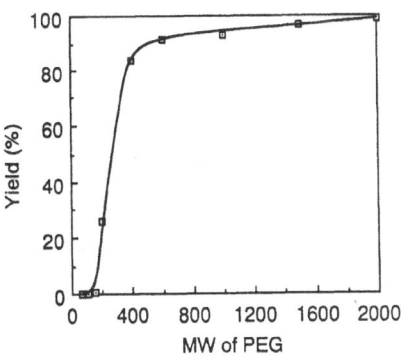

Figure 1. Yields of the complexes of α-CD with PEG as a function of the molecular weight of PEG.

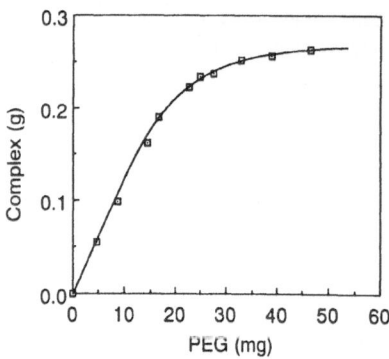

Figure 2. Amount of α-CD-PEG complexes as a function of added PEG (MW=600). A total amount of 2 mL of saturated aqueous solution of a-CD was used.

the case of the complex formation of α-CD with PEG. This is due to the fact that PPG is more hydrophobic than PEG.

β-CD and γ-CD formed complexes with polyisobutylene (PIB). The yields of the complexes of β-CD with PIB decreased with increase in the molecular weight of PIB. In contrast, the yields of the complexes of γ-CD with PIB increased with increase in the molecular weight of PIB. The chain-length selectivity between β-CD and γ-CD is totally reversed.

2.2. Stoichiometries of the Complexes

The complex formation of α-CD with PEG was studied quantitatively. Figure 2 shows the yields of the complexes of α-CD with PEG of average molecular weight 600 as a function of added PEG. The yields increased linearly when the amount of PEG added is small and leveled off at the molar ratio of 2:1 (ethylene glycol unit : CD). These results indicate that the complex formation is stoichiometric. The saturation values show that more than 90 % of the α-CD was consumed by the complex formation with PEG. The continuous variation plots for the complexation between α-CD and PEG also suggest that the stoichiometries of the complexes are 2:1. The stoichiometries were confirmed by the ¹H NMR spectrum. Figure 3 shows the ¹H NMR spectrum of the complex of PEG-600 with α-CD. It should be noted that the stoichiometries of the complexes are always 2:1 even if α-CD and PEG are mixed in any ratio.

PPG-β-CD complexes are also 2:1 (monomer unit:CD). γ-CD-PMVE complex, γ-CD-PIB complex, and α-CD-oligoethylene complex are 3:1 (monomer unit:CD). The length of two ethylene glycol units (or propylene glycol units) or three MVE (or PIB) units corresponds to the depth of the CD cavities (6.7 A).

2.3. Complex Formation of α-Cyclodextrin with Monodisperse Oligo(ethylene glycol)s[15]

We have used so far commercially available PEGs, which are polydisperse. Therefore, the complexes obtained were polydisperse and heterogeneous. We also found that α-CD did not form complexes with low molecular weight analogs, such as ethylene glycol and bis(ethylene glycol). In order to make clear the chain-length selectivity and obtain pure monodisperse complexes, we prepared monodisperse oligo(ethylene glycol)s (OEG) and studied the interactions between α-CD and the pure oligo(ethylene glycol)s.

Oligo(ethylene glycol)s HO(-CH₂CH₂O)n-OH (n=8,12, 28, 20, 28, 36, 44) were prepared by stepwise reactions starting from α,ω-tetrakis(ethylene glycol) ditosylate and the

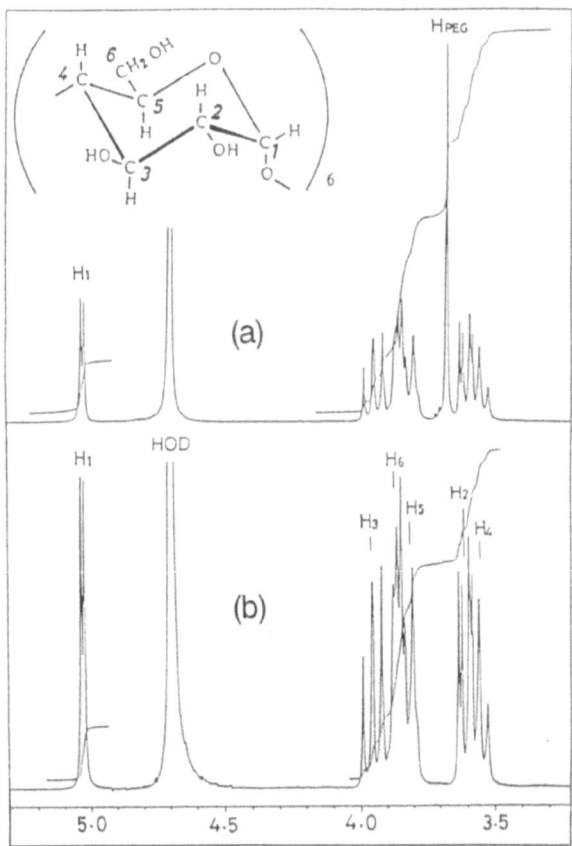

Figure 3. ¹H NMR spectra of the complexes of PEG with a-CD (a) and α-CD (b) in D₂O.

monosodium tosylate. The products were purified by preparative size-exclusion chromatography (SEC) repeatedly.

Figure 4 shows the yields of the complexes of α-CD with OEG as a function of the degree of polymerization of OEG. The yields are calculated on the basis of 2:1 stoichiometry. α-CD did not form complexes with ethylene glycol, bis(ethylene glycol), and tris(ethylene glycol). α-CD formed complexes with tetrakis(ethylene glycol) (TEG) and larger OEG. The yields increase sharply with an increase in the degree of polymerization from 5 to 12. The complexes were obtained almost quantitatively with eicosakis(ethylene glycol) and larger OEG. β-CD did not form complexes with any OEG. The stoichiometries of the complexes are 2:1 (two ethylene glycol units and one α-CD) when the degree of polymerization is higher than 6. The stoichiometries of the complexes of α-CD with TEG and pentakis(ethylene glycol) (PEG) are 2:1 (CD:OEG). This finding that a minimum PEG length is required for the formation of stable cyclodextrin complexes shows the importance of cooperativity in complexation and is similar to the formation of PEG complexes with hydrogen-donor polymers such as poly(acrylic acid).

Table III shows the results of complex formation between α-CD and cyclic oligomers of ethylene glycol together with those of linear OEG for comparison. It is interesting that the yields of the complexes of α-CD with cyclic OEG decreased with an increase in the size

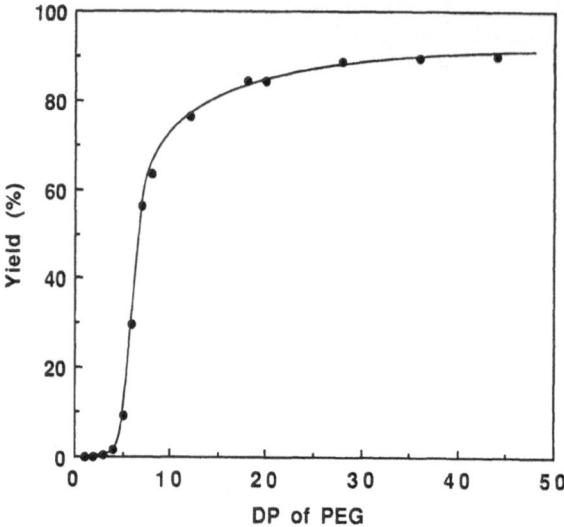

Figure 4. Yields of the complexes of α-CD with oligo(ethylene glycol) as a function of the degree of polymerization.

of the guest, and those of the complexes of α-CD with linear OEG increased with an increase in the chain length. Cyclic oligomers of ethylene glycol (crown ethers, 15-crown-5 and 18-crown-6) did not form complexes with α-CD, except for 12-crown-4 which gave complexes with α-CD in low yield. These crown ethers are too large to fit in the CD cavity and α-CDs are not able to penetrate the chain due to the absence of the chain ends. These results indicate that end groups are required for the complex formation.

2.4. Properties of the Complexes

The complex of α-CD with PEG, that of β-CD with PPG, and that of γ-CD with PMVE of low molecular weights (1000) are soluble in water. The complexes of higher

Table III. Complex formation between CD and PEG with various end groups

R(CH$_2$CH$_2$O)$_n$CH$_2$CH$_2$R'			Yield (%)	
R	R'	MW	α-CD	β-CD
–OH	–OH	1000	90	0
–NH$_2$	–NH$_2$	1450	90	0
–OCH$_3$	–OCH$_3$	1000	93	0
-OC(O)-C$_6$H$_3$(NO$_2$)$_2$	-OC(O)-C$_6$H$_3$(NO$_2$)$_2$	900	0	0
-OC(O)-C$_6$H$_3$(NO$_2$)$_2$	–OCH$_3$	900	77	10
-NH-C$_6$H$_3$(NO$_2$)$_2$	-NH-C$_6$H$_3$(NO$_2$)$_2$	3700	0	0

molecular weight can be dissolved in water by heating. The addition of an excess amount of low molecular weight guests, such as benzoic acid, propionic acid, and propanol, to the suspension of the complex resulted in solubilization of the complex when the molecular weight is low (1000). The formation of the complex is reversible. In solution, complexes are in equilibrium between the complex and its component. The addition of salts, such as NaCl and KCl, did not cause any change in the solubility of the complexes. This result indicates that there are no ionic interactions between CDs and the polymers. The addition of urea, which is thought to affect hydrogen bonds, results in solubilization of the complexes. The results indicate that hydrogen bonding plays an important role in forming the complexes between these polymers and CDs.

The complexes of α-CD with oligoethylene and those of γ-CD with PIB are sparingly soluble in water. Hydrophobic interactions plays an important role in stabilizing the complexes.

The decomposition point of the complexes is a little higher than that of the cyclodextrin. The complex of α-CD with PEG-1000 decomposes above 300 °C, whereas α-CD melts and decomposes below 300 °C. Thus, poly(ethylene glycol) stabilizes α-CD.

2.5. Inclusion Modes

Figure 5 shows the X-ray powder patterns of the complex of α-CD with oligoethylene and those with other low molecular weight compounds. Saenger et al. reported that the structures of the inclusion complexes of CDs with low molecular weight compounds can be classified by two groups; one is "cage type", and the other is "channel type"[16]. The X-ray powder pattern of the α-CD-oligoethylene complex shows that the complexes are crystalline, and the patterns are very similar to those of the complex of α-CD with valeric acid or octanol, which have been reported to have extended column structure, and totally different from those of the complexes with small molecules, such as acetic acid, propionic acid, and propanol, which have a cage type structure. These results indicate that the complexes of α-CD and oligoethylene are isomorphos with those of channel type structure rather than the so-called "cage type" structure. Similar results were obtained with α-CD-PEG complexes. The X-ray powder pattern of β-CD-PPG complex is similar to that of β-CD-p-iodoaniline complex, which has been reported to have channel structures by means of single crystal X-ray crystallography.

Molecular models show that PEG and oligoethylene chains are able to penetrate α-CD cavities, while the PPG and PMVE chains cannot pass through the α-CD cavity. These views are in accordance with our results that α-CD formed complexes with PEG and oligoehtylene but not with PPG and PMVE. β-CD did not form complexes with PEG. A PEG chain is too slim to fit in the β-CD cavity. However, β-CD forms complexes with PPG. Model studies further indicate that the single cavity (depth 6.7Å) accommodates two ethylene glycol (or propylene glycol) units (6.6 Å) or three methyl vinyl ether units (or ethylene units) when chains assume a planar zigzag conformation.

Figure 6 shows the ^{13}C CP/MAS NMR spectra of α-CD and the α-CD-oligoethylene complex. α-CD assumes a less symmetrical conformation in the crystal when it does not include a guest in the cavity. In this case, the spectrum shows resolved C-1 and C-4 resonances from each of the six α-1,4-linked glucose residues. Especially C-1 and C-4 adjacent to a conformationally strained glycosidic linkage are observed at 80 and 98 ppm, respectively. On the other hand, in the spectrum of the α-CD-oligoethylene complex the peaks at 80 and 98 ppm disappeared. Each carbon of glucose can be observed in a single peak. These results indicate that α-CD adopts a symmetrical conformation and each glucose unit of CD is in a similar environment. The X-ray studies of single crystals showed that α-CD

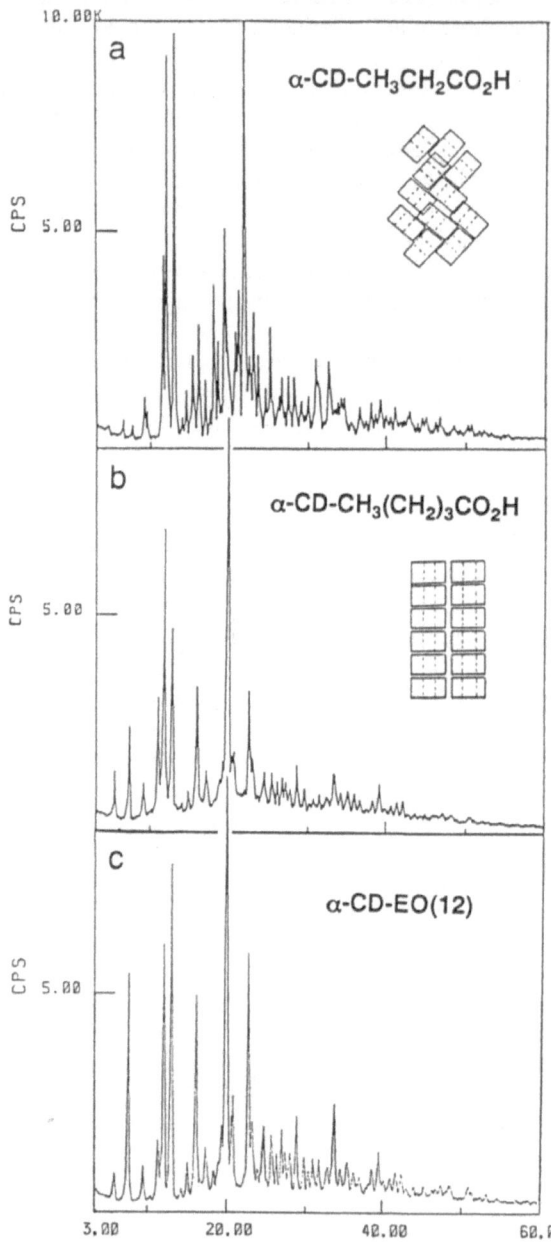

Figure 5. X-ray diffraction patterns for α-CD complexes, α-CD-propionic acid (a), α-CD-valeric acid (b), and α-CD-oligoethylene.

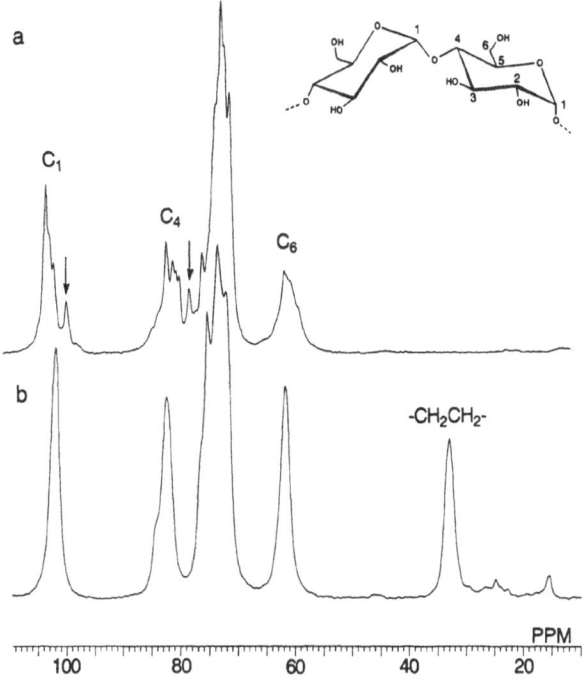

Figure 6. ^{13}C CP/MAS NMR spectra of α-CD (a) and the α-CD-oligoethylene complex (b).

assumes a less symmetrical conformation when it does not include guests in the cavity and α-CD adopts a symmetrical conformation when it includes guests in the cavities. CP/MAS NMR spectra of complexed and uncomplexed CDs are consistent with the results by X-ray. So an oligoethylene chain is thought to be included in the cavities. Similar results were obtained to those with α-CD-PEG complexes.

2.6. Complex Formation between Cyclodextrin and Poly(ethylene glycol) Derivatives

Table IV shows results of complex formation between α-CD and PEG with various end groups. First, PEGs with small end groups, such as methyl, dimethyl, and amino groups, form complexes. The yields are rather higher than unmodified PEG. These results indicate that interactions (hydrogen bonds) between OH groups of OEG and OH groups of α-CD are not the driving force for complex formation.

PEG carrying bulky substituents such as a 3,5-dinitrobenzoyl group and a 2,4-dinitrophenyl group at both ends of the PEG, which do not fit or pass through the α-CD cavity, did not form any complexes with α-CD.

Figure 7 shows a proposed structure of the complex of poly(ethylene glycol) with α-CD. The inclusion complex formation of PEG in the α-CD channel is entropically unfavorable, However, formation of the complexes is thought to be promoted by hydrogen bond formation between neighboring cyclodextrins. Therefore, head-to-head and tail-to-tail arrangement is thought to be the most probable structure.

Table IV. Yields for the complex formation between oligo(ethylene glycol) and α-CD

	Number of -CH$_2$CH$_2$O-							
	2	3	4	5	6	7	8	12
Linear (%)	0	0	2	9	30	56	64	76
Cyclic (%)	21	–	9	0	0	–	–	–

3. SYNTHESIS OF MOLECULAR NECKLACE (POLYROTAXANES)[17]

Recently, much attention has been focused on supramolecular science, science of noncovalent assembly, because of the recognition of the importance of specific noncovalent interactions in biological systems and in chemical processes. Rotaxanes are one of the classical classes of molecules consisting of a noncovalent entities, a "rotor" and an "axle" in a single molecule[18], and have been synthesized in a statistical way; thereby, the yields were very low[19]. More recently, rotaxanes have attracted renewed interests in the field of supramolecular chemistry because of their unique structures and properties. Rotaxanes can be prepared by closing the end groups of "axle" by large groups within the ordered environments of the noncovalent templating forces in such a way as to retain the order originally imposed by the weak interactions[20]. By this method complexes containing methylated β-CD and thread[21] have been synthesized. Both symmetric[22] and asymmetric[23] ionic rotaxanes containing α-CD have been reported.

3.1. Synthesis of Polyrotaxanes from Polydisperse PEG

We have succeeded in preparing compounds in which many cyclodextrins are threaded on a single PEG chain and are trapped by capping the chain ends with bulky groups as shown in **Scheme 1**. This is the first example that many rotors are imprisoned in a single molecule. We named this molecule "molecular necklace". Wenz et al. also reported a rotaxane with many α-CD[24].

The inclusion complexes of α-CD with PEG bisamine (PEG-BA) were prepared by adding an aqueous solution of PEG-BA to a saturated aqueous solution of α-CD at room temperature, using a method similar to that used to prepare complexes of α-CD and PEG. The resulting complex was allowed to react with an excess of 2, 4-dinitrofluorobenzene, which is bulky enough to prevent dethreading. The product was purified by column chromatography on Sephadex G-50 by using dimethylsulfoxide (DMSO) as solvent.

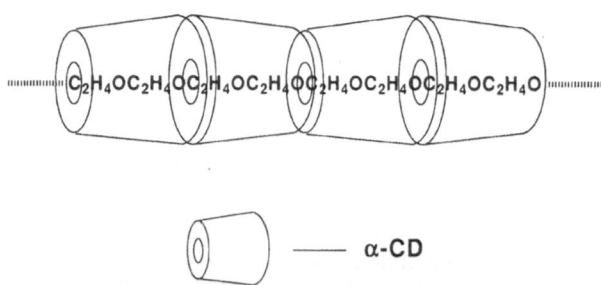

Figure 7. Proposed structure of the α-CD-PEG complex.

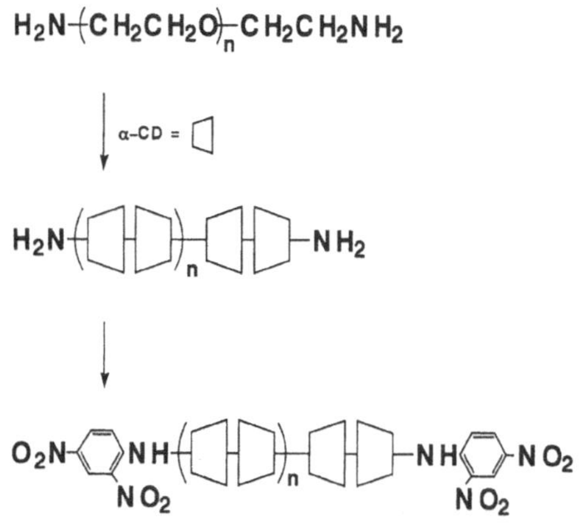

$$H_2N\text{-}(CH_2CH_2O)_n\text{-}CH_2CH_2NH_2$$

$$\downarrow \quad \alpha\text{-CD} = \square$$

$$H_2N\text{-}(\square\square)_n\text{-}\square\square\text{-}NH_2$$

$$\downarrow$$

$$O_2N\text{-}\langle\rangle\text{-}NH\text{-}(\square\square)_n\text{-}\square\square\text{-}NH\text{-}\langle\rangle\text{-}NO_2$$
$$\qquad NO_2 \qquad\qquad\qquad\qquad\qquad NO_2$$

Scheme 1

The products are insoluble in water and dimethylformamide, but they are soluble in DMSO and in 0.1 N NaOH. The products were characterized by UV-vis, X-ray diffraction, ^{1}H NMR, ^{13}C NMR ^{13}C CP/MAS NMR and 2D NOESY NMR spectra. The ^{1}H NMR spectra of the product shows that the product is composed of α-CD, PEG-BA, and dinitrophenyl groups and the peaks of CD, PEG, and dinitrophenyl groups are broadened, suggesting that α-CDs are difficult to move on a PEG chain. 2D NOESY NMR spectra show that the signals of H-3 and H-5 protons of α-CD, which are directed toward the inside of the cavity, correlate with the resonance of the CH_2 of PEG, but the H-1, H-2, and H-4 protons, which are located outside the cavity, do not correlate with PEG. These results indicate that a PEG chain is included in α-CD cavities.

Table V shows the results of the preparation of polyrotaxanes of various molecular weights. The number of CDs increases with an increase in the molecular weight. MN-3350, which was prepared from PEG (MW=3350), has 20-23 CDs on a PEG chain. This corresponds to the molar ratio of ethylene glycol units to α-CDs of 3.9. More than half of the polymer chain is covered with α-CDs. MN-1450 has 15 α-CDs on a PEG chain. The molar ratio of ethylene glycol units to α-CD is 2.3. The ratio indicates that the molar ratio is almost stoichiometric; that is, CDs are almost close-packed from end-to-end of the polymer chain.

3.2. Synthesis of a Polyrotaxane Containing Monodisperse Oligo(ethylene glycol)[25]

Preparation of polyrotaxanes containing many α-CDs are described in Section 3.1. However, in these cases, polymers used are polydisperse and the number of CDs in a polymer chain is also polydisperse. Thereby the rotaxanes obtained are highly heterogeneous.

Table V. Polyrotaxanes prepared from PEGBA with various molecular weights

Polyrotaxane	Molecular weight[a]	Number of ethylene glycol units	Number of α-CD included[a]	Molar ratio between ethylene glycol units and α-CD
MN-3350	23,500	77	20	3.9
MN-2000	20,000	45	18	2.5
MN*-2001[b]	19,000	45	17	2.6
MN-1450	16,500	35	15	2.3

[a] Calcd. from UV-Vis spectra and ^{1}H NMR spectra. [b] Prepared from JED-2001.

Moreover, in both rotaxanes, only the part of the polymer chain is covered with cyclodex-trins. In order to obtain homogeneous polyrotaxanes, we have prepared monodisperse PEGs (28mer, MW=1248) because PEG of molecular weight 1000-1500 were found to be most favorable for complex formation. We have succeeded in preparing complexes between α-CDs and monodisperse diamino-PEG and imprisoning 12 α-CDs on monodisperse diamino-PEG by capping PEG chain ends with bulky substituents. It is an important step toward the "molecular abacus".

The ^{13}C NMR spectrum of the polyrotaxane show that the C-4 and C-6 peaks were doublets; a broad peak at a higher magnetic field and a sharper peak at a lower field, respectively. The broad peaks can be assigned to C-6 and C-4 in the rotaxane, which are difficult to move due to hydrogen bonds between CDs. The sharp peaks at lower magnetic field can be assigned to C-6 and C-4 of the cyclodextrins at both ends because they are not involved in hydrogen bonds and are more flexible than the others and they are susceptible to the effects of the dinitrophenyl groups at the ends of the rotaxane.

The bulky end groups (dinitrophenyl groups) were removed by cleaving the C-N bond with strong base, and CDs were recovered. The number of cyclodextrins in the polyrotaxane can be estimated from the ^{1}H NMR spectra, optical rotation, and UV-vis spectra. Twelve α-CDs were found to be included in the polyrotaxane.

4. SYNTHESIS OF TUBULAR POLYMERS[26]

Tubular polymers have been prepared as shown in **Scheme 2.** Polyrotaxanes were prepared as described previously[17]. In this case poly(ethylene glycol) of molecular weight 1,450 was used, because α-CD forms complexes with PEG of molecular weight 600-2000 most efficiently. The complexes are nearly stoichiomeric (2 ethylene glycol units:a single CD), that is , α-CDs are almost closed-packed from end to end of the polymer chain. The polyrotaxane(22.5 mg) was dissolved in 10 % NaOH solution and epichlorohydrin (3.84 mmol) was added to the solution. The solution was stirred at room temperature for 36 hours. The reaction mixture was neutralized by the addition of HCl. The yellow solid was precipitated from ethanol. In order to remove the bulky stoppers at both ends, the product was treated with strong base(25% NaOH) at 45 °C for 24 hours. The reaction mixture was cooled and then neutralized with HCl. Figure 8 shows the elution diagram of the reaction mixture of polyrotaxane with 25 % NaOH (a), that of the crosslinked polyrotaxane(b), and that of the reaction mixture of the crosslinked polyrotaxane with 25 % NaOH (c). The crosslinked product was eluted at the void volume(b), the same as the polyrotaxane. After the polyrotaxane was treated with 25 % NaOH and the solution was neutralized before crosslinking, two peaks were observed: one could be detected only by UV(360 nm) , which is assigned to dinitrophenyl group(DNP), and the other, which could be detected only by optical rotation, is identified as dethreaded α-CDs. After the crosslinked product was treated with strong base, two peaks were observed(c): one at the void volume, which is detected only by optical rotation, is identified as the product, the molecular tube. The second peak, that was detected only by UV(360 nm), is assigned as a dinitrophenyl group(DNP). The molecular weight of the molecular tube was estimated by GPC on Sephadex G-100 using dextran as standard. The molecular tube was eluted just after dextran(MW=20,000), indicat-ing that the average molecular weight of the molecular tube is less than 20,000. This value is consistent with the fact that the molecular weight of the molecular tube prepared from PEG of molecular weight 1,450 is about 17,000. Therefore intra chain crosslinking took place predominantly. The yield of the final product is 92 %.

The product was soluble in water, DMF, and dimethylsulfoxide (DMSO), although polyrotaxanes are insoluble in water and DMF and soluble in DMSO. The product was

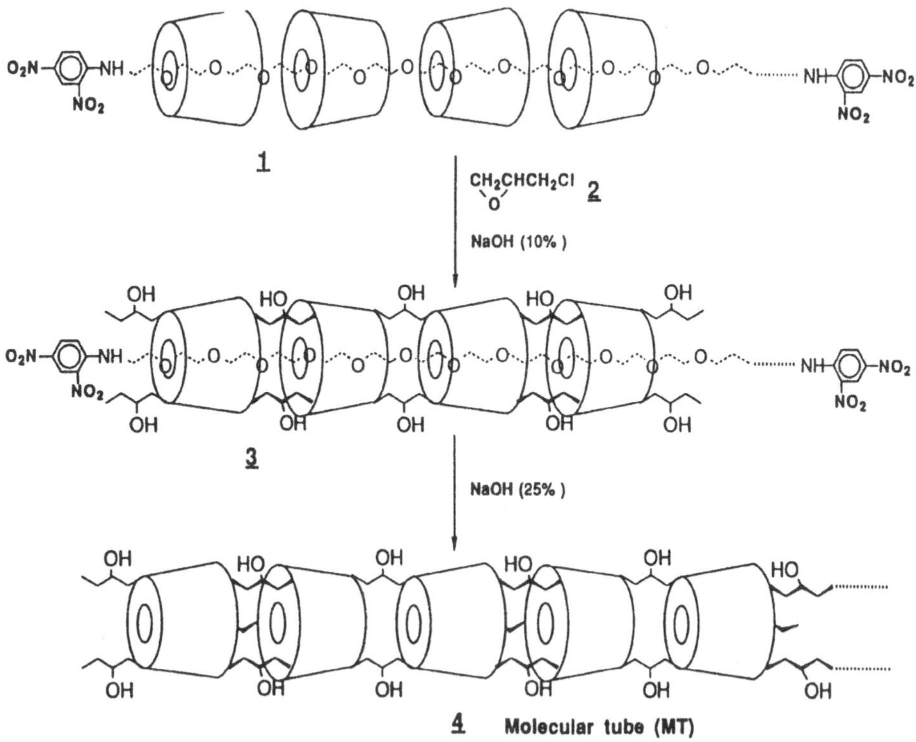

Scheme 2

characterized by ^{1}H NMR, ^{13}C NMR, IR, and UV spectra and GPC. The ^{1}H NMR spectra of the molecular tube in D_2O and in DMSO d_6. and ^{13}C NMR spectra show that both CDand the bridge can be observed. All the peaks of the ^{1}H NMR spectrum are broadened, indicating the product is polymeric.

When the solution of the molecular tube was added to a KI-I_2 solution(pale yellow), the solution turned deep red instantaneously, although the addition of an α-CD solution to KI-I_2 solution caused nothing. The absorption spectra of KI-I_2 solution in the presence and absence of α-CD and molecular tube show that the position of the absorption maximum changed little with some increase in the absorption on addition of α-CD. On addition of the molecular tube, the absorption maximum shifted to longer wavelength and a large tailing over 500 nm was observed. On addition of randomly crosslinked α-CD with epichlorohydrin, no visible changes took place. These results indicate that I_3^- ions arrange linearly in the tube. The shift is not so large compared with that of amylose and iodide, but is similar to that of amylopectin and polyiodide. The change of the spectra was found to be the maximum at one to one (CD unit and I_3^-).

5. CONCLUSION

We found that cyclodextrins form complexes with various polymers with high selectivities and that polymer chains are included in the channels formed by CDs. Polyrotaxanes consisting of cyclodextrins and PEG were prepared by capping the end groups of the complexes between CDs and PEG bisamine with bulky groups. Tubular polymers were

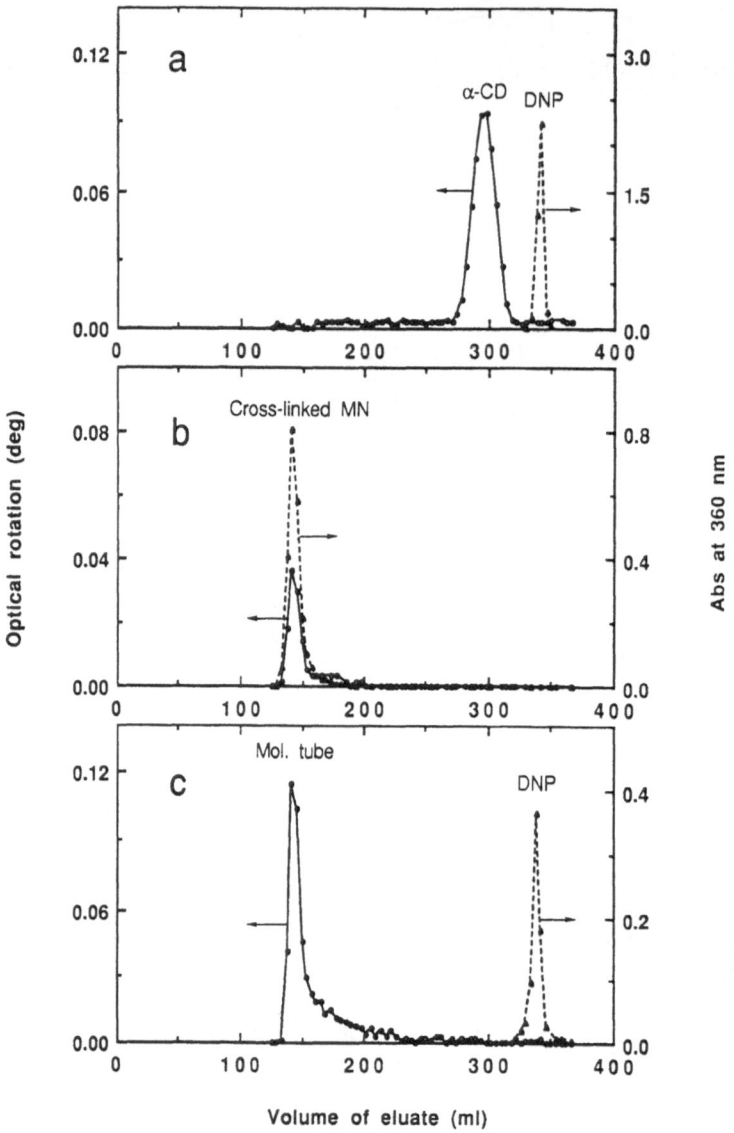

Figure 8. Elution diagrams of the reaction mixture of polyrotaxane with 25% NaOH (a), that of the crosslinked polyrotaxane (b), and that of the reaction mixture of the crosslinked polyrotaxane with 25% NaOH (c). A Sephadex G-25 column (2.2 x 93 cm) with water eluate was used.

obtained by crosslinking neighboring CDs threaded on a PEG chain in a polyrotaxane, followed by removing bulky end groups and the polymer chain. These findings indicate that such template synthesis provides a new approach to construct new nano-structures[26].

REFERENCES

1. Y. Tsukahara, K. Tsutsumi, Y. Yamashita, S. Shimada, *Macromolecules, 23* (1990) 5201.
2. T. W. Bell, H. Jousselin, *Nature, 367* (1994) 441.

3. H-B. Mekelburger, W. Jaworek, F. Vogtle, *Angew. Chem. Int. Ed. Engl. 31* (1992), 1571.

4. S. Asakuma, H. Okada, T. Kunitake, *J. Am. Chem. Soc., 113* (1991) 1749.

5. P. M. Ajayan, S. Iijima, Nature, 361 (1993) 333.

6. (a) M. L. Bender and M. Komiyama, *Cyclodextrin Chemistry,* Springer-Verlag, Berlin, 1978, (b) J. Szejtli, *Cyclodextrins and Their Inclusion Complexes*, Akademiai Kiado, Budapest, 1982.

7. (a) A. Harada, J. Li, and M. Kamachi, *Proc. Jpn. Acad.*, 69, Ser. B (1993) 39, (b) A. Harada, *Polym. News,* 18 (1993) 358.

8. A. Harada, J. Li, and M. Kamachi, *Ordering in Macromolecular Systems,* 69, Springer-Verlag, Berlin, 1994.

9. A. Harada, J. Li, and M. Kamachi, *Nature, 364* (1993) 516.

10. (a) A. Harada and M. Kamachi, *Macromolecules*, 23 (1990) 2821, (b) A. Harada, J. Li, and M. Kamachi, *Macromolecules, 26* (1993) 5698.

11. A. Harada and M. Kamachi, *J. Chem. Soc., Chem. Commun.*, 1990, 1322.

12. A. Harada, J. Li, and M. Kamachi, *Chem. Lett.*, 1993, 237.

13. J. Li, A. Harada, and M. Kamachi, *Bull. Chem. Soc., Jpn.*, 67 (1994) 2808.

14. A. Harada, J. Li, S. Suzuki, and M. Kamachi, *Macromolecules, 26* (1993) 5267.

15. A. Harada, J. Li, and M. Kamachi, *Macromolecules, 27* (1994) 4538.

16. W. Saenger, *Jerusalem Symp. Quantum Chem. Biochem.* Ed. E. B. Pullman, D. Reidel Co., Dordrecht, 1976.

17. (a) A. Harada, J. Li, and M. Kamachi, *Nature 356* (1992) 325. (b) A. Harada, J. Li, and M. Kamachi, *Carbohydr. and Carbohydr. Poly.*, 25 (1993) 266. (c) A. Harada, T. Nakamitsu, J. Li, and M. Kamachi, *J. Org/ Chem.*, *58* (1993) 7524.

18. G. Schill, *Catenanes, Rotaxanes, and Knots*, Academic Press, New York, 1971.

19. G. Agam, D. Graiver, and A. Zilkha, *J. Am. Chem. Soc., 98* (1976) 5206.

20. P.L. Annelli, P. R. Ashton, R. Ballardini, V. Balazani, M. Delgado, M. T. Gandolfi, T. T. Goodnow, A. E. Kaifer, D. Philip, M. Pietraszkiewicz, L. Prodi, M. V. Reddington, A. M. Z. Slawin, N. Spencer, J. F. Stoddart, C.Vicent, D. J. Williams, *J. Am. Chem. Soc., 112* (1990) 2440.

21. (a) J. S. Manka and D. S. Lawrence, *J. Am. Chem. Soc.,* 112 (1990) 2440, (b) T. V. S. Rao and D. S. Lawrence, *J. Am. Chem. Soc.,* 112 (1990) 3614.

22. (a) H. Ogino, *J. Am. Chem. Soc., 103* (1981) 1303, (b) R. S. Wylie and D. H. Macartney, *J. Am. Chem. Soc.,* 114 (1992) 3138.

23. R. Ishnin and A. E. Kaifer, *J. Am. Chem. Soc.,* 113 (1991) 8118.

24. G. Wenz and B. Keller, *Angew. Chem., Int. Ed. Engl.,* 31 (1992) 197.

25. A. Harada, J. Li, and M. Kamachi, *J. Am. Chem. Soc., 116* (1994) 3192.

26. A. Harada, J. Li, and M. Kamachi, *Nature 356* (1994) 325.

10

MULTI-COMPONENT POLYMERS CONTAINING POLYISOBUTYLENE VIA MULTI-MODE POLYMERIZATION

Munmaya K. Mishra

Polymers Research
Texaco, Inc., Research and Development
P.O. Box 509
Beacon, New York 12508

INTRODUCTION

Recently, the trend in polymer research has been geared toward producing advanced materials via macromolecular engineering [1]. It is known that a desired combination of physical properties could be achieved by designing tailor made block and graft copolymers. Various methods including polycondensation reactions (using telechelic oligomers) or living polymerization techniques have been used for the synthesis of multi-component polymers. Although telechelic oligomers can be made by a wide variety of techniques, their use in block copolymerization suffer from a number of disadvantages. It is known that sequential monomer addition (SMA) technique in living ionic polymerization [2] is a convenient way to prepare block copolymers possessing well-defined and predetermined structures. In addition to well established living anionic polymerization, cationic living polymerization of isobutylene has been developed during the past few years [3-8]. Well defined block copolymers are prepared by these living systems following the common strategy of SMA [7]. It should be pointed out that there are various drawbacks which quite often retards the practical application of living systems to prepare block copolymers. These drawbacks essentially relate to limitation of the method to certain monomers and exclude monomers that polymerize by other mechanisms.

Because of these practical limitations however, multi-mode polymerization seems to be promising and various block copolymers have been synthesized by this method [9]. "Multi-mode polymerization" involve synthesis in which two (or more) mutually exclusive polymerization mechanisms are sequentially combined [10-14]. Since the work of Richards and his coworkers [11, 15] on the preparation of block copolymers by combining different polymerization mechanisms, the range of possible monomer combinations in block copolymers have expanded and many examples have been reported [9, 14, 16, 17]. There are generally two types of transformation processes i.e., direct and indirect transformation [17].

Macromolecular Engineering, Edited by M.K. Mishra et al.
Plenum Press, New York, 1995

In the case of direct transformation, propagating active centers are transformed directly to another active center with different polarity by an electron transfer process. However, the indirect transformation involves the combination of various polymerization modes and the design of multi-component polymers have been well documented by this method. Although indirect transformation involves several multi-step paths leading to the transformation of active centers, it is much more convenient to achieve and can be applied to a wide range of monomer couples. In this process, suitable functionality is introduced in one or two step(s) by appropriate initiation and termination reactions in the polymerization of the first monomer. Subsequently, a second monomer is polymerized using the first generation polymer as the macroinitiator to produce block/graft/multi-component polymers. Polyisobutylene based multi-component polymers can be synthesized by three types of combinations i.e., cationic with anionic, radical and group transfer (GTP) polymerizations. This chapter is a review of various multi-component polymers containing polyisobutylene via multi-mode polymerization including some of our recent work.

1. COMBINATION OF CATIONIC AND ANIONIC POLYMERIZATION

Cationic to anionic transformation is one of the well studied system reported in the literature. The particular advantage of these transformations is that both cationic and anionic blocks can be prepared under living polymerization conditions. About a decade ago the examples of such conversions were reported by Richards et al. [18, 19]. With the ready availability of functional polymers by new polymerization techniques, efforts have been made to prepare block copolymers by cation to anion transformation. This field has been recently reviewed [9, 17]. Various types of polyisobutylene based block copolymers produced by this method have been summarized in the Table 1.

In recent past the synthesis of PIB-siloxane di- and multi-block copolymers [22] has been demonstrated using combination of cationic and anionic polymerization techniques. The key toward syntheses was the definition of conditions for the initiation of living anionic polymerization of hexamethylcyclotrisiloxane (D_3) at the -CH_2OLi termini of well defined telechelic PIB sequences. PIB and PDMS exhibit a combination of desirable physical properties and are valuable commercial materials for a great variety of applications [7, 26, 27]. By combining these polymers into PIB-PDMS block copolymers, one can create new materials that exhibit a combination of the properties of the original two components.

The macroinitiator [22] for the living anionic polymerization of D_3 was prepared by converting the hydroxyl terminated telechelic PIB to the alkoxide with use of butyl lithium. Well-defined PDMS-b-PIB-b-PDMS triblocks and -(-PDMS-b-PIB-b-PDMS-)-n multiblocks were obtained by quenching the living D_3 polymerization by $(CH_3)_3SiCl$ and by

Table 1. Multi-component polymers containing PIB via cationic to anionic transformation

Block Copolymer	Reference
Poly(isobutylene-b-butadiene)	[20]
Poly(isobutylene-b-methyl methacrylate)	[21]
Poly(isobutylene-b-siloxane)	[22]
Poly(isobutylene-b-methyl methacrylate)	[23]
Poly(isobutylene-b-ε-caprolactone)	[24]
Poly(isobutylene-co-methylstyrene-g-pivalactone)	[25]

Table 2. Syntheses of PDMS-b-PIB-b-PDMS and (-PDMS-b-PIB-b-PDMS-)-n^A

Expt.#	LiO-PIB-OLi^B mmol/L	D$_3$ mol/L	Conv.%	PDMS-b-PIB-b-PDMS exp. Mn^C	(-PDMS-b-PIB-b-PDMS-)-n Mn
1	10	0.99	30	9,800	30,800
2	10	0.99	71	18,500	—
3	5	0.52	47	15,400	47,200
4	5	0.22	65	9,600	44,800
5	4.8	0.08	90	6,800	27,100

A. In THF, 26°C
B. M_n = 4200
C. By VPO for Mn < 10,000; otherwise by membrane osmometry

stoichiometric quantities of (CH$_3$)$_2$SiCl$_2$, respectively. We found that Li alcoholate-ended PIB can induce the rapid and quantitative initiation of living D$_3$ polymerization.

The macroinitiator was synthesized by adding equimolar amount of n-BuLi relative to the hydroxyl group of HO-PIB-OH solution in THF. The polymerization of D$_3$ was initiated by adding D$_3$ solution in THF at ambient condition (~26°C). The conversion of D$_3$ was followed by GC. Samples were withdrawn by syringes and deactivated with either (CH$_3$)$_3$SiCl (50% molar excess relative to ~OLi) or (CH$_3$)$_2$SiCl$_2$ (equimolar amounts of Cl-Si- relative to the ~OLi groups). Few representative results are presented in the following Table 2.

Figure 1 represents the GPC scans of the starting material and block copolymers. GPC scans show that the elution volumes of the block copolymers are much lower than that of the HO-PIB-OH prepolymer, which indicates an increase of the hydrodynamic volume on account of "blocking" and subsequent extension.

Hydrolytic Stability of PDMS-b-PIB-PDMS

The hydrolytic stability of a representative triblock copolymer (M_n = 18,500) was estimated by comparing the GPC UV traces of the triblock sample before and after

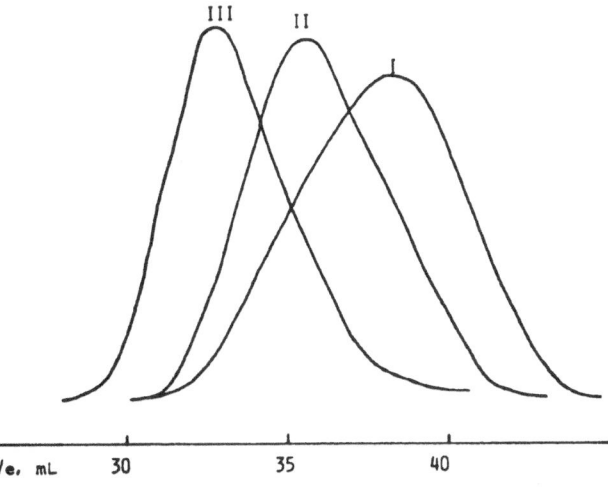

Figure 1. Normalized RI GPC traces of (I) HO-PIB-OH; (II) (CH$_3$)$_3$Si-PDMS-b-PIB-b-PDMS-Si(CH$_3$)$_3$; (III) -(-PDMS-$_b$-PIB-$_b$-PDMS-SiMe$_2$-)-$_n$.

hydrolysis. The triblock was stable in refluxing wet THF for 18 hours. Similarly heterogeneous systems (i.e., in bulk or in blend with PDMS) or hydrophobic (hexanes) solutions of the triblock were resistant to aqueous HCl. In contrast, the triblock in THF solution degraded in the presence of HCl. Detail results have been published earlier. By analyzing these results, it has been concluded that the triblock copolymer exhibited enhanced resistance toward hydrolysis compared with low molecular weight alkoxysilanes.

2. COMBINATION OF CATIONIC AND RADICAL POLYMERIZATION

Cationic to radical transformation reactions has become attractive because of the recent discovery of living cationic polymerization systems involving vinylethers, styrene and isobutylene. Most of the reported cation to radical transformation involved polytetrahydrofuran Poly(THF) as cationic segment [9, 17].

General synthetic approach in this particular transformation involves introduction of a common radical initiator functional group, such as peroxide or azo groups, into the center of the macromolecular chain which is suitable for the subsequent radical generation. The first example of this type of transformation was the introduction of a common radical initiator group, such as peroxide into the center of polyTHF. Expectedly, subsequent thermolysis would yield two polymeric radicals. Nuyken et al. [28, 29] prepared polyvinyl ether possessing a central azo group by using azo-divinylether and adding sequentially HI, coinitiator, and isobutyl vinylether as monomer. AB type block copolymers were formed upon thermolysis in the presence of methacrylonitrile since the main mode of termination is combination in this particular polymerization.

After the discovery of the truly living polymerization [3-8] of isobutylene, research has been continued into the exploration of the scope and significance of this discovery. Although cationic to radical transformation (CRT) [9, 17] is one promising method to design various other copolymers, the synthesis of polyisobutylene (PIB) based block copolymers using CRT was unknown [30]. Preparatory to a detailed presentation of a large body of accumulated information, representative portions have been selected to provide an overview of our finding.

As described elsewhere [9, 17, 23] block copolymers of PIB and poly(methyl methacrylate) (PMMA) can be made by a cationic to anionic transformation. The process involved the synthesis of PIB by cationic polymerization. Then the functional group of PIB was converted to be used as an anionic initiator for the polymerization of MMA. However, the method has advantages and limitations and involves many reaction steps.

The main thrust of our study was to eliminate several reaction steps and utilize a direct method for the synthesis of macroinitiator followed by the polymerization of vinyl monomers using the same. To achieve this objective a new strategy was devised [30] and described as follows.

The first synthesis of a PIB with central azo group (PIB-N=N-PIB) and its application as a macroinitiator for the synthesis of block copolymers via cationic to radical transformation (multi-mode polymerization) are described. The synthesis strategy (Figure 2) involved a two step process. At first isobutylene was polymerized cationically using a difunctional organic tertiary ether (having azo-group) and a Lewis acid initiator combination to produce azo group containing PIB. In the second step the functional PIB was used as a macroinitiator for the radical polymerization of vinyl monomers such as styrene, methyl methacrylate, etc. and the results are presented in Table 3.

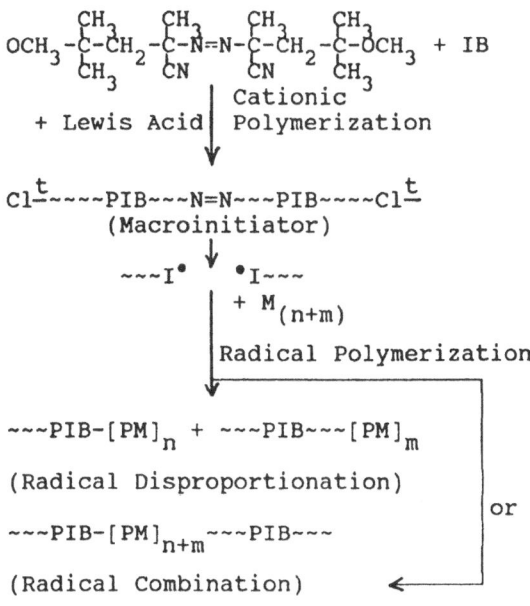

Figure 2. Reaction scheme of block copolymer synthesis

3. COMBINATION OF CATIONIC AND GROUP TRANSFER POLYMERIZATION

The mechanism of group transfer polymerization [31, 32] implies that the initiating silyl keten acetal moiety remains attached to the propagating chain. Using this concept, in a recent report Ruth et al. [33] have described the block copolymer synthesis of IB and methyl methacrylate (MMA) by combining cationic and group transfer polymerization. Thus, α,ω-dihydroxy PIB prepared by living cationic polymerization was converted to the diester which is sequentially treated with lithium diisopropylamide (LDA) and chlorotrimethylsilane to afford the silyl keten acetal macroinitiator. Reaction of the polymeric silyl keten acetal with methacrylate in the presence of tetrabutyl ammonium dibenzoate (TBABB) produced a block copolymer. The methylmethacrylate conversion was quantitative, however significant amount of unreacted polyisobutylene was reported by the authors. Also, the synthesis of PMMA-*g*-PIB copolymers have been reported recently [34] by the combination of living

Table 3. Polymerization of MMA and STY Initiated with PIBMI*

Expt#	PIBMI		Monomer		Conv.%	M_n§	M_w/M_n
	M_n	[-N=N-] (mole x 10^3)	Type	mole x 10^3			
1	19,850	7.0	MMA	46.7	22.8	36,700	1.4
2	19,350	7.0	MMA	46.7	20.5	25,370	1.2
3	6,900	4.0	MMA	28.0	14.2	9,000	2.9
4	6,900	4.0	STY	26.0	14.6	7,150	2.8
5	19,850	7.0	STY	43.6	17.6	28,750	1.6

* Solvent = Toluene (50 ml), Temperature = 65^0C, Time = 7 hr;
§ Molecular weights were determined by GPC using a PIB broad standard calibration.

cationic and group transfer polymerization using polyisobutylene methacrylate macro-monomers.

4. CONCLUSION

Using this multi-mode polymerization, i.e., combining various polymerization mechanisms, novel polymeric materials may be synthesized from new and classical mono-mers. As this chapter provides an overview of various multi-component polymers containing polyisobutylene, it is hoped that it will serve as a reference for the future developments along these lines to produce novel materials for various applications.

REFERENCES

1. Mishra, M.K., Ed., 1994, *"Macromolecular Design: Concept & Practice" Polymer Frontiers* Int'l., NY, ISBN: 0-9639138-0-8 and chapters therein
2. Van Beylen, M. and Scwarz, M., *1993, Ionic Polynmerization and Living Systems*, Chapman and Hall, New York
3. Mishra, M. K. and Osman, A., 1990, US Patent, 4,908,421
4. Mishra, M. K. and Osman, A., 1990, US Patent, 4,943,616
5. Kennedy, J.P. and Mishra, M. K.,1990, US Patent, 4,929,683
6. Kennedy, J.P. and Faust, R., 1990, US Patent, 4,910,321
7. Kennedy, J.P. & Ivan, B.,*1991, "Designed Polymers by Carbocationic Macromolecular Engineering," Hanser, Munich*
8. Mishra, M.K.,*1993, Ind. J. Technol, V31: 197*
9. Yagci, Y and Mishra, M., 1994, chapter X of Reference 1
10. Richards, D.H.,1979, *Dev. Polym.*, 1: 1
11. Richards, D.H.,1980, *Br. Polym. J.*, 12: 89
12. Quirk, R.P., Kinning, D.J., and Fetters, L.J.,1989, in Comprehensive Polymer Sci., V 7 (G. Allen and J. C. Bevington, eds.), Pergamon, Oxford, Chapter 1
13. Rempp, P., Franta, E., and Herz,*1988, J.-E., Adv. Polym. Sci.*, 86: 145
14. Schue, F.,1989, in Comprehensive Polymer Sci., vol 7 (G. Allen and J. C. Bevington, eds.), Pergamon, Oxford, Chapter 10
15. Burgess, F., Cunliffe, A.V., Richards, D.H., and Sherrington,*1979, D.C., J.Polym.Sci., Polym.Lett.Ed.*, 14: 471
16. Steward, M. J.,1991, *New Methods of Polymer Synthesis*, Blackie & Sons, New York, chapter IV.
17. Yagci, Y. and Mishra, M. K., 1995, in Polymeric Materials Encyclopedia, Salomone, J. Ed, CRC Press, inpress
18. Abadie, M.J.M., Schue, F., Souel, Th., and Richards, D.H.,1982, *Polymer,* 23: 445
19. Cohen, P., Abadie, M.J.M., Schue, F., and Richards, D.H., *1982,Polymer*, 23: 1105
20. Nemes, S. and Kennedy, J.P.,*1991, J.Macromol.Sci.Chem.*, A28: 311
21. Kennedy, J. P. and Price, J.,*1991, Polym.Mat. Sci.Eng*, 64: 40
22. Wilczek, L., Mishra, M.K., and Kennedy, J.P.,*1987, J.Macromol.Sci.Chem.*, A24: 1033
23. Kitayama, T., Nishiura, T., and Hatada, *K, 1991, Polym.Bull.*,26: 513
24. Wondraczek, R and Kennedy, J.P., *1982, J.Polym,Polym.Chem.Ed*, 20: 173
25. Harris, Jr.,J.F. and Sharkey, *W.H.,1986, Macromolecules*, 19: 2903
26. Noll, W.,1968, Chemistry and Technology of Silicones, Academic, New York, p.437
27. Lichtenwalner, H. K. and Sprung., H.S.,1973, "Silicones," in Encyclopedia of Polymer Science and Technology, V12, Wiley , New York, p. 79
28. Nuyken, O, Kröner, H., and Aechtner, S.,*1988, Makromol.Chem.,Rapid Commun.*,9: 671
29. Nuyken, O., Kröner, H., and Aechtner,S.,1990, *Polym.Bull.*, 24: 513
30. Mishra, M.K.,*1994, Abstract Book of MacroAkron'94 IUPAC Symposium, Akron, July 11- 15 (1994)*, p 38; Preprints of 5th SPSJ Int'l Polymer Conference, Osaka, 1994, p. 35
31. Webster, O.W., Hertler, W.R., Sogah, D.Y., Farnham, W.B. and Rajanbabu,T.V., *1983, J. Am. Chem. Soc.*, 105: 5706

32. For a Recent Review See. Eastmond, G.C., Webster, O.W.,1991, in " New Methods of Polymer Synthesis", (Ed. J. R. Ebdon) Blackie & Sons, New York, p. 22

33. Ruth, W.G., Moore, C.G., Brittain, W.J., Si, J., and Kennedy, J.P., *1993, Polymer Preprints*, ACS, 34 (1): 479

34. Takacs, A. and Faust, R.,1994 , *Abstract Book of MacroAkron '94 IUPAC Symposium, Akron, July 11- 15,* p 99

11

MACROPHOTOINITIATORS

Synthesis and Their Use in Block Copolymerization

Yusuf Yaġci

Istanbul Technical University
Department of Chemistry
Maslak 80626
Istanbul, Turkey

INTRODUCTION

There has been growing interest in the preparation of polymeric photoinitiators [1]. They present several atractive aspects over low molecular weight counterparts including low volatility and migration . As far as migration is concerned, their value in UV curing associated with food packing will be more important since future legislation will require photoinitiating systems with zero migration. Besides UV curing applications, block and graft copolymers can be prepared by using main and side chain polymeric photoinitiators, respectively [2]. These systems consists of a photochemical reactions by which active sites are produced at the chain ends or side chains, which themselves initiates the polymerization of a second monomer.

Low temperature conditions, usually room temperature, prevents the side reaction leading to the formation of homopolymers and high block and graft yields are attained. On the other hand, molecular weight distribution of the resultant polymers shows characteristics of free radical polymerization and homogeneity of the living polymers are not observed.

From the synthetic point of view, there are many ways to incorporate light sensitive groups into a polymer chain and these methods have been reviewed recently by Yagci and Schnabel [3].

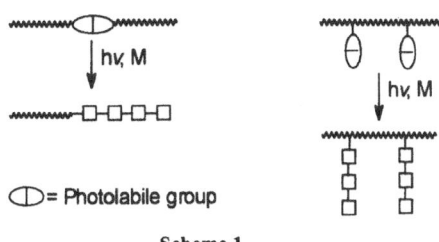

= Photolabile group

Scheme 1

Macromolecular Engineering, Edited by M.K. Mishra et al.
Plenum Press, New York, 1995

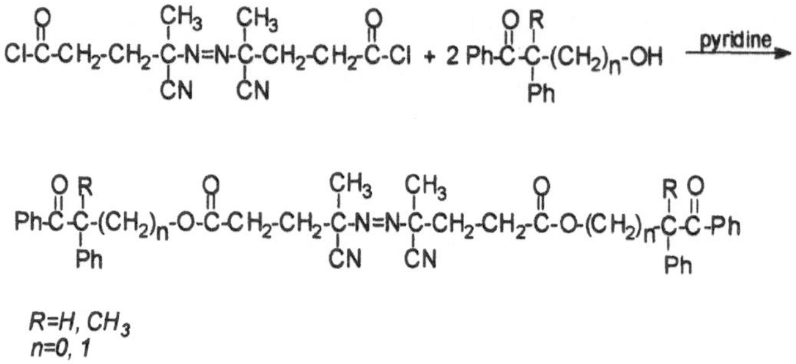

R=H, CH$_3$
n=0, 1

Scheme 2

   This paper is concerned with block copolymerization by using macrophotoinitiators, which recently were carried out in our laboratory. These investigations revealed that macrophotoinitiators are readily prepared by functional azo-initiators, cationic polymerization and activated monomer polymerization.

BLOCK COPOLYMERIZATION BY MACROPHOTOINITIATORS PREPARED BY FUNCTIONAL AZO-INITIATORS

Azo-Benzoin Initiators

   Azo-benzoin initiators were synthesized [4,5] by reactions of benzoin or α-methylolbenzoin methyl ether and the acid chloride as shown in scheme 2.
   These initiators were used as free radical initiators for the polymerization of styrene (St) and exhibit initiator properties similar to common azo initiators except their participation in transfer reaction. The initiation of polymerization by means of azo-benzoin initiators yields polymers with one or two benzoin end groups according to the termination mode of particular monomer involved. UV irradiation of the resulting prepolymers caused α- scission and yielded benzoyl radicals and polymer bound radicals. Either or both of these radicals may then initiate the polymerization of the second monomer [6]. The overall process is presented in scheme 3.
   Together with block copolymer, homopolymers are also formed. Homopolystyrene (PSt) formation may be due to the primary radical combination of polymeric alkoxy benzil radicals. PSt molecules, which for any reason failed to acquire benzoin end groups during the synthetic procedure in the first step, are also automatically included in the nonblock PSt component. GPC analysis clearly indicates the formation of block copolymer. Figure 1 shows

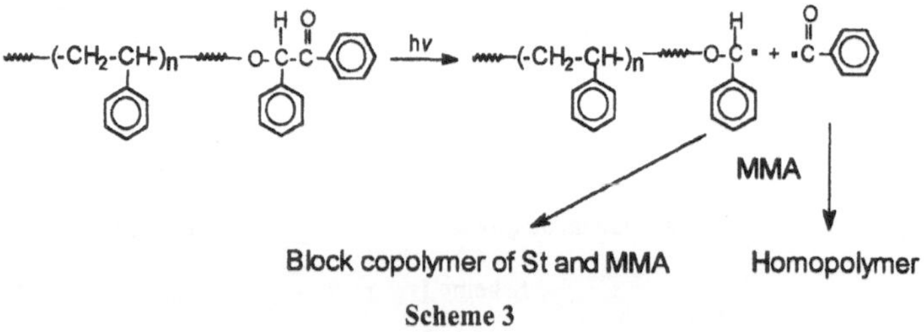

Block copolymer of St and MMA   Homopolymer

Scheme 3

the chromatograms recorded with prepolymer, block copolymerization mixture and extraction products. The block copolymer structure was also assigned by means of spectral measurements.

The use of photoactive polymers in blocking reactions are not limited to free radical process. Electron donating polymeric radicals, formed according to scheme 3, may conveniently be oxidized to polymeric carbocations to promote cationic polymerization of cyclic ethers and alkyl vinyl ethers. We have demonstrated [7] that irradiation of benzoin terminated polymers in conjuction with pyrdinium salts as oxidants in the presence of cyclohexene oxide (CHO) makes it possible to synthesize block copolymers of different chemical nature.

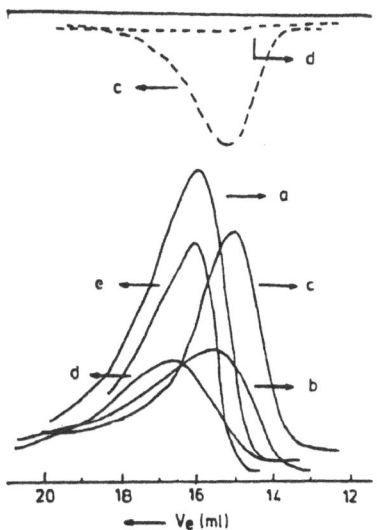

Styrene-cyclohexene oxide block copolymer

Typical results concerning block copolymerization by promoted cationic polymerization are presented in Table 1. As can be seen both molecular weight and the concentration of photoactive PSt affect the conversion, block yield, and composition of the block copolymer.

· For the practical value of this process, we have prepared [8] block copolymers of crystalline and liquid crystalline segments of the following structure by promoted cationic polymerization.

Figure 1. Gel permeation chromatograms recorded with the products formed on block copolymerization. (a) Prepolystyrene, (b) Crude block copolymer, (c) Pure block copolymer after extraction, (d) Homopoly(methyl methacrylate, Homopolystyrene.

Table 1. Promoted Block Copolymerization of Cyclohexene Oxide (CHO) with Polystyrene Using Benzoin Terminated Prepolymer and Pyridinum Salt $(5 \times 10^{-3}$ mol/l) at $\lambda=350$ nm in CH_2Cl_2, [CHO] = 5.9 mol/l

PSt (g/l)	Mn (g/mol)	Conversion (%)	PSt-b-PCHO (%)	PSt in Block Copolymer (%)
PSt1 (13.3)	35400	19.6	17.1	81.6
PSt2 (13.3)	94100	6.2	38.9	83.2
PSt3 (13.3)	130000	8.3	23.5	85.3
PSt1 (26.6)	35400	8.5	46.8	88.1

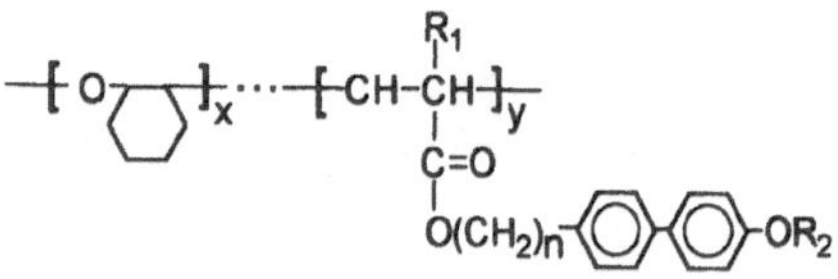

Azo-Acyloxime Ester Initiators

Acyloxime esters have been proven [9] to be highly effective in photogenerating free radicals. the photolysis of acyloxime esters has been proposed to occur by clevage of N-O bond followed by a secondary fragmentation step. The mechanism is depicted in scheme 4.

Polymers containing acyloxime moieties in the main chain can act as as photochemical macroinitiator for block copolymerization. It seemed, therefore, appropriate to use functional azo-initiator approach for the incorporation of acyl oxime ester groups to polymers in a manner similar to that described above [10]. The thermal initiation of polymerization using azo-acyloxime ester initiator yields polymers with photoscissable groups in the main chain (scheme 5) as was demonstrated by optical absorption measurements. Recent studies [11] concerning their photolysis in the presence of a second monomer revealed that acyl oxime groups may also degrade during the thermal treatment in the first step and give relatively low block yields.

Macrophotoinitiators obtained by using the above described functional azo-initiatots participate directly in genareting macromoleculer radicals for the subsequent block copolymerization. In the following section, we will describe another type of functional azo-initiators which can be used to yield polymers capable of participating indirectly in the photochemical radical forming process. These polymers are transarent to UV light but readily react with the excited states of certain compounds.

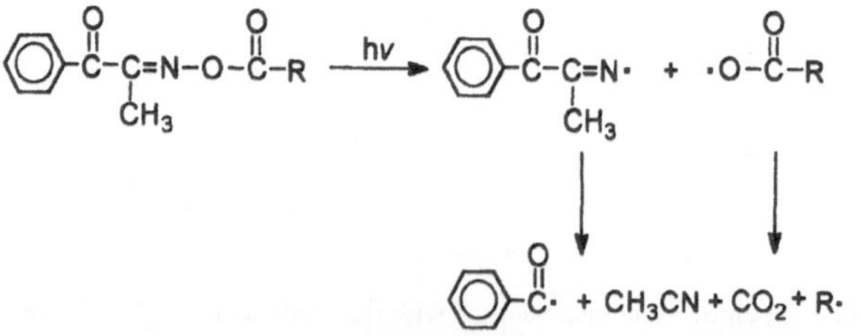

Scheme 4

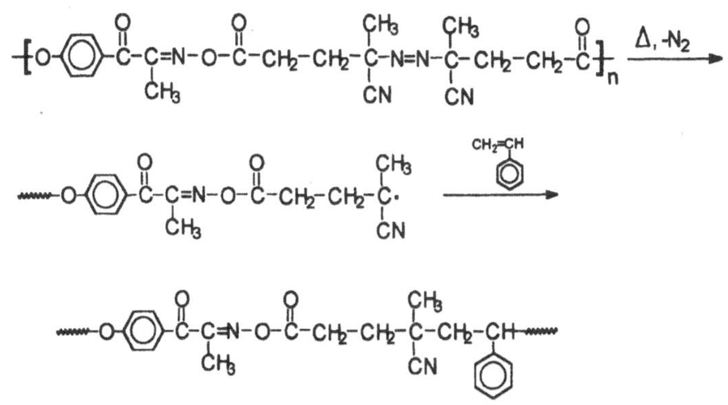

Scheme 5

Amino-Azo Initiator

Selective and efficient formation of block copolymers may be accomplished by UV irradiation of appropriate amine containing prepolymers in the presence of aromatic ketones and vinyl monomers [12]. Amino azo-initiator, 4,4'-azobis(N,N-dimetylamino ethyl-4-cyanopentanoate), facilitates the preparation of polymers with tertiary amine end groups [13].

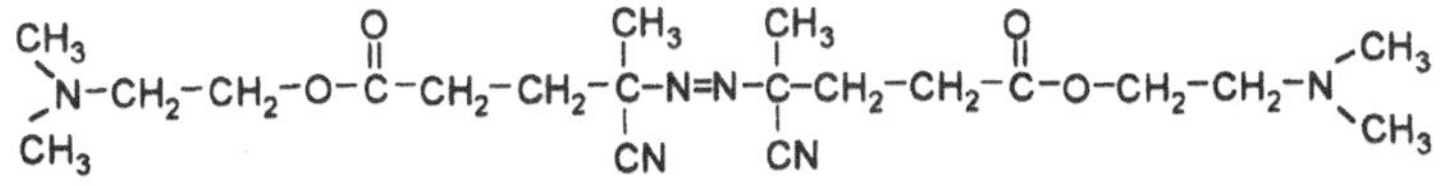

Polystyrene functionalized this way was used in photoinduced polymerization of MMA in the presence of 9-fluorenone. The resulting polymer was presumed to contain a copolymer of poly(St-*b*-MMA) and was completly soluble in acetonitrile which is non-solvent for homoPSt. Semipinacol type radicals produced by reduction of aromatic ketones are not noticebaly active in initiation of polymerization of MMA at room temperature as observed by Ledwith [14]. This behaviour reduces homopolymer formation in the block copolymerization via photoreduction of 9-fluorenone.

Trichloroacetyl-Azo Initiator

According to Bamford et al. [15] terminal carbon centered macroradicals are formed upon irradiation of polymers having -CX$_3$, -CHX$_2$, -CH$_2$X end groups (X: halogen atom) in the

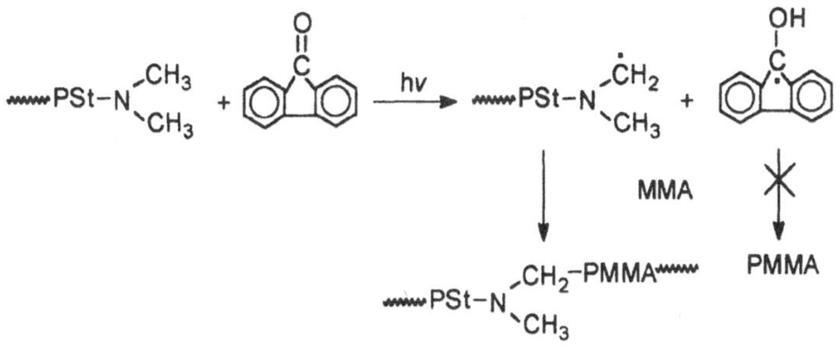

Scheme 6

presence of carbonyl compounds of transition metals, preferably the manganese and rhenium carbonyls, $Mn_2(CO)_{10}$ and $Re_2(CO)_{10}$. These compounds absorb light at rather long wavelengths. The principal radical-generating reaction is an electron transfer process from transtion metal to halide, the former assuming a low oxidation stage (presumably the zero state)

$$M_0 + X_3C\text{-}R \longrightarrow M_1X + X_2\dot{C}\text{-}R$$

Scheme 7

Polymeric initiators with terminal halogen groups can be prepared by using CCl_4 as chain transfer agent in free radical polymerization or terminating living polymerization by halogen containing compounds [15]. We have recently developed [16] another bifunctional initiator, 4,4'-azobis(4-cyanopentane-trichloroacetylamide), to obtain polymers containing CCl_3 groups,

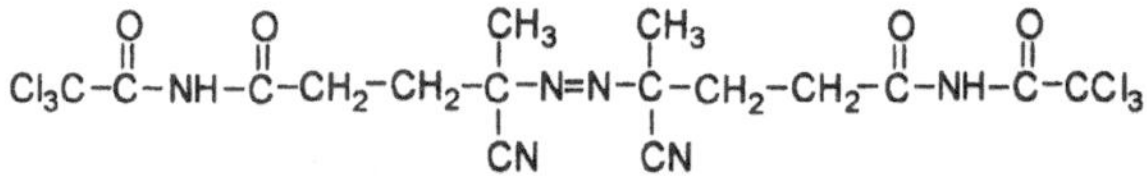

$$Cl_3C\text{-}\overset{O}{\overset{\|}{C}}\text{-}NH\text{-}\overset{O}{\overset{\|}{C}}\text{-}CH_2\text{-}CH_2\text{-}\overset{CH_3}{\underset{CN}{\overset{|}{C}}}\text{-}N{=}N\text{-}\overset{CH_3}{\underset{CN}{\overset{|}{C}}}\text{-}CH_2\text{-}CH_2\text{-}\overset{O}{\overset{\|}{C}}\text{-}NH\text{-}\overset{O}{\overset{\|}{C}}\text{-}CCl_3$$

With the aid of this polymer, block copolymers can be obtained by applying a method proposed by Bamford: the polymer is irradiated with UV light in the presence of manganes carbonyl and appropriate monomer. Upon absorption of light, $Mn_2(CO)_{10}$ decomposes into $Mn(CO)_5$. The latter reacts with terminal CCl_3 groups yielding macroradicals which can start the polymerization of a monomer contained in the system. The overall mechanism is shown in scheme 8.

$$Mn_2(CO)_{10} \xrightarrow{h\nu} 2\,Mn(CO)_5$$

$$\text{\textasciitilde\textasciitilde PSt-}CCl_3 + Mn(CO)_5 \longrightarrow \text{\textasciitilde\textasciitilde PSt-}\dot{C}Cl_2 + Mn(CO)_5Cl$$

$$\text{\textasciitilde\textasciitilde PSt-}\dot{C}Cl_2 + MMA \longrightarrow \text{\textasciitilde\textasciitilde PSt-}CCl_2\text{-}PMMA\text{\textasciitilde\textasciitilde}$$

Scheme 8

Typical results obtain from PSt/MMA system are shown in Table 2. Notably, Rp, the rate of MMA conversion increase with increasing concentration of CCl_3. The increase, Δm, in total mass of polymer reflects the formation of block copolymer. The formation of block copolymer was clearly shown by gpc analysis as can be seen from Figure 2, where chromatograms recorded with the prepolymer and with the polymeric products obtained after irradiation times of 45, 90 and 180 minutes, respectively, are presented. One can see the development of a new peak at higher molecular weight, which correlates with decrease in the prepolymer peak with increasing irradiation time. The new peak is ascribed to the block copolymer.

It is clear that block copolymerization via photofunctional azo-initiators provides a versatile two-stage method applicable to vinyl monomers. In the first stage, photoactive prepolymers were synthesized by using these initiators. The resulting polymers can act directly or indirectly as macrophotoinitiators in a broad spectral range. The direct initiation is based on the generation of macroradicals via photoiniduced cleavage of of benzoin and acyl oxime ester groups incorporated to polymers. Regarding indirect action, excited states of aromatic carbonyl compounds and photodecomposition products of transition metal carbonyl compounds readily react with amine and halogen terminated polymers, respectively.

Table 2. Photopolymerization of MMA initiated by Cl_3C-PSt-CCl_3/$Mn_2(CO)_{10}$ at λ=436 nm. PSt1, M_n = 14500, PSt2, M_n = 5800

PSt (g/l)	$Mn_2(CO)_{10}$ (mol/l)	Irradiation Time (min.)	Δm^a (%)	f_{block}^b
PSt1(70)	1.0×10^{-3}	150	42	0.49
PSt2(40)	0.8×10^{-3}	150	133	0.67
PSt1(40)	0.8×10^{-3}	180	55	0.38

[a]Increase in mass of polymer
[b]Weight fraction of block copolymer in product

BLOCK COPOLYMERIZATION BY MACROPHOTOINITIATORS PREPARED BY ACTIVATED MONOMER POLYMERIZATION

It has been shown by Penczek et al. [17] that oxiranes can be polymerized in the presence of hydroxyl containing compounds by the activated monomer (AM) mechanism. The propagation in these systems can be represented by the following reaction.

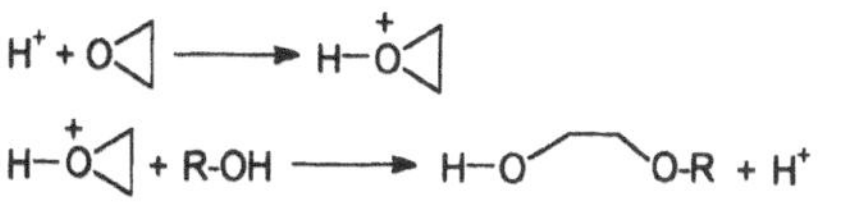

$$H^+ + O\triangleleft \longrightarrow H-\overset{+}{O}\triangleleft$$

$$H-\overset{+}{O}\triangleleft + R\text{-}OH \longrightarrow H\text{-}O\frown O\text{-}R + H^+$$

Scheme 9

By its nature, this process may be adapted so that hydroxy containing functional molecules, monomers and polymers are used as initiators to yield telechelic [18], macro-monomer and block copolymers[19], respectively. We have recently demonstrated [20] that the following benzoin type photoinitiators may also be used in AM polymerization of epichlorohydine (ECH) to produce polymers with terminal photoactive groups.

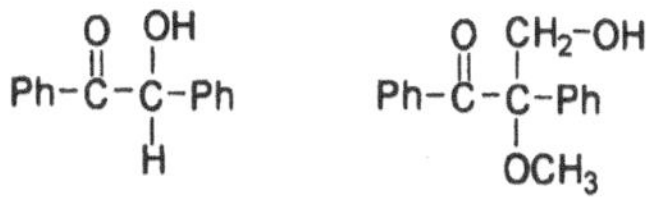

$$\underset{\textbf{(B)}}{Ph-\overset{O}{\overset{\|}{C}}-\overset{OH}{\overset{|}{\underset{|}{C}}}-Ph}\qquad\underset{\textbf{(MBME)}}{Ph-\overset{O}{\overset{\|}{C}}-\overset{CH_2-OH}{\overset{|}{\underset{|}{C}}}-Ph}$$

In both cases reaction was carried out by slow addition of ECH to the solution of B or MBME in methylene chloride, in the presence of the catalyst. At these conditions, both initiators were consumed at the early stages of the reaction, as evidenced by the decrease of corresponding signals in GPC chromatograms of the reaction mixtures at different conversions (Figure 3).

After termination of of the polymerization, the products were extracted with methanol in order to remove the possibly present traces of unreacted initiators. The resultant

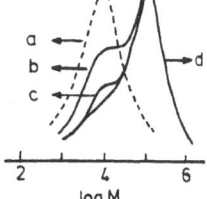

Figure 2. Gel permeation chromatograms recorded with products formed by irradiation of Cl_3C-PSt-CCl_3 in the presence of MMA and $Mn_2(CO)_{10}$. (a) prepolystyrene (b) after 45 min. (c) after 90 min. (d) after 180 min.

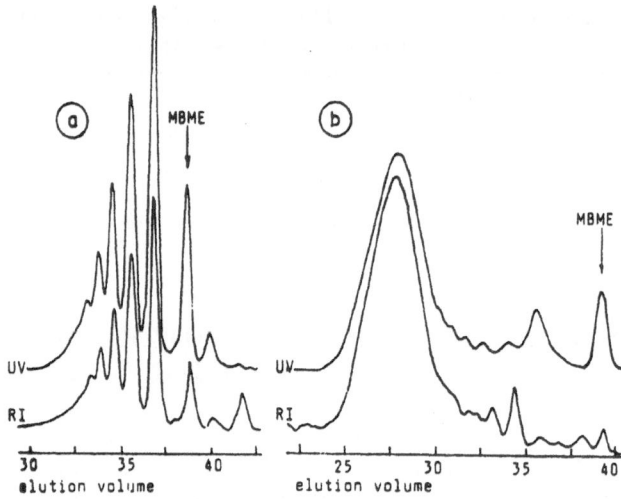

Figure 3. Gel permeation chromatograms of the reaction mixture (with MBME as initiator) at different conversions. (a) [ECH] added / $[MBME]_o$ = 5, (b) [ECH] added / $[MBME]_o$ = 25.

polymers exhibited absorption characteristics of precursor benzoin compounds and may undergo clevage upon irradiation to produce polymeric radicals. Block copolymerization via polyepichlorohydin macrophotoinitiator are now being investigated and will be reported elsewhere.

BLOCK COPOLYMERIZATION BY PYRIDINIUM SALT MACROPHOTOINITIATORS

It has been well established that certain N-alkoxy pyridinium and N-alkoxy isoquinolinium salts are capable of acting as photoinitiators for the cationic polymerization. Initiation is based on the photolytic generation of nitrogen centered radical cations as illustrated for the case of N-ethoxy isoquinolinium ions [21].

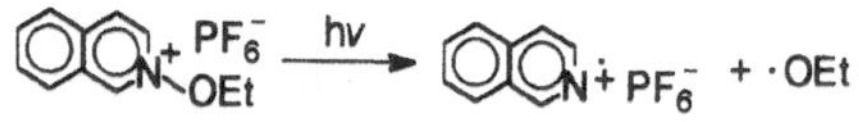

Scheme 10

Reactive radical cations may also be generated indirectly via the reaction of with electronically excited states of certain sensitizers [22]. This way spectral sensitivity of the initiating systems are shifted longer wavelengths which has importance in uv curing applications.

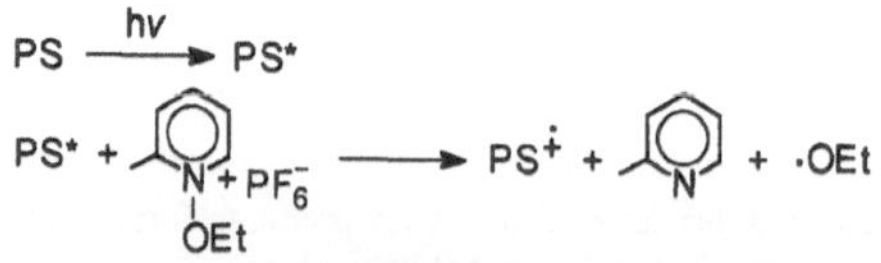

Scheme 11

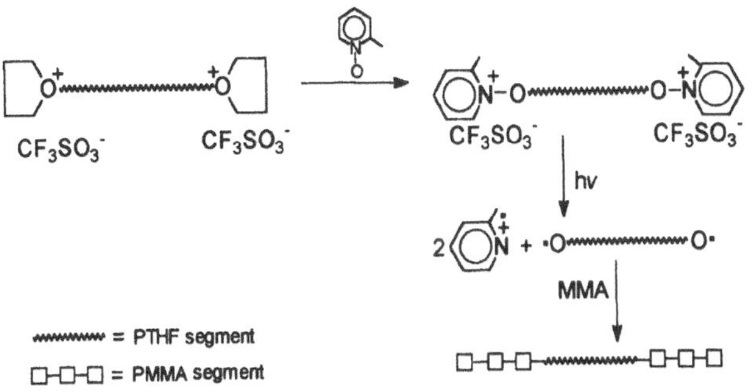

Scheme 12

Ethoxy radicals, which are also formed in both direct and sensitized decomposition of pyridinium ions, can inititiate free radical polymerization of appropriate monomers [23]. Obviously, polytetrahydrofurans (PTHF), terminated by *N*-alkoxy pyridinium ions, can act macrophotoinitiators for the polymerization of monomers such as MMA that readily polymerize by a free radical mechanism. PTHF macrohotoinitiatots were prepared by termination of living polymerization of THF by the corresponding *N*-oxides [24]. The well-defined macrophotoinitiators with exact functionalities, confirmed by ¹H-nmr, uv-vis and g.p.c. analysis, were obtained. Upon irradiation of macrophotoinitiators at suitable wavelengths, polymeric alkoxy radicals are produced. The overall process is shown for the pyridinium macrophotoinitiator in scheme 12. Photolysis in the presence of MMA gives quantitative yields of PMMA-*b*-PTHF-*b*-PMMA block copolymers. Block copolymers with various MMA segment lengths were obtained by precipitating at room temperature and -20 °C.

Indirect photolysis using anthracene sensitizer in the presence of MMA gives rise to the formation of the same block copolymers. Again the reaction products are almost pure block copolymers.

BLOCK COPOLYMERS BY POLYSILANE MACROPHOTOINITIATORS

High molecular weight, soluble and formable polsilanes have found application as photoresist materials, photoconductors and photoinitiators in radical polymerization [25]. It was assumed that the initiating process consists of silyl type radicals, formed by rapid degradation of polysilanes upon photolysis in the 300-350 nm region, with vinyl monomers [26]. Recently, it was shown that in conjuction with pyridinum [27]or iodonium salts [28], polysilanes can be used as photoinitiators for the cationic polymerization of cyclic ethers and vinyl ethers. More recently, we included this rather novel class of polymers in our current program concerning block copolymerization by using them as multifunctional initiators [29]. Both cationic and radical routes are employed.

Block Copolymers by Free Radical Route

Photoactive PMMA was prepared by photoinitiated free radical polymaerization of MMA initiated poly(methylphenylsilane) (PMPSi). The segment length of polysilane incorporated in PMMA was controlled by irradiation time. The possible reaction mechanism is presented in Scheme 13. The primary silyl radicals from polysilane degradation, initiate and

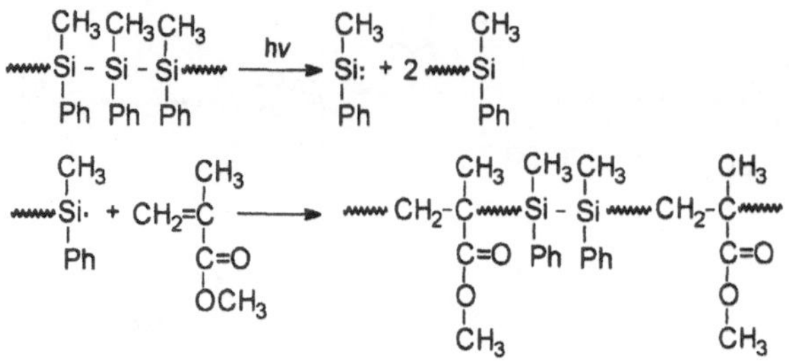

Scheme 13

propagate with MMA yielding Si-Si bonds in the polymer chain. The resulting PMMA exhibits the polysilane absorption band. Conversion and molecular weight of PMMA and the silicon content in the PMMA backbone can be adjusted by irradiation time.

Photoactive PMMA was used to initiate the polymerization of St. Obviously, the initiating radicals are derived from silyl radicals as a result of photodegradation of polysilane moieties in the PMMA backbone (scheme 14). By adjusting the amount and position of polysilane segments in the initially formed PMMA, higher block yields may be obtained.

Scheme 14

Block Bopolymers by Cationic Route

Previously, we have shown [27] that the cationic polymerization of CHO is initiated upon UV irradiation of metylene chloride solutions containing N-ethoxy-2-methylpyridinium hexafluorophosphate and polysilane. A plausible explanation mechanism involves oxidation of silyl biradicals and silyl radicals by pyridinium ions to yield reactive cations capable of initiating the polymerization of CHO. Provided silylenium ions formed would add to CHO, the resulting oxonium ion would initiate chain propagation as shown below.

The same reaction principle was applied for the polymerization of CHO using remainig polysilane units in PMMA and pyridinium ions (scheme 15). Although block copolymers are formed, a considerable amount of homopolymers was also formed which may be due to extensive degradation of polysilane units during irradiation. Moreover, the character of the initiation mechanism, which needs further investigation, may contribute to the formation of non-blocked component.

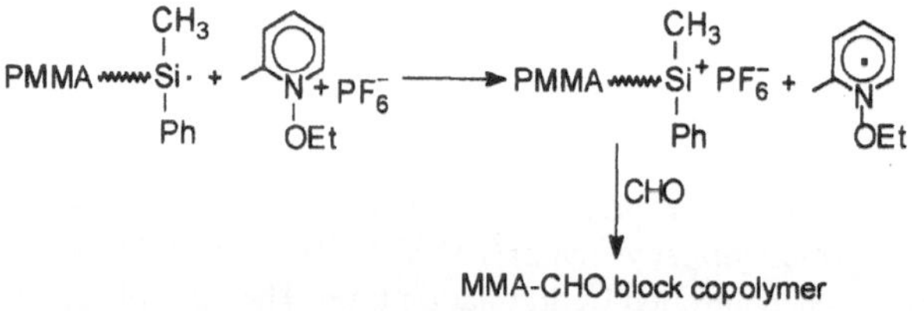

Scheme 15

REFERENCES

1. R.S.Davidson, *J.Photochem.Photobiol. A, Chem*, 69, 263 (1993)
2. Y.Yagci and M.K.Mishra, *"Macromeculer Design: Concept and Practice"*, Polymer Front.Int'l Inc., (M.K.Mishra, ed.) New York, Chapter 10, (1994)
3. Y.Yagci and W.Schnabel, *Prog.Polym.Sci.*, 15, 551 (1990)
4. A.Onen and Y.Yagci, *J.Macromol.Sci.Chem.*,A27, 743 (1990)
5. A.Onen and Y.Yagci, *Angew.Makromol.Chem.Chem.*, 181, 191 (1990)
6. Y.Yagci and A.Onen, *J.Macromol.Chem.Sci.Chem.*, A28, 129 (1991)
7. Y.Yagci, A.Onen and W.Schnabel, *Macromolecules*, 24, 4620 (1991)
8. I.E.Serhatli, E.Chiellini, G.Galli and Y.Yagci, *Polym.Bull.*, in press
9. P.Baas and H.Cerfontain, *J.Chem.Soc., Perkin Trans II*, 1963 (1979)
10. A.Onen, S.Denizligil and Y.Yagci, *Angew.Makromol.Chem.*, 217, 79 (1994)
11. T.Imamoglu, A.Onen and Y.Yagci, *Angew.Makromol.Chem.*, in press
12. A.Ledwith, *J.Oil Colour Chem.Assoc.*, 59, 157 (1976)
13. Y.Yagci, G.Hizal and U.Tunca, *Polymer Commun.* 31, 7 (1990)
14. H.Block, A.Ledwith and A.R.Taylor, *Polymer*, 19, 354 (1978)
15. C.H.Bamford, *"New Trends in the Photochemistry of Polymers"*(Eds.N.S.Allen and J.F.Rabek) Elsevier, London, p.129, (1985)
16. Y.Yagci, M.Müller, and W.Schnabel, *Macromol.Reports*, A28 (Suppl.1) 37 (1991)
17. S.Penczek, P.Kubisa and R.Szymanski, *Makromol.Chem., Macromol.Symp.*, 3, 203 (1986)
18. P.Kubisa, *Makromol.Chem.,Macromol.Symp.*,13/14, 203 (1986)
19. Y.Yagci, I.E.Serhatli, P.Kubisa and T.Biedron, *Macromolecules*, 26, 2397 (1993)
20. Y.Yagci, Y.Hepuzer, A.Onen, I.E.Serhatli, P.Kubisa and T.Biedron, *Polym.Bull.*, 33, 411 (1994)
21. Y.Yagci, A.Kornowski and W.Schnabel, *J.Polym.Sci.,Polym.Ed.*,30, 1987 (1992)
22. Y.Yagci, I.Lukac and W.Schnabel, *Polymer*, 34, 1130 (1993)
23. N.Kayaman, A.Onen, Y.Yagci and W.Schnabel, Polym.Bull., 32, 589 (1994)
24. G.Hizal, Y.Yagci and W.Schnabel, *Polymer*, 35, 4443 (1994)
25. R.D.Miller and J.Michl, *Chem.Rev.*, 89, 1359 (1989)
26. A.R.Wolf and R.West, *Appl.Organomet.Chem.*, 1, 7 (1987)
27. Y.Yagci, I.Kminek and W.Schnabel, *Eur.Polym.J.*, 28, 387 (1992)
28. Y.Yagci, I.Kminek and W.Schnabel, *Polymer*, 34, 426 (1993)
29. D.Yucesan, H.Hostoygar, S.Denizligil and Y.Yagci, *Angew.Makromol.Chem.*, 221, 207 (1994)

12

AMPHIPHILIC POLYMER NETWORKS BY COPOLYMERIZATION OF BIS-MACROMONOMERS

Peiwen Tan, Saskia R. Walraedt, Jan M.M. Geeraert and Eric J. Goethals[*]

Department of Organic Chemistry
Polymer Chemistry Division
University of Ghent
9000 Ghent
Belgium

INTRODUCTION

Polymers carrying a polymerizable end group, generally known as macromonomers, have frequently been utilized for the synthesis of graft copolymers (1)(2). When a polymer contains polymerizable groups at *both* chain ends, -a bis-macromonomer-, polymerization or copolymerization leads to the corresponding networks or segmented copolymer networks respectively. The physicochemical properties of the latter may be expected to be the result of a combination of the properties of segmented polymers at the one hand and of polymer networks at the other. For example, it should be interesting to investigate the morphology of such compounds if they are build up of two incompatible chain segments. The theory of the segmented copolymers would predict a phase separation, the morphology of which is determined by the relative lengths of the two constituting polymers. However, in the network form, phase separation is inhibited due to the restricted chain mobilities and therefore a compatibilization of two incompatible polymers may be expected. As a consequence, new materials with formerly unreachable physicochemical properties could arise.

In this paper, the synthesis and some properties of such networks, in which one of the polymer segments is hydrophilic and the other is hydrophobic, will be presented. Two approaches can be used to achieve these polymer structures. In a first approach, a hydrophobic bis-macromonomer is copolymerized with a hydrophilic monomer, in the second one, a hydrophilic bis-macromonomer is copolymerized with a hydrophobic monomer. In a previous paper, block-copolymer networks derived from polylactide-bis-macromonomers have been reported (3). Other authors have recently described segmented amphiphilic polymer networks obtained from bis-macromonomers derived from the hydrophobic polyisobutylene

---

[*] Address correspondence to Dr. Goethals.

Macromolecular Engineering, Edited by M.K. Mishra et al.
Plenum Press, New York, 1995

and hydrophilic monomers (4)(5)(6). In the present review emphasis will be put on the combination hydrophilic bis-macromonomer - hydrophobic comonomer.

CHOICE OF THE HYDROPHILIC POLYMERS

The following hydrophilic polymer chains were selected for our study : poly(ethylene oxide) (PEO), poly(1,3-dioxolane) (PDXL) and poly(N-tert.butylaziridine) (PTBA). The structures of these polymers are described below:

PEO   PDXL   PTBA

The last polymer, PTBA, is not hydrophilic as such but becomes hydrophilic by quaternization of the amino functions.

As polymerizable end groups, acrylate and methacrylate esters were selected for free radical initiated copolymerizations.

SYNTHESIS OF THE BIS-MACROMONOMERS

Three methods have been used for the synthesis of bis-macromonomers : end group modification, end-capping of living polymerizations, and polymerizations carried out in the presence of functional transfer agents.

The first method was used for the synthesis of PEO α,ω-bis-methacrylates. The starting material was a commercial poly(ethylene glycol) of molecular weight 4000. The esterification of the hydroxyl groups was performed with methacryloyl chloride in the presence of triethylamine.

End-capping was used for the synthesis of PTBA α,ω-bis-acrylates. The cationic polymerization of TBA has a high living character and the polymer can therefore be prepared in a controlled way by the proper choice of initiator and reaction conditions(7). Bifunctionally growing living polyTBA is obtained by initiation with a bifunctionally growing living poly(tetrahydrofurane), which in turn is obtained by initiation of the THF polymerization with trifluoromethane sulfonic acid anhydride (8). End-capping of the active species with acrylic acid leads to the bis-acrylate ester (9).

PDXL-bis-methacrylate was obtained by cationic ring-opening polymerization of DXL in the presence of the formal of hydroxyethyl methacrylate which acts as a transfer agent (10).The molecular weight of the polymers can be controlled by the ratio [monomer]/[transfer agent]. For this study, a polymer with a molecular weight (Mn) of 4200 was selected.

COPOLYMERIZATIONS

The bis-macromonomers were copolymerized with methyl methacrylate (MMA), in bulk or in toluene solution at 70°C with AIBN as initiator. In all cases the end products were, as expected, insoluble materials. The appearance of the networks depended on the constitution of the network. Generally speaking, high MMA contents lead to transparent strong materials. As the concentration of the bis-macromonomer was increased, phase separation

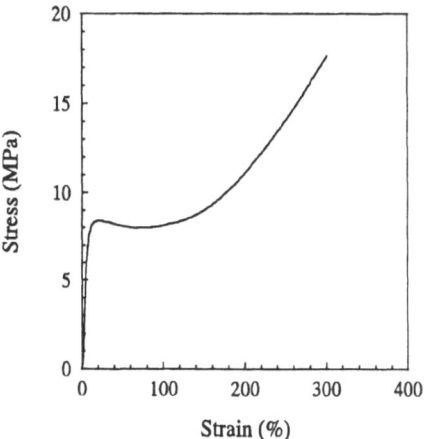

Figure 1. Stress-strain curve of a PMMA$_{(50)}$-cross-PEO$_{(50)}$(4000) network at 25 °C.

during the polymerization became apparent. Generally this occurred at bis-macromonomer contents of over 50%.

PROPERTIES OF SEGMENTED COPOLYMER NETWORKS

In this paragraph, some properties of the segmented copolymer networks, as they were known at the time of the redaction of this paper, are reported.

For the composition of the segmented copolymer networks a nomenclature indicating the fractions of both components as well as the molecular weight (Mn) of the bis-macromonomers was used. In the following example, the network consisted of 20 wt%PMMA and 80 wt% of PDXL and the Mn of the PDXL bis-macromonomer was 4200:

$$\text{PMMA}_{(20)}\text{-cross-PDXL}_{(80)}(4200).$$

PEO-PMMA Networks

These networks form transparent or slightly opaque, strong materials. In Fig.1, a stress-strain curve of a network composed of 50/50 PEO/PMMA, measured at 25°C, is shown. It can be seen that it behaves as a tough material rather than a glass. Fig.2 shows four dynamic mechanical thermal analyses (DMTA) of networks with varying PEO/PMMA ratios.

It can be seen that curves A, B and C show one continuous decrease of the E' modulus over a relatively broad range of temperature, indicating that the materials consist of one phase with a glass transition situated in between the Tg's of the two pure components. Compound D, containing the highest PEO fraction, shows two decreases of E' but, according to DSC analysis, material D does not contain a measurable crystalline fraction. Therefore, the DMTA results can not be attributed to the presence of crystalline PEO but rather to the presence of two amorphous phases with different compositions. These results are in agreement with a recent publication describing the compatibility and crystallization in linear PEO-crosslinked PMMA semi-interpenetrating networks (11), where it was also concluded that two PEO-PMMA phases differing in PEO content and having different Tg's, were

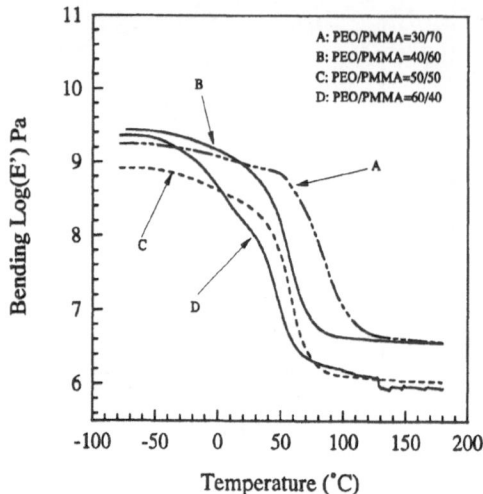

Figure 2. DMTA of PMMA-cross-PEO(4000) networks with varying PEO/PMMA ratios.

present. However, for semiIPNs with a PEO content of 30% or higher, a crystalline fraction, increasing with the content of PEO, was found. In contrast, our segmented copolymer networks did not show any sign of crystallinity with PEO fractions up to 60%.

PDXL - PMMA Networks

PDXL is a low Tg (-50°C), crystallizable (Tm 55°C), moderately hydrophilic polymer. The phase diagram of PDXL-water mixtures has been described (12). The segmented copolymers obtained by copolymerization of PDXL-bis-macromonomer with MMA are strong to soft transparent materials for polymers containing up to 45% of PDXL and

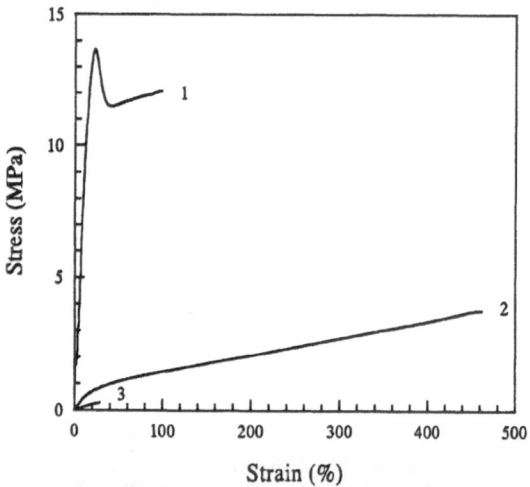

Figure 3. Stress-elongation curves of PMMA-cross-PDXL networks. 1. PMMA(72)-cross-PDXL(28)(4700); 2. PMMA(55)-cross-PDXL(45)(4700); 3. PMMA(24)-cross-PDXL(76)(4700).

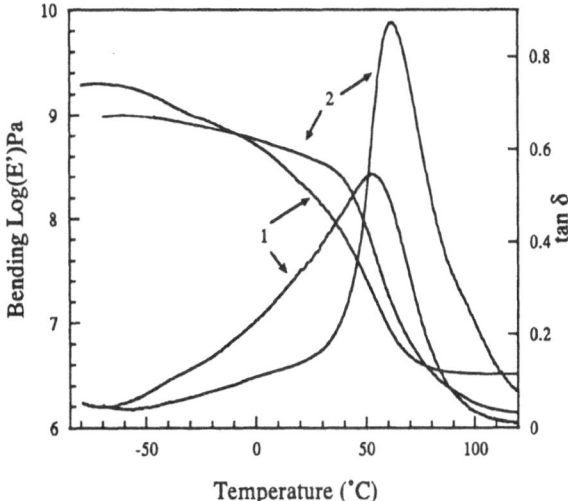

Figure 4. DMTA of a PMMA(54)-cross-PDXL(46)4200 (1) and a PMMA(72)-cross-PDXL(28)(4200) network (2).

opaque, brittle materials when the PDXL content is higher. In Fig.3 the stress-strain curves of three networks with different compositions are compared.

DMTA of the materials (Fig.4) reveals that at low PDXL content, the material behaves as a one phase amorphous material with a Tg (as measured by the maximum of the tan δ) in between the two Tg's of the constituents.

When the PDXL content is raised to appr. 50%, the top of the tan δ curve is further shifted to lower values but the curve becomes broad, indicating the presence of a variety of compositions with a correspondingly varying array of Tg's.

As indicated in Table 1, networks containing small fractions of PDXL show no melting endotherm in DSC measurements. At a 1/1 composition, a very small fraction of crystalline material can be detected. When PDXL is the major component of the network, a considerable crystallinity is observed, which, at 20/80 ratio, amounts to appr. 50% of the crystallinity of cross-linked homoPDXL obtained from a bis-macromonomer of similar molecular weight.

In order to evaluate the influence of the network structure on the physical properties of the PDXL-PMMA systems, a series of blends of linear PDXL and PMMA were studied. It was found that the two polymers showed a very limited compatibility and that PDXL phase-separates as soon as its content exceeds 10%.

It can be concluded that the incorporation of the two polymer segments in a single network forces the two otherwise incompatible polymers to form more or less homogeneous blends and inhibits the crystallization of the PDXL to a great extent. Consequently, a series

Table 1. Melting behaviour of PMMA-cross-PDXL(4200) networks with varying PDXL/PMMA ratio

	Tm (°C)	Heat of fusion (J/g)
PMMA$_{80}$-cross-PDXL$_{20}$(4200) —		—
PMMA$_{50}$.cross-PDXL$_{50}$(4200) 28		0.25
PMMA$_{20}$-cross-PDXL$_{80}$(4200) 24		20.7
PMMA$_{0}$-cross-PDXL$_{100}$(4200) 22		49

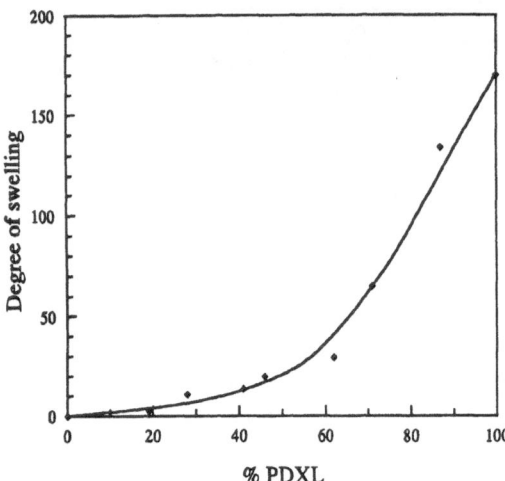

Figure 5. Degree of swelling (expressed as $100(w_{swollen}-w_{dry})/w_{dry}$) in water at 50 °C of PMMA-cross-PDXL(4200) networks as a function of the amount of PDXL.

of materials exhibiting a continuous change from hydrophobic to hydrophilic properties can be prepared by variation of the composition of the network. This is visualised in Fig.5 in which the swelling behavior of various networks in water is traced as a function of the PDXL content. It can be seen that the hydrophilicity increases continuously, a behavior which would be expected for random copolymers (which obviously cannot be synthesized from the present two monomers) but not for block-copolymers, where a rather sudden change in properties is expected when the morphology changes from hydrophobic matrix to hydrophilic matrix by increasing the fraction of hydrophilic segments.

PTBA-PMMA Networks

PTBA is a crystalline polymer with a melting point of 142°C and a glass transition temperature of 24°C. Due to its crystallinity, the solubility of the polymer is low in most solvents, including water, at room temperature. Crystallization of pure PTBA occurs only from solution and not from the melt. This is attributed to the limited flexibility of the sterically hindered polymer chain in the condensed phase. The PTBA bis-acrylate used in the present study had the structure PTBA2000-PTHF1200-PTBA2000 . Three segmented networks were prepared by copolymerization with different amounts of MMA. The end products ranged from slightly opaque (for 75% of PTBA) to transparent materials (for 50 and 25% PTBA). PTBA is insoluble in water but becomes soluble in acidic medium due to the quaternization of the amines to the corresponding ammonium salts. Therefore, the degree of swelling of the networks was expected to depend on the pH of the medium. Table 2 shows the results of swelling experiments for the three segmented networks.

It can be seen that the swelling increases dramatically by shifting from pure water to 0.1 N HCl and that this increase is the highest for the network with the highest TBA fraction. DSC analysis showed that the network with 75% of TBA, contained some crystalline fraction. Apparently, phase separation of the two components took place during the network formation. It is also interesting to note that networks containing 50% of TBA show a higher degree of swelling in pure water than those containing 75%. This is probably to be attributed to differences in degree of crystallinity of the PTBA fractions in these two networks.

Table 2. Degree of swelling [a] of PTBA-PMMA segmented copolymers in different solvents as a function of composition

Solvent	PMMA/PTBA		
	75/25	50/50	25/75
CHCl$_3$	—[b]	1300	1140
H$_2$O	7	55	20
0.1 N HCl	15	350	980
0.1 N NaOH	0.5	0.7	2

a: degree of swelling expressed as $100(w_{swollen}-w_{dry})/(w_{dry}\rho_{solvent})$
b: network was broken due to the high degree of swelling of the PMMA segments.

CONCLUSIONS

Three hydrophilic α,ω-bis-methacrylate prepolymers derived from PEO, PDXL and PTBA have been synthesized and copolymerized with MMA to form the corresponding segmented copolymer networks. Some physical properties of the end products as a function of their composition have been studied.. The general trend observed in the thus obtained materials is that phase separation of the two constituting polymers is strongly inhibited by the incorporation of cross-links. Physical properties such as glass transition temperature and swelling behavior can be shifted in a continuous manner by variation of the ratio of the two components, indicating a good compatibility over a broad range of compositions. The copolymer networks obtained from PTBA show a pH-dependent swelling behavior.

REFERENCES

1. Rempp P.F., Franta E., 1984, Adv. Polym. Sci, 58:1
2. Chujo, Y., Yamashita, 1989, Telechelic Polymers: Synthesis and Applications, p 163, CRC press; Boca Raton, FL, Ed: Goethals, E.J..
3. Barakat I., Dubois Ph., Jérome R., Teyssié Ph. and Goethals E.J., 1994, J. Polym. Sci, Part A, Polym. Chem., 32: 2099.
4. Ivan B., Kennedy J.P., Mackey P.W., 1990, Polymer Preprints, 31 (2):215.
5. Ivan B., Kennedy J.P., Mackey P.W., 1990, Polymer Preprints, 31(2):217.
6. Ivan B., Kennedy J.P., Mackey P.W.,1991, Polymer Drugs and Delivery Systems, ACS Symp. Book Series, 469;194 and 203, Eds.: Dunn R.L. and Ottenbrite R.M..
7. Munir A., Goethals E.J., 1981, J. of Polym. Sci., 19:165.
8. Vandevelde M., Goethals E.J., 1986, Makromol. Chem., Macromol. Symp., 6:271.
9. Goethals E.J., Vlegels M.A.,1981, Polymer Bulletin, 4:521.
10. De Clercq R.R., Goethals E.J., 1992, Macromolecules, 25:1109.
11. Ahn S.H., An J.H., Lee D.S., Kim S.C., 1993, J. Polym. Sci, 31:1627.
12. Benkhira A., Franta E., François J.,1992, Macromolecules, 25:5697.

13

STEREOSPECIFIC POLYMERIZATION AND COPOLYMERIZATION OF STEREOREGULAR PMMA MACROMONOMERS

Koichi Hatada, Tatsuki Kitayama, Osamu Nakagawa and
Takafumi Nishiura

Department of Chemistry
Faculty of Engineering Science
Osaka University
Toyonaka, Osaka 560
Japan

1. INTRODUCTION

It is one of the most effective methods for the structural control of comblike polymer and graft polymer to utilize macromonomers. In fact, many papers have been published on utilization of macromonomers for the preparation of comblike polymers and graft polymers with controlled structures[1-17]. However, few investigations have been reported on the preparation of comblike polymer and graft polymer with high stereoregularity. We have reported the preparation of isotactic (*it*-) and syndiotactic (*st*-) poly(methyl methacrylate) (PMMA) macromonomers with styrene end group, polymerizations of which give comblike polymers with stereoregular PMMA branches[6-11]. We have also reported the stereospecific anionic polymerization of polyisobutylene macromonomer having methacryloyl end-function, which afforded comblike polymers with stereoregular main chain[12]. The reason why methacryloyl group was selected as a polymerizable function in this case is that the stereoregularity of polymethacrylate can be controlled in a wide range by selecting a proper initiator and polymerization conditions[18,19].

Stereospecific polymerization of stereoregular macromonomer should give a comblike polymer or graft polymer with controlled stereoregularities in main chain as well as in side chains. In this work, we carried out the synthesis of stereoregular PMMA macromonomers having methacryloyl function and their polymerizations[13,14] and copolymerizations[11,15-17] with methacrylates by anionic initiators to obtain the stereoregular comblike polymers and graft polymers. Polymers of totally deuterated PMMA macromonomers with undeuterated methacryloyl function was prepared to analyze the main-chain stereoregularities of the comblike polymers. Some of the tacticity dependent properties of the resulting branched PMMAs were also discussed.

Macromolecular Engineering, Edited by M.K. Mishra et al.
Plenum Press, New York, 1995

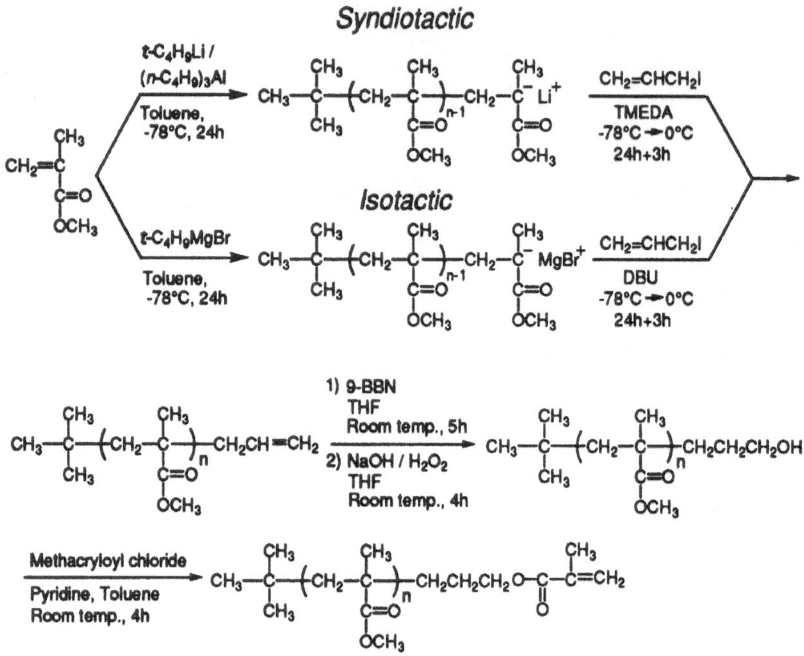

Scheme 1 Preparation of stereoregular macromonomer

2. SYNTHESIS OF STEREOREGULAR PMMA MACROMONOMER HAVING METHACRYLOYL FUNCTION

Stereoregular PMMA macromonomers having methacryloyl function was prepared by three-step reactions from *it*- and *st*-PMMA anions as shown in Scheme 1[11,13-17].

2.1. Preparation of Stereoregular PMMA-CH$_2$-CH=CH$_2$

To obtain highly *it*- and *st*-PMMA macromonomers having methacryloyl function, we utilized the living polymerizations of methyl methacrylate (MMA) with *t*-C$_4$H$_9$MgBr[20] and *t*-C$_4$H$_9$Li/(*n*-C$_4$H$_9$)$_3$Al (1/3)[21], respectively, which have already been applied to the preparation of highly stereoregular PMMA macromonomers with styrene end group[8]. From the experience in the synthesis of the styrene-type PMMA macromonomer, 1,8-Diazabicyclo[5.4.0]undec-7-ene (DBU) and N,N,N',N'-tetramethylethylenediamine (TMEDA) were used, to facilitate the coupling reaction between allyl iodide and *it*- or *st*-PMMA living anions, respectively.

^{1}H NMR spectra of *st*-PMMA-CH$_2$-CH=CH$_2$ are shown in Figure 1a as an example. The number average molecular weight (Mn) was determined as 3100 from the intensity ratio of the signals due to the protons of *t*-C$_4$H$_9$ (0.82ppm) and OCH$_3$ (3.61ppm) groups. The functionality of allyl end group was determined as 0.98 from the intensity ratio of the signals due to *t*-C$_4$H$_9$ and olefinic protons in allyl group (-CH=: 5.54-5.67ppm, =CH$_2$: 4.92-4.98ppm)[15].

Functionality and Mn of *it*-PMMA-CH$_2$-CH=CH$_2$ were similarly determined as 0.95 and 2910, respectively[17], from its ^{1}H NMR spectrum.

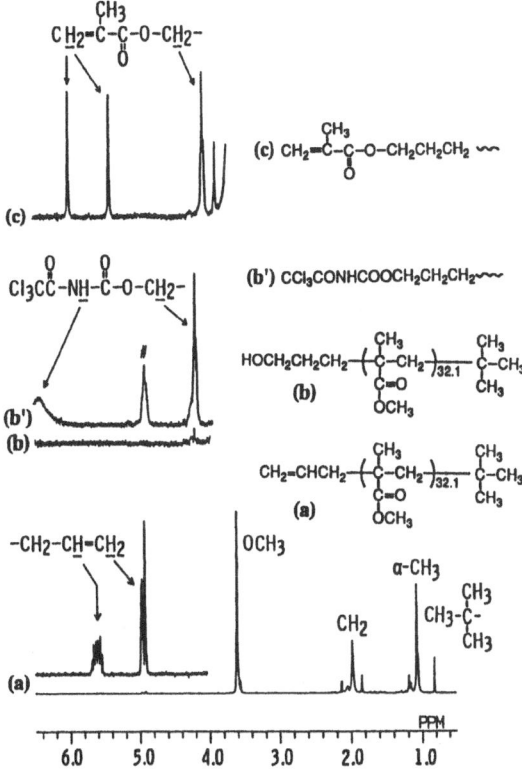

Figure 1. 270MHz ^{1}H NMR spectra measured in nitrobenzene-d_5 at 110°C of st-PMMA-CH$_2$-CH=CH$_2$ (a), st-PMMA-(CH$_2$)$_3$-OH (b), st-PMMA-(CH$_2$)$_3$-O-CONH-CO-CCl$_3$ (b') and st-PMMA macromonomer(c) # Signal due to the reaction product of trichloroacetyl isocyanate and 1,5-cyclooctanediol derived from 9-BBN.

2.2. Preparation of Stereoregular PMMA-(CH$_2$)$_3$-OH

PMMA-(CH$_2$)$_3$-OH was prepared from the PMMA-CH$_2$-CH=CH$_2$ through hydroboration. The stereoregular PMMA with allyl end group was dissolved in THF under dry nitrogen and 9-borabicyclo[3.3.1]nonane (9-BBN) in THF was added to the solution at room temperature, and then aqueous NaOH and aqueous H$_2$O$_2$ were added successively. When aqueous H$_2$O$_2$ was added, the temperature of the mixture was kept below 30°C in order to prevent decomposition of H$_2$O$_2$.

The olefinic proton signals of allyl group disappeared completely in the ^{1}H NMR spectrum of the product as shown in Figure 1b. In order to determine the hydroxyl functionality, an excess amount of trichloroacetyl isocyanate was reacted with the st-PMMA-(CH$_2$)$_3$-OH in nitrobenzene-d_5 in an NMR sample tube to form st-PMMA-(CH$_2$)$_3$-O-CONH-CO-CCl$_3$ and the resulting solution was subjected to ^{1}H NMR measurement. The same method has been successfully utilized for the analysis of hydroxyl end groups of poly(ethylene oxide)[22] and poly(propylene oxide)[23]. In the ^{1}H NMR spectrum of the st-PMMA-(CH$_2$)$_3$-O-CONH-CO-CCl$_3$ shown in Figure 1b', a new signal ascribable to the methylene protons of -CH$_2$-O-CONH-CO-CCl$_3$ was observed at 4.17ppm. Another new signal appeared at 4.91ppm was assigned to the reaction product of trichloroacetyl isocyanate and 1,5-cyclooctanediol derived from 9-BBN. The functionality of hydroxyl end group was determined as 0.96 from the intensity ratio of the signals due to protons of t-C$_4$H$_9$ and -CH$_2$-O-CONH-

CO-CCl$_3$ (4.17ppm) (Figure 1b')[15]. The conversion from allyl group to hydroxyl group was 98%.

The functionality of hydroxyl end group of it-PMMA-(CH$_2$)$_3$-OH was similarly determined as 0.90[17] from its ^{1}H NMR spectrum. In this case the conversion from allyl group to hydroxyl group was 95%.

2.3. Preparation of Stereoregular PMMA-(CH$_2$)$_3$-OCOC(CH$_3$)=CH$_2$

PMMA macromonomer was prepared from the stereoregular PMMA-(CH$_2$)$_3$-OH and methacryloyl chloride in the presence of pyridine.

In the ^{1}H NMR spectrum of the st-PMMA macromonomer, new signals ascribable to vinylidene protons of methacryloyl end group were observed at 5.40ppm and 5.99ppm (Figure 1c). The functionality of methacryloyl end group of the macromonomer was determined as 0.92 from the intensity ratio of the signals due to protons of t-C$_4$H$_9$ and vinylidene group. The conversion from hydroxyl group to methacryloyl group was 96%. The Mn of the macromonomer was determined as 3530 from the ^{1}H NMR signal intensities of OCH$_3$ and t-C$_4$H$_9$ protons.

The values of functionality and Mn of methacryloyl group of it-PMMA macro-monomer could be determined as 0.89 and 3140, respectively[17], from its ^{1}H NMR spectrum. The conversion from hydroxyl group to methacryloyl group was 99%.

The tacticities of the PMMA macromonomers were determined from the α-CH$_3$ proton signals.

3 SYNTHESIS OF STEREOREGULAR COMBLIKE POLYMER[13,14]

3.1. Stereospecific Polymerization of Stereoregular PMMA Macromonomer

The stereoregular PMMA macromonomer was purified by reprecipitation and dried under vacuum at room temperature. The macromonomer was dissolved in benzene and the solution was dried over calcium dihydride (CaH$_2$). The CaH$_2$ was removed by filtration under dry nitrogen, and the macromonomer was recovered by freeze-drying without being exposed to air. The polymerization was carried out under dry nitrogen in glass ampules. Polymeriza-tions of stereoregular PMMA macromonomers were carried out with t-C$_4$H$_9$MgBr in toluene at -78°C and with 1,1-diphenylhexyllithium (DPHLi) in THF at -78°C, respectively. The results are summarized in Table 1.

GPC chromatogram of the product from the polymerization of st-PMMA macro-monomer with DPHLi in THF is shown in Figure 2a as a typical example. The chromatogram indicates the presence of low molecular weight fraction. From the peak intensities of the higher and lower molecular weight fractions, the conversion of the macromonomer could be determined as 87% by taking the functionality of the starting macromonomer into account. The conversion of the macromonomer was also determined by the ^{1}H NMR spectroscopic analysis for the vinylidene proton signals of the polymerization products. The value deter-mined by ^{1}H NMR was 96% and larger than those obtained from the GPC chromatogram (Table 1), indicating that the lower molecular weight peak in the GPC chromatogram contains the contribution by the unimer which failed to propagate further.

The conversions of it- and st-macromonomers in the polymerization with t-C$_4$H$_9$MgBr in toluene were 78% and 47%, respectively, suggesting the higher reactivity of it-PMMA macromonomer. We have reported that the styrene-type it-PMMA macromonomer

Table 1. Anionic polymerization of stereoregular PMMA macromonomer (Mac.)[a]

No.	Mac.(mmol)	Time day	Conversion[b](%) GPC	Conversion[b](%) NMR	$\bar{M}$n LALLS	$\bar{M}$n NMR	DP of main chain	$\bar{M}$w[c] $\bar{M}$n
Initiator: DPHLi, Solvent: THF								
1	st-[d] (0.174)	7	87	96	43300	44800	13.1[h]	1.17
2	it-[e] (0.122)	7	72	100	29500	24500	7.73[h]	1.35
Initiator: t-C₄H₉MgBr, Solvent: Toluene								
3	st-[f] (0.292)	23	47	48	21100	–	5.95[c]	1.23
4	it-[g] (0.253)	23	78	100	35700	–	9.65[c]	1.56

[a] $[M]_0 = 0.05$ (mol/l), $[M]_0/[I]_0 = 5.0$, Temperature -78°C.
[b] Conversion of the macromonomer was based on the functionality.
[c] Determined by GPC-LALLS.
[d] $\bar{M}$n = 3400, DP = 32.1, mm:mr:rr = 1:10:89, functionality = 0.89.
[e] $\bar{M}$n = 3140, DP = 29.5, mm:mr:rr = 94: 3: 3, functionality = 0.89.
[f] $\bar{M}$n = 3530, DP = 33.4, mm:mr:rr = 1:11:88, functionality = 0.92.
[g] $\bar{M}$n = 3690, DP = 35.0, mm:mr:rr = 93: 5: 2, functionality = 0.84.
[h] Determined by ¹H NMR.

showed higher reactivity than the corresponding st-PMMA macromonomer in radical polymerization, which were ascribed to the higher segmental mobility of it-PMMA chain[10]. These results confirm the tacticity dependence of the reactivity of PMMA macromonomers.

The polymacromonomers formed were fractionated from the reaction product by repeated precipitation in toluene-hexane mixtures. The GPC curve of the polymacro-monomer thus isolated is shown in Figure 2b. The Mn value and weight average molecular weight (Mw) of the polymacromonomers were determined by GPC-LALLS, and number average degree of polymerization (DP) of the main chain, that is, number of branches per molecule could be determined from the Mn values of the polymacromonomer and starting macromonomer (Table 1). In the case of the polymacromonomers prepared with DPHLi in THF, Mn's were also determined from the ¹H NMR signal intensities of phenyl protons of the initiator fragment and t-C₄H₉ protons of the branch end of the macromonomer unit, and the values agreed well with those determined by GPC-LALLS (Table 1).

3.2. Stereoregularity of Main Chain of the Comblike Polymer

The side-chain tacticity of the polymacromonomer should be identical to that of the starting macromonomer which was already determined by ¹H NMR analysis. However, the

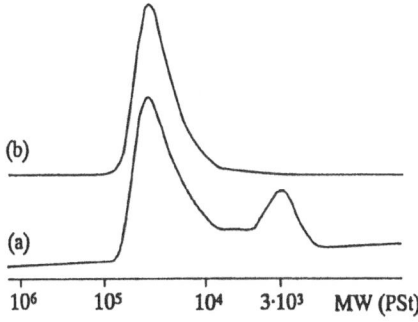

Figure 2. GPC curves of the polymerization mixtures of st-PMMA macromonomer with DPHLi in THF (a) and the polymacromonomer fractionated therefrom (b).

direct determination of main-chain stereoregularity from ^{1}H NMR spectra of the polymacro-monomers is rather difficult due to the overlap of the signals of the main-chain and side-chain methacrylate units. In order to avoid this difficulty, that is, to avoid the contribution of the side-chain signals, the *it*- and *st*-macromonomers were prepared from totally deuterated MMA (MMA-d_8) [Eq. 1] and polymerized under the same conditions as those for the non-deuterated macromonomer[14]. The degree of deuteration for α-CH$_3$, CH$_2$= and OCH$_3$ groups of MMA-d_8 were 99.07, 99.12 and 99.81, respectively.

$$
CH_3-\underset{\underset{CH_3}{|}}{\overset{\overset{CH_3}{|}}{C}}\left(CD_2-\underset{\underset{\underset{OCD_3}{|}}{C=O}}{\overset{\overset{CD_3}{|}}{C}}\right)_n CH_2CH_2CH_2O-\underset{\underset{O}{\|}}{\overset{\overset{CH_3}{|}}{C}}-\overset{|}{C}=CH_2
$$

[1]

Figure 3 shows ^{1}H NMR spectra of polymers of deuterated *st*-macromonomer (polymacromonomer-*d*) prepared with DPHLi in THF and with *t*-C$_4$H$_9$MgBr in toluene. The signals in the regions of α-CH$_3$ and main-chain CH$_2$ protons should be mostly due to those for the backbone of the polymacromonomers since the PMMA side chain are almost totally deuterated. The signal patterns of these regions clearly indicated that the polymacromonomer prepared with DPHLi in THF is syndiotactic (Figure 3a) and that prepared with *t*-C$_4$H$_9$MgBr in toluene is isotactic (Figure 3b).

Quantification of the main-chain stereoregularity of the *st*-polymacromonomer-*d* was made from the ^{1}H NMR spectrum (Figure 3a) as follows[14]. Signal intensity of α-CH$_3$ protons in *rr* triad, [α-CH$_3$(*rr*)], and that of CH$_2$ protons in *r* diad, [CH$_2$(*r*)], can be easily determined from the signals at 1.24ppm and 2.10ppm, respectively. Signal intensity of [α-CH$_3$(*mr*)] was

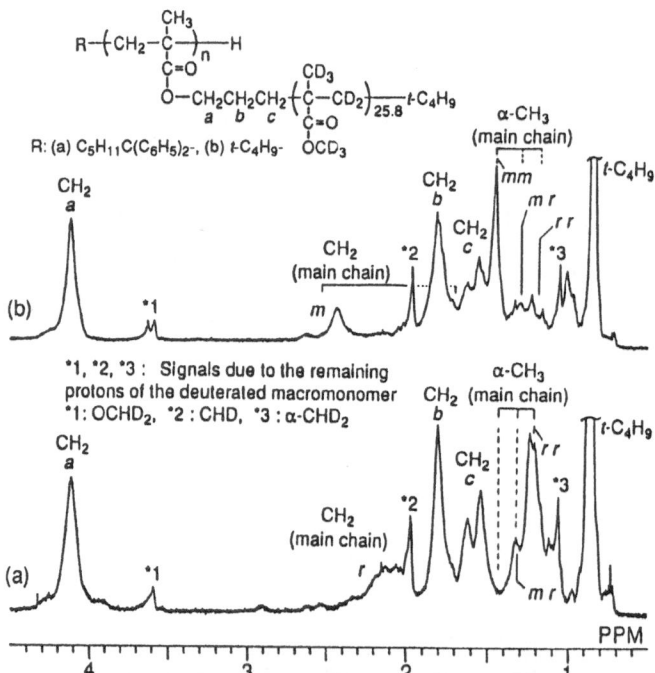

Figure 3. 500MHz ^{1}H NMR spectra measured in nitrobenzene-d_5 at 140°C of polymers of *st*-PMMA macro-monomer-*d* prepared with DPHLi in THF (a) and with *t*-C$_4$H$_9$MgBr in toluene (b).

Table 2. Characterization of polymers of deuterated PMMA
macromonomers (Mac-*d*) prepared with DPHLi in
THF [A] and with *t*-C₄H₉MgBr in toluene [B]

Mac-*d*	DP		Tacticity (%)^c					
[Polymn]	Main^a chain	Side^b chain	Main chain			Side chain		
			mm	*mr*	*r*	*mm*	*mr*	*r*
st- [A]	26.3	25.8	1	21	78	0	13	87
it- [A]	37.4	26.7	8	17	75	93	6	1
st- [B]	15.0	25.8	69	21	10	0	13	87
it- [B]	34.6	26.7	69	22	9	93	6	1

^{a} Determined by GPC-LALLS.
^{b} Determined by GPC. ^{c} Determined by ^{1}H NMR.

calculated from the following equation [2] since the α-CH₃ signal due to *mr* triad is obscured by overlap of the peaks.

$$[\alpha\text{-CH}_3(mr)]=2\{1.5[\text{CH}_2(r)]\text{-}[\alpha\text{-CH}_3(rr)]\} \qquad [2]$$

The signal of α-CH₃ protons in *mm* triad could not be observed probably due to the peak overlapping and thus the intensity was estimated by the equation [3].

$$[\alpha\text{-CH}_3(mm)]=1.5[\text{OCH}_2]\text{-}\{[\alpha\text{-CH}_3(rr)]+[\alpha\text{-CH}_3(mr)]\} \qquad [3]$$

Thus the main-chain tacticity of the *st*-polymacromonomer-*d* could be determined as shown in Table 2.

The main-chain tacticity of the *it*-polymacromonomer-*d* was determined similarly based on the signal intensities of [α-CH₃(*mm*)] and [CH₂(*m*)] and shown in Table 2.

Figure 4 shows the structural characteristics of the polymacromonomers obtained by the anionic polymerizations under the conditions shown in Table 1. The values of the main-chain tacticity are the values for the polymacromonomers-*d* obtained under the same conditions. The stereoregularity of the PMMA side chains is well controlled through the stereospecific synthesis of PMMA macromonomers *via* stereospecific living polymerizations[13,14]. The main-chain tacticities of the polymacromonomers prepared with DPHLi in THF and *t*-C₄H₉MgBr in toluene are predominantly syndiotactic and isotactic, respectively, though the extent of stereoregularity are slightly lower than those for PMMAs obtained under similar conditions. Thus the knowledge on stereospecificity of polymerization of methacrylate may be applicable to the polymerization of the methacrylate-type macromonomers.

The DP values of polymacromonomers are larger than those expected from feed ratio of the macromonomer to the initiator and the conversion. Moreover, the conversion determined from GPC chromatogram was smaller than that determined by ^{1}H NMR as described in the previous section. These results suggest that a part of propagating species was quenched by impurities. Even with the low initiator efficiency, probably arising from the difficulty of purification of the macromonomer, the anionic synthesis of polymacromonomer provides us a superior way to attain the structural control of comblike polymers such as tacticity, DP, and end group.

3.3. Solution Viscosity of the Stereoregular Comblike Polymer

Solution viscosities of the four types of stereoregular polymacromonomers were measured in chloroform and in acetone using an Ubbelohde type viscometer[14]. These values are summarized in Table 3. Though the different *M*n's of these polymacromonomers do not

(a) *st*-main chain and *st*-side chain [Table 1-No.1]

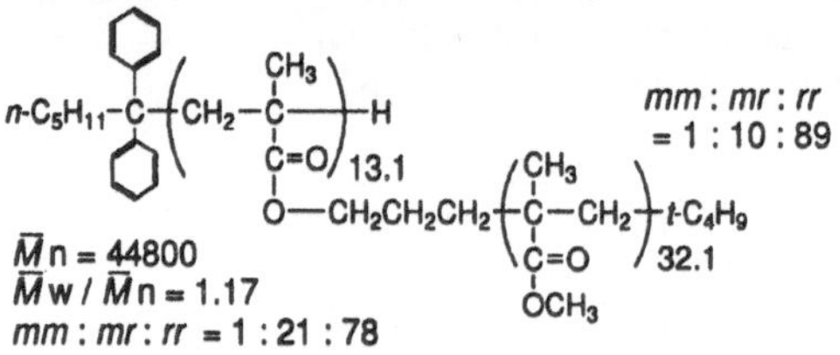

$$mm : mr : rr = 1 : 10 : 89$$

$\overline{M}n = 44800$
$\overline{M}w / \overline{M}n = 1.17$
mm : *mr* : *rr* = 1 : 21 : 78

(b) *st*-main chain and *it*-side chain [Table 1-No.2]

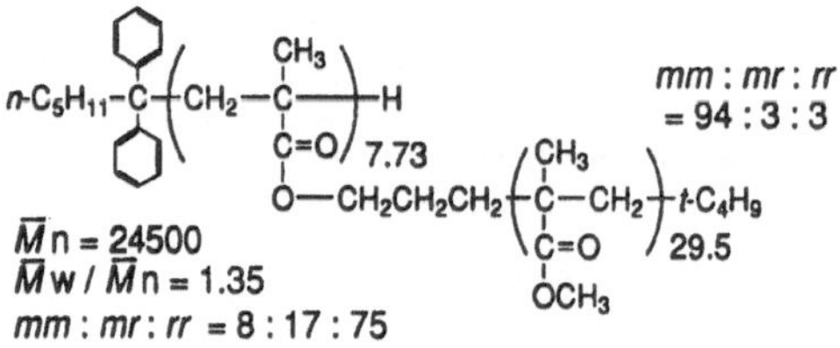

$$mm : mr : rr = 94 : 3 : 3$$

$\overline{M}n = 24500$
$\overline{M}w / \overline{M}n = 1.35$
mm : *mr* : *rr* = 8 : 17 : 75

(c) *it*-main chain and *st*-side chain [Table 1-No.3]

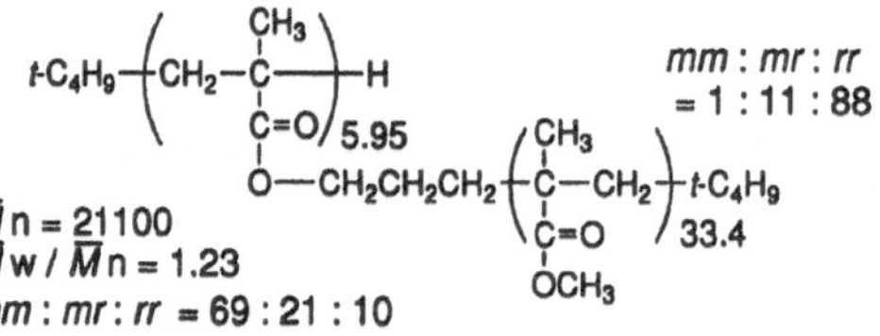

$$mm : mr : rr = 1 : 11 : 88$$

$\overline{M}n = 21100$
$\overline{M}w / \overline{M}n = 1.23$
mm : *mr* : *rr* = 69 : 21 : 10

(d) *it*-main chain and *it*-side chain [Table 1-No.4]

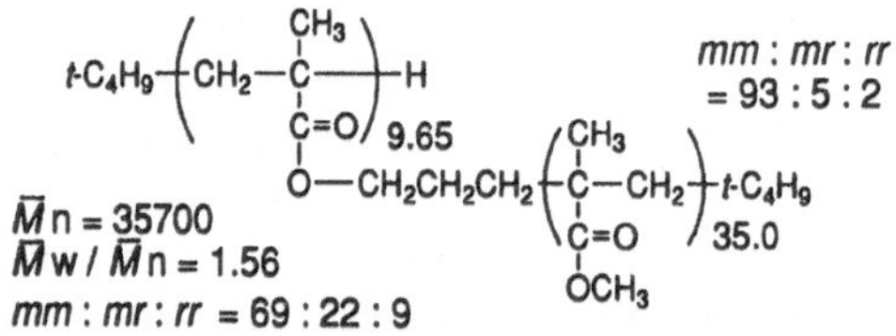

$$mm : mr : rr = 93 : 5 : 2$$

$\overline{M}n = 35700$
$\overline{M}w / \overline{M}n = 1.56$
mm : *mr* : *rr* = 69 : 22 : 9

Figure 4. Structures of polymers of *st*-PMMA macromonomer (a) and *it*-PMMA macromonomer (b) prepared with DPHLi in THF and those of polymers of *st*-PMMA macromonomer (c) and *it*-PMMA macromonomer (d) prepared with *t*-C₄H₉MgBr in toluene.

allow us exact comparison of [η] values, [η]'s of the polymers of *st*-macromonomer are generally smaller than those of the polymers of *it*-macromonomer except for the data of the polymacromonomer with *st*-main chain and *it*-side chains measured in chloroform. These results may show that solution viscosity of the polymacromonomer is mainly affected by side-chain stereoregularity.

To obtain more detailed information which reflect the influence of main-chain stereoregularity, solution viscosity of the polymacromonomer was determined by GPC-dif-

Table 3. Intrinsic viscosity [η] and Huggins' coefficient (k_H) of the stereoregular polymacromonomers measured in chloroform and in acetone at 35°C

Polymer	Main chain	Side chain	Chloroform		Chloroform	
			[η]	K_H	[η]	K_H
Fig.3a	*st-*	*st-*	0.105	1.49	0.063	1.95
Fig.3b	*st-*	*it-*	0.104	1.23	0.071	1.53
Fig.3c	*it-*	*st-*	0.085	1.51	0.060	1.79
Fig.3d	*it-*	*it-*	0.107	0.92	0.079	0.82

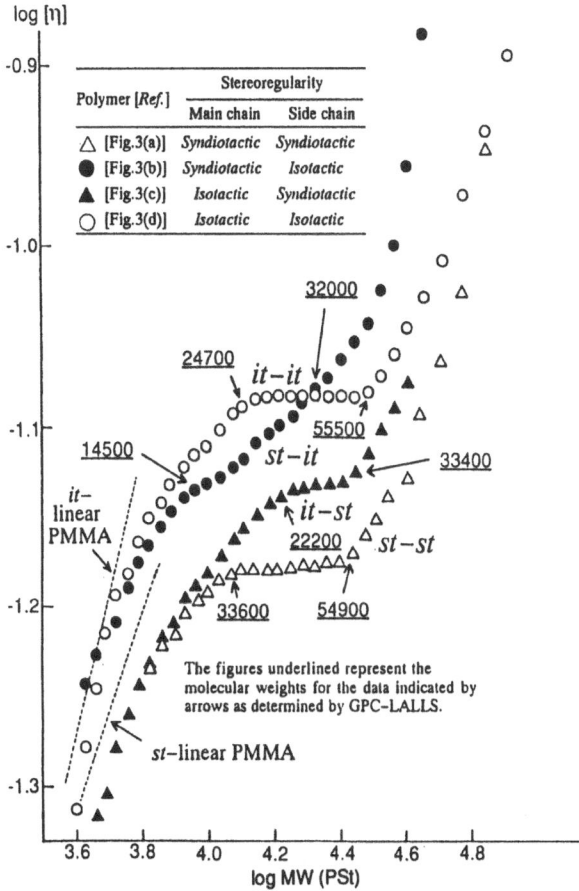

Figure 5. log[η]-logMW relationships of the stereoregular comblike polymers determined by GPC-VIS in THF at 40°C.

ferential viscometer (GPC-VIS) in THF at 40°C[13,14]. The relationship between [η] and molecular weight (MW) for the four types of stereoregular polymacromonomers shown in Figure 5 were obtained.

Though log[η]-logMW plot for linear polymer is usually linear as expected from Mark-Houwink-Sakurada's equation, the plot for these polymacromonomers shows peculiar patterns, that is, deviation from the linearity. At low molecular weight range, log[η] of polymacromonomer increases linearly with logMW similarly to that of linear PMMA. However, the plot deviates from the linear relationship around apparent MW of 10,000, and the slope of the plot decreases in the range of apparent MW range from 10,000 to 30,000, and then the plot goes up again linearly with further increase in MW. These results reflect the change in shapes of the polymacromonomers in solution from linear shape, through star-like shape, to comblike shape with an increase in MW.

The value of [η] is expressed as [η]=φ/ρ where φ and ρ represent shape factor and value of segmental density, respectively[24]. The value of φ depends on the shape of a molecule and increases as the shape of the molecule deviates from a spherical one. In the case of a linear polymer, mean-square radius of gyration, $<S^2>^{1/2}$, increases proportionally to $DP^{1/2}$ and thus the volume of a polymer in solution (hydrodynamic volume) is in proportion to

$DP^{3/2}$. The value of ρ is defined as segment density, expressed in number per unit volume, and proportional to $DP^{-1/2}$ [25]. Therefore, [η] of linear polymer increases with increasing DP as far as the polymer molecules in solution take spherical shapes regardless of DP.

In this low MW range, the polymacromonomers have only two or three side-chain branches (DP=2-3) and the shape of polymacromonomer molecule would be little affected by branching and almost the same as that of the linear polymer.

In the middle MW range, the plots for the polymacromonomer with *st*-main chain and *it*-side chains deviate a little from the linearity and the plots of the other polymacromonomers showed a plateau. In this range, the main-chain skeleton is forced to collapse to a core with the branches radiating outside from the core (star-like). Thus, the addition of one more branch to the star-like polymer does scarcely affect the shape and size of the molecule in solution, while the segmental density (ρ) of the polymacromonomer increases with an increase in MW. As a result, the slope of the plot decreased as expected for linear polymer.

In the higher MW range, a further increase in the number of the PMMA branches along the main chain no longer allows the collapsed core but forces the polymer molecule to expand along the main chain, leading to comblike shape. Since the molecule can not take a spherical form and begins to expand in solution, the value of ϕ increases and the value of ρ may decrease. As a result, the [η] value begins to increase again with an increases in MW. Tsukahara and his coworkers have reported partly similar behavior for the log[η]-logMW relationship for polymers of polystyrene macromonomer [26].

Figure 5 clearly shows that the log[η]-logMW relationship is affected by the stereoregularities of main chain and side chain. It is particularly interesting that polymacromonomers which have the same types of stereoregularity in the main chain and in the side chains (*it-/it-* or *st-/st-*) exhibit the plateau regions more clearly. The plots of these two types of polymacromonomers and that for the polymacromonomer with *it*-main chain and *st*-side chains went up again from the plateau around apparent MW of 30,000 (logMW=4.5). The MW value of the horizontal axis in Figure 5 is based on the calibration with linear polystyrenes and thus reflects hydrodynamic volume. At the points where the plots begin to go up again, the MW values determined by GPC-LALLS, as shown in Figure 5 by figures underlined, are different from each other, though the apparent MW values (the horizontal axis), which should correspond to the volume of the polymacromonomers in solution, are almost the same. These results suggest that the segmental density of the polymacromonomer molecule in THF depends on its tacticity of the main chain and side chain [13,14].

Currently we are studying the behavior of solution viscosity of polymacromonomer in detail by using a polymer of uniform macromonomer. Since polymers of the uniform macromonomer has side chains with the same chain lengths, it is expected that more detailed information on the relationship between [η] and DP of the main chain of polymacromonomer will be obtained.

4. SYNTHESIS OF STEREOREGULAR GRAFT POLYMER [11,15-17]

4.1. Stereospecific Copolymerization of Stereoregular PMMA Macromonomer

Anionic copolymerizations of stereoregular PMMA macromonomers with MMA or ethyl methacrylate (EMA) were carried out under the same conditions as the homopolymerizations. The results are summarized in Table 4 and Table 5.

The conversion of macromonomer determined by ^{1}H NMR analysis of the reaction product indicates that in the copolymerization with MMA by t-C_4H_9MgBr in toluene,

Table 4. Anionic polymerization of stereoregular PMMA
macromonomer (Mac.) and methacrylate monomer (RMA)
with t-C_4H_9MgBr in toluene at -78°C

No.	Mac. (mmol)	Comonomer RMA (mmol)	RMA Mac.	Initiator mmol	Time day	Conversion[a] (%) Mac.	RMA
5	st-[b] (0.13)	MMA (1.40)	10.8	0.060	16	65	100
6	st-[b] (0.13)	EMA (1.32)	10.0	0.058	16	83	98
7	it-[c] (0.15)	MMA (1.50)	9.90	0.083	14	89	100
8	it-[c] (0.12)	EMA (1.14)	9.40	0.063	14	83	100

[a] determined by ^{1}H NMR spectroscopy.
[b] Syndiotactic PMMA macromonomer: $\bar{M}n$= 3400, DP=32.1, mm:mr:rr= 1:10:89,
functionality= 0.89, [Mac]$_0$= 0.033(mol/l).
[c] Isotactic PMMA macromonomer: $\bar{M}n$= 3000, DP=28.1, mm:mr:rr= 94:3:3,
functionality= 0.84, [Mac]$_0$= 0.036(mol/l).

it-PMMA macromonomer has larger reactivity than st-macromonomer as in the case of the homopolymerization. We have reported that it-PMMA macromonomer with styrene-type end group showed higher reactivity than st-one in radical polymerization and copolymerization with styrene[8,10,11], and Ziegler polymerization[11], which was ascribed to the higher segmental mobility of it-PMMA chain. These results mean that it-PMMA macromonomers showed higher reactivity than st-ones in polymerizations with ionic mechanisms as well as radical one.

The copolymer products were fractionated by repeated precipitation in toluene-hexane or chloroform-hexane mixtures to remove the unreacted macromonomer. The graft polymers prepared with t-C_4H_9MgBr in toluene showed bimodal molecular weight distribution. Among them, the graft polymer derived from st-PMMA macromonomer and MMA was further fractionated into higher and lower molecular weight fractions and characterized (*cf.* Table 6, Nos. 5a, 5b).

Figure 6 and Figure 7 show ^{1}H NMR spectra of the graft polymers of st-PMMA macromonomer. The composition of the graft polymer derived from st-PMMA macromonomer and MMA could be determined by comparing peak intensities of OCH_2 protons of the macromonomer units (4.04ppm) and OCH_3 of MMA units (3.61ppm) (Figure 6a and Figure 7a). Though MMA units in the main chain and those in the side chains are indistinguishable by the spectra, the contribution of the side-chain MMA unit could be subtracted

Table 5. Anionic polymerization of stereoregular PMMA
macromonomer (Mac.) and methacrylate monomer (RMA)
with 1,1-diphenylhexyllithium (DPHLi) in THF at -78°C

No.	Mac. (mmol)	Comonomer RMA (mmol)	RMA Mac.	DPHLi mmol	Time day	Conversion[a] (%) Mac.	RMA
9	st-[b] (0.13)	MMA (1.60)	12.3	0.060	16	100	100
10	st-[b] (0.13)	EMA (1.40)	10.8	0.058	16	100	96
11	it-[c] (0.12)	MMA (1.20)	10.0	0.066	14	100	100
12	it-[c] (0.11)	EMA (1.23)	11.2	0.061	14	100	100

[a] determined by ^{1}H NMR spectroscopy.
[b] Syndiotactic PMMA macromonomer: $\bar{M}n$= 3400, DP=32.1, mm:mr:rr= 1:10:89,
functionality= 0.89, [Mac]$_0$= 0.033(mol/l).
[c] Isotactic PMMA macromonomer: $\bar{M}n$= 3140, DP=29.5, mm:mr:rr= 94:3:3,
functionality= 0.89, [Mac]$_0$= 0.035(mol/l).

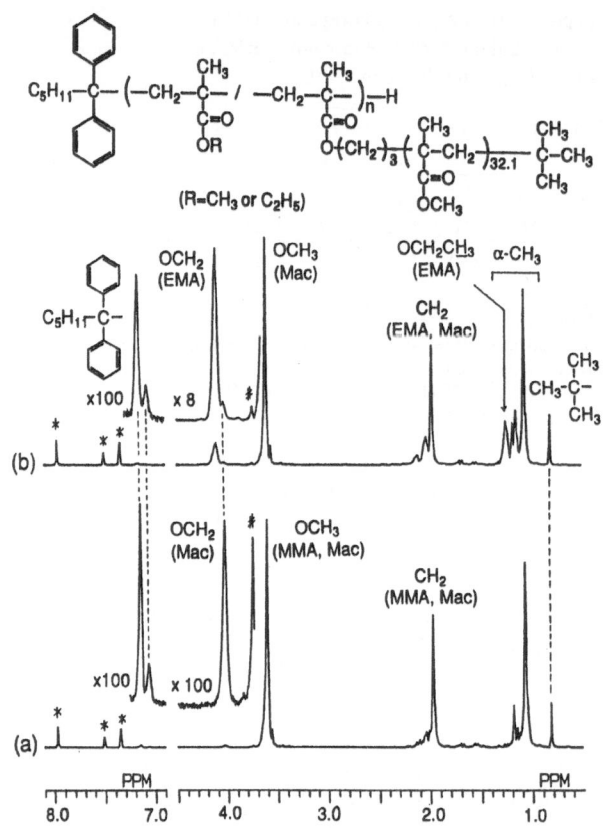

Figure 6. 500MHz ^{1}H NMR spectra measured in nitrobenzene-d_5 at 110°C of the graft polymers of st-PMMA macromonomer with MMA (a) and with EMA (b) prepared with DPHLi in THF * Signals due to the remaining protons in nitrobenzene-d_5. # ^{13}C satellite signal of OCH$_3$ signal.

by assuming the DP of the macromonomer units in the graft polymers to be equal to that of the starting macromonomer. The ratios of MMA unit to the macromonomer unit ([MMA]/[macromonomer]), thus determined from the spectra shown in Figure 6a and Figure 7a, were 14.7 and 23.9, respectively[11,15].

The graft polymer formed with DPHLi has one 1,1-diphenylhexyl group at the initiating end and t-C$_4$H$_9$ groups equivalent to the number of branches (N$_{br}$). Thus N$_{br}$ could be determined as 6.62 directly from the relative signal intensities of the phenyl protons and t-C$_4$H$_9$ protons of the spectrum shown in Figure 6a[11,15].

In the case of the graft polymer prepared with t-C$_4$H$_9$MgBr, however, direct determination of N$_{br}$ from end-group analysis by ^{1}H NMR is difficult, since the t-C$_4$H$_9$ group at the initiating chain end of the graft polymer and those at the terminal of the branches show almost identical chemical shift (Figure 7a). Thus, the peak intensity for the t-C$_4$H$_9$ group at the initiating chain end of the graft polymer was determined by subtracting the contribution from the t-C$_4$H$_9$ groups at the branch ends, which was estimated from the signal intensity of OCH$_2$ protons of the macromonomer unit (N$_{br}$=2.45)[15].

The composition and the N$_{br}$ of the graft polymers prepared with it-PMMA macromonomer and MMA were determined by the same procedures and shown in Table 6[17].

Figure 6b and Figure 7b show the ^{1}H NMR spectra of the graft polymers derived from st-PMMA macromonomer and EMA. The compositions of the graft polymers with EMA main chain were determined from ^{1}H NMR spectra by comparing peak intensities of OCH$_2$ (EMA) at 4.12ppm and OCH$_3$ (macromonomer) at 3.61ppm, based on the assumption that the Mn of the PMMA branches in the graft polymer was same as that of the starting

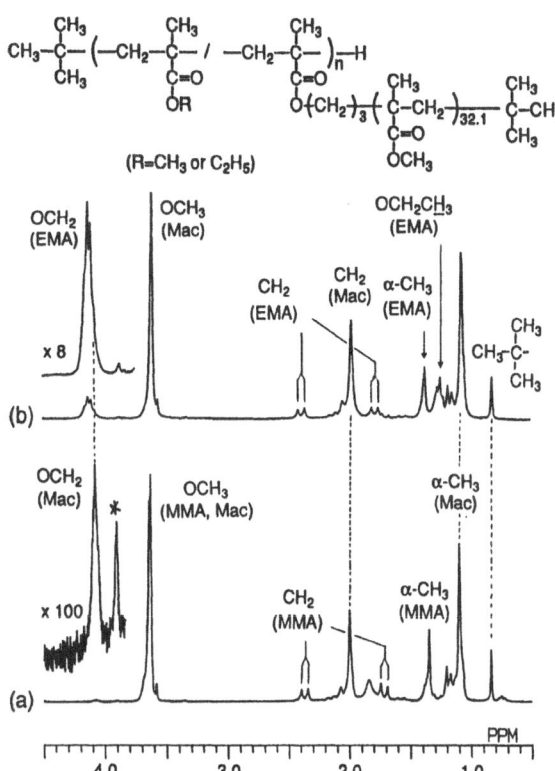

Figure 7. 270MHz ^{1}H NMR spectra measured in nitrobenzene-d_5 at 110°C of the graft polymers of *st*-PMMA macromonomer with MMA (a) (the lower molecular weight fraction: see Table 6. 5a) and with EMA (b) prepared with *t*-C$_4$H$_9$MgBr in toluene. * ^{13}C satellite signal of OCH$_3$ signal.

macromonomer (Table 6). As OCH$_2$ protons of the macromonomer unit resonated very near the OCH$_2$ (EMA) signal and thus the peak intensity could not be determined separately, its contribution was subtracted based on the intensity of OCH$_3$ signal by taking the DP of the macromonomer into account. The ratios of EMA unit to the macromonomer unit ([EMA]/[macromonomer]), thus determined from the spectra shown in Figure 6b and Figure 7b, were 8.9 and 9.10, respectively[11,15].

The N_{br} of the graft polymer with EMA main chain prepared with DPHLi in THF was determined as 7.89 from the spectrum shown in Figure 6b by the same procedure as the graft polymers with MMA main chain prepared with DPHLi in THF[11,15].

In the case of the graft polymer with EMA main chain prepared with *t*-C$_4$H$_9$MgBr in toluene, both of the initiating chain end of the main chain and the terminal of the side chain are *t*-C$_4$H$_9$ group, and thus direct determination of N_{br} by the end-group analysis was not possible. In the case of the graft polymer with MMA main chain prepared with *t*-C$_4$H$_9$MgBr, the OCH$_2$ signals of the macromonomer unit could be used to estimate the contribution of *t*-C$_4$H$_9$ group of the macromonomer units (Figure 7a). However, the corresponding signals of the graft polymer with EMA main chain overlap with those of EMA unit (Figure 7b). Thus the contribution of the branch-end *t*-C$_4$H$_9$ group was estimated from the intensity of OCH$_3$ and the DP of the macromonomer, and was subtracted from the total *t*-C$_4$H$_9$ signal intensity to obtain the signal intensity of the *t*-C$_4$H$_9$ group at the initiating chain end. The N_{br} values determined by this method was found to be affected severely by only a small error in the estimation of the number of branch-end *t*-C$_4$H$_9$ group. Therefore, the N_{br} of the graft polymer was estimated as 9.51 based on the Mn determined by GPC and the composition[15].

The composition and the N_{br} of the graft polymers prepared from *it*-PMMA macromonomer and EMA were determined by the same procedures[17].

From the values of the composition (C) and N_{br} of the graft polymers thus determined, *Mn*'s of the graft polymers could be determined according to equation 4[11,15-17].

$$Mn = N_{br} \times (Mn \text{ of macromonomer}) + C \times N_{br} \times (MW \text{ of comonomer}) \\ + (MW \text{ of initiator fragment}) + 1 \qquad [4]$$

4.2. Stereoregularity of the Main Chain of the Resulting Graft Polymer

From the methylene proton signals in the ^1H NMR spectra shown in Figure 6 and Figure 7, it is evident that main-chain stereoregularity of the graft polymers prepared with DPHLi in THF was predominantly syndiotactic and that with t-C_4H_9MgBr in toluene was predominantly isotactic. For the quantitative determination of the tacticity, however, ^{13}C NMR analyses of the graft polymers were necessary[15-17].

Figure 8 illustrates the carbonyl carbon spectra of the graft polymers prepared from *st*-PMMA macromonomer and MMA with DPHLi in THF and with t-C_4H_9MgBr in toluene, along with that of the macromonomer. Commonly observed signals in these spectra are mostly due to MMA units in the macromonomer chains. While the spectrum of the graft polymer prepared with DPHLi in THF (Figure 8b) is almost the same as that of the *st*-macromonomer, the spectrum of the graft polymer formed with t-C_4H_9MgBr in toluene (Figure 8c) is clearly different from others, particularly in the range of around 177.5ppm. This signal is ascribable to *mm* triad of MMA sequence in the main chain of the graft polymer. An average tacticity of the main chain was determined by subtraction of the signals due to MMA units in the macromonomer chains as follows[15]. The sharp resonance at 178.2 ppm observed in the spectrum of the macromonomer (Figure 8a) can be assigned to the carbonyl carbons in the first and second MMA units from the t-C_4H_9 chain end (α_1 and α_2 units)[27]. The same signal could be observed in the spectra of the graft polymers and thus used as an intensity standard for the estimation of the contributions of the macromonomer unit signals. The quantitativeness of the peak intensity should be fulfilled, since the DP (32.2) of the

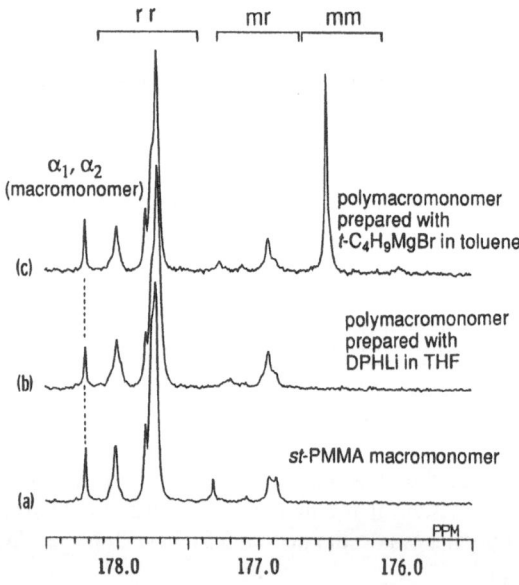

Figure 8. 67.8MHz ^{13}C NMR spectra measured in chloroform-*d* at 55°C of carbonyl carbons of *st*-PMMA macromonomer (a), and the graft polymers of *st*-PMMA macromonomer and MMA prepared with DPHLi in THF (b) and with t-C_4H_9MgBr in toluene (c).

macromonomer estimated from signal intensities of the α_1 and α_2 carbonyl carbons relative to those of all the carbonyl carbons agreed well with that determined from the ^{1}H NMR spectrum (32.1).

The tacticities of the graft polymer derived from st-PMMA macromonomer and EMA were determined by the same procedure, and are shown in Table 6.

In the case of the graft polymers prepared from it-PMMA macromonomer, the carbonyl carbon signals due to α_1 or α_2 unit were observed in mm and mr triad regions with a complicated splitting owing to lower stereoregularity at the initiating end, and thus not useful as an intensity standard[27]. Thus, the tacticities of the main chain were determined from the quaternary carbon signals by the following procedure[17]. The triad tacticities determined from the quaternary carbon signals should be expressed as a sum of those of the main chain and side chains.

$$mm = x_{\mathrm{m}}mm_{\mathrm{m}} + x_{\mathrm{s}}mm_{\mathrm{s}} \ [5]$$

$$mr = x_{\mathrm{m}}mr_{\mathrm{m}} + x_{\mathrm{s}}mr_{\mathrm{s}} \ [6]$$

$$rr = x_{\mathrm{m}}rr_{\mathrm{m}} + x_{\mathrm{s}}rr_{\mathrm{s}} \ [7]$$

mm, mr, rr: fractions of triad tacticity obtained from the quaternary carbon.

mm_{m}, mr_{m}, rr_{m}: triad tacticities in the main chain.

mm_{s}, mr_{s}, rr_{s}: triad tacticities in the side chains.

x_{m}, x_{s}: mole fractions of methacrylate units in the main chain and in the side chains. $x_{\mathrm{m}} + x_{\mathrm{s}} = 1$

The values of mm_{s}, mr_{s} and rr_{s} should be those of the starting macromonomer, and x_{m} and x_{s} can be estimated from the composition of graft polymer and DP of the macromonomer. Then the tacticity of main chain can be calculated from the equations [5], [6] and [7].

Control of stereoregularity in the side chain of the graft polymer could be achieved by using the highly stereoregular macromonomers. Copolymerizations of the macromonomers with t-C$_4$H$_9$MgBr in toluene (Nos. 5-8) and with DPHLi in THF (Nos. 9-12) gave the graft polymers with it- and st-main chains, respectively. Stereoregularities in the MMA and EMA sequences in main chain were close to those for PMMA and poly(EMA), respectively, formed under the same conditions[15,20,28]. Thus, the tacticity in the main chain of the graft polymer could be controlled by the stereospecific anionic copolymerization of the stereoregular macromonomer.

The graft polymers prepared with t-C$_4$H$_9$MgBr in toluene (Nos.5-8) have broad and bimodal molecular weight distribution. The results suggest that there exist at least two types of active species which are different in their reactivities. The higher and lower molecular weight fractions of the graft polymer derived from st-PMMA macromonomer and MMA (No.5a and 5b) had almost the same composition and main-chain tacticity, indicating that the two active species have the same stereoselectivity and monomer selectivity. A similar phenomenon has been observed for the copolymerization of MMA and EMA with t-C$_4$H$_9$MgBr in toluene at low temperature[28].

The results of the structural analyses of the graft polymer described in this chapter are summarized in Table 6.

The values of Mn and N_{br} of the graft polymers were much larger than those expected from the feed ratios of the total of the macromonomer and comonomer to the initiator. These results mean that the initiator efficiencies are low as in the case of homopolymerization, and thus further improvement of purification of macromonomer is required.

Since the graft polymers were prepared by the polymerization systems which are effective for living polymerization of methacrylates, the chain-end structures of the graft

Table 6. Characterization of the stereoregular graft polymers obtained by the copolymerization of stereoregular PMMA macromonomer (Mac.) and methacrylate monomer (RMA)

No[a] RMA	$\frac{RMA^b}{Mac.}$	$\overline{M}n$	$\frac{\overline{M}w^c}{\overline{M}n}$	N_{br}	Main chain				Side chain			
					DP	mm	mr	rr	DP	mm	mr	rr
9 MMA	14.7	32500	1.35	6.62	103.9	0	22	78	32.1	1	10	89
11	9.41	12000	1.30	2.89	30.1	3	19	78	29.5	94	3	3
5a	23.9	14200[d]	1.27	2.45	61.0	96	4	0	32.1	1	10	89
5b	22.2	101800[e]	2.63	18.1	419.9	99	1	0	32.1	1	10	89
7	13.1	23600	1.69	5.47	77.1	100	0	0	28.1	94	3	3
10 EMA	8.90	35100	1.39	7.89	78.1	0	12	88	32.1	1	10	89
12	7.70	14600	1.69	3.57	31.1	6	23	71	29.5	94	3	3
6	9.10	41400[c]	3.38	9.51[f]	96.1	85	7	8	32.1	1	10	89
8	8.90	32700[c]	4.37	8.13[f]	80.6	92	8	0	28.1	94	3	3

[a] Polymerization conditions are summarized in Table 4 or Table 5.
[b] Composition.of graft polymer.   [c] Determined by GPC.
[d] The lower molecular weight fraction of the graft polymer.
[e] The higher molecular weight fraction of the graft polymer.
[f] Estimated from the $\overline{M}n$ determined by GPC and the composition.

polymers are well predictable. The end-group analysis by ^{1}H NMR spectroscopy provides the detailed information on the structure of the graft polymers. For example, N_{br} value of the graft polymer prepared with DPHLi was determined directly by ^{1}H NMR, which is more reliable than N_{br} value obtained by viscometry. Characteristics of graft polymer in solution will be investigated systematically by using graft polymers with well defined structure including stereoregularity.

4.3. Solution Viscosity of the Stereoregular Graft Polymers

Table 7 summarizes the [η] and k_H values of the stereoregular graft polymers of MMA measured in chloroform and in acetone at 35°C, together with those of linear st-PMMA and a mixture of it- and st-PMMAs as references. The graft polymer with st-main chain and st-side chains showed lower viscosities than linear st-PMMA of a similar Mn as commonly observed for branched polymers[29]. The k_H values of the st-graft polymer and the st-PMMA did not differ from each other significantly, though homopolymers of it- and st-PMMA macromonomers exhibited larger k_H values than the corresponding linear PMMAs[8]. This may be due to the smaller number of branches of the graft copolymer than those of the polymacromonomers.

The graft polymer with it-main chain and st-side chains (it/st-graft polymer) shows large k_H value (4.99) in acetone as compared with other graft polymers (Table 7). The k_H value for this graft polymer in chloroform is not so large and similar to those of other graft polymers (Table 7). The large k_H value in acetone, or strong concentration dependence of reduced viscosity, suggests the possibility of association in acetone through stereocomplex formation between it- and st-PMMA chain segments[30] in the graft polymer. ^{1}H NMR spectrum of this graft polymer in acetone-d_6 (1 w/v%) at 35°C showed the peaks with intensities of 61% of the expected value, indicating that 39% of the polymer did not show the signals due to association through stereocomplex formation[15]. The DSC curve of the graft polymer in the solid state showed an endothermic peak due to the melting of the stereocomplex, when it was recovered from acetone solution by evaporating the solvent[15].

Table 7. Intrinsic viscosity [η] and Huggins' coefficient (k_H) of the stereoregular graft polymers of MMA and PMMA measured in chloroform and in acetone at 35°C

No.[a]	Main chain (DP)	Side chain (DP)	Mn	$\dfrac{Mw}{Mn}$	N_{br}	Chloroform		Acetone	
						[η]	k_H	[η]	k_H
9	st- (103.9)	st- (32.1)	32500	1.35	6.62	0.139	0.86	0.086	1.57
5a	it- (61.0)	st- (32.1)	14200	1.27	2.45	0.111	1.23	0.072	4.99
11	st- (30.1)	it- (29.5)	12000	1.30	2.89	0.103	1.04	0.075	1.36
13[b]	st- (198.6)	it- (30.4)	32700	1.45	4.02	0.256	0.31	0.172	2.15
st-PMMA[c]			22700	1.26		0.220	0.88	0.118	1.35
it-PMMA[d] + st-PMMA macromonomer[e] (42:58)						0.070	1.23	0.057	1.82

[a] See Table 6.   [b] [MMA]/[Mac.] in feed was 30.   [c] $mm:mr:rr$= 0:11:89.
[d] $\overline{M}n$= 5370, $\overline{M}w/\overline{M}n$= 1.10, $mm:mr:rr$= 97:2:1.
[e] $\overline{M}n$= 3400, $\overline{M}w/\overline{M}n$= 1.07, $mm:mr:rr$= 1:10:89.

it-PMMA of DP=53.1 and the st-PMMA macromonomer (Table 7) mixed in acetone-d_6 were found by ^{1}H NMR to associate to a similar extent to that for the it/st-graft polymer[15]. The k_H value for the graft polymer (4.99) was almost three times as large as that for the mixture (1.82), although the k_H value of 1.82 is still larger than that for linear polymer (Table 7). The it/st-graft polymer may undergo, at least in part, intramolecular association between it-main chain and st-PMMA side chain, and thus exist as shrinked particles. This might be one of the reasons for the large k_H value, since spherical particles in solution usually exhibit large k_H values[29]. The occurrence of the intramolecular association may weaken the intermolecular association of the it/st-graft polymer, which is inherently sensitive to concentration. Such a weakened intermolecular association may cause enhancement of concentration dependence of reduced viscosity, i.e., larger k_H value.

The graft polymer with st-main chain and it-side chains, whose DP along the main chain was 30.1, showed a smaller k_H value (1.36) in acetone (Table 7) and no endothermic peak in DSC measurement[17]. The results indicate that this graft polymer could not form a stereocomplex in the solid state as well as in solution. On the other hand, the same type of the graft polymer with a larger DP (198.6) of main chain formed a stereocomplex in the solid state as confirmed by DSC measurement and showed a large k_H value (2.15) in acetone (Table 7), although it was smaller than the k_H value for the it/st-graft polymer. Challa and his coworkers proposed a double-stranded helix structure for PMMA stereocomplex, in which it-PMMA helical chain is surrounded by st-PMMA helix with a larger radius with the ester functions pointing outward of the double helix[31]. The structure model is consistent with our findings that a large variety of st-polymethacrylates with various ester groups form stereocomplexes with it-PMMA and that no it-polymethacrylates other than it-PMMA form stereocomplex with st-PMMA[32-34]. If the double-stranded helix model is adopted for the intramolecular complex formation in st/it-graft PMMA, main-chain st-PMMA segment should surround it-PMMA side chains. The process should be much more difficult than that in it/st-graft PMMA. In the case that the st-main chain segment is short, particularly, the complex formation becomes impossible as described above. Therefore, geometrical arrangement of it- and st-PMMA segments in the graft polymers is an important factor for intramolecular stereocomplex formation in stereoregular graft PMMAs.

REFERENCES

1. R. Milkovich and M. T. Chiang, U. S. Patent, 3, 786, 116 (1974); R. Milkovich, *Polym. Prepr., Polym. Chem. Div., Am. Chem. Soc.,* **21**, 40 (1980).
2. P. F. Rempp and E. Franta, *Adv. Polym. Sci.,* **58**, 1 (1984).
3. Y. Yamashita, *J. Appl. Polym. Sci., Appl. Polym. Symp.,* **36**, 193 (1981).
4. R. Asami and M. Takaki, *Makromol. Chem., Rapid Commun.,* **12**, 163 (1985).
5. K. Ito, *Kobunshi Kako (Polymer Applications),* **35**, 262 (1986).
6. K. Hatada, H. Nakanishi, K. Ute and T. Kitayama, *Polym. J.,* **18**, 581 (1986).
7. K. Hatada, T. Shinozaki, K. Ute and T. Kitayama, *Polym. Bull.,* **19**, 231 (1988).
8. K. Hatada, T. Kitayama, K. Ute, E. Masuda, T. Shinozaki and M. Yamamoto, *Polym. Bull.,* **21**, 165 (1989).
9. K. Hatada, T. Kitayama, E. Masuda and M. Kamachi, *Makromol. Chem., Rapid Commun.,* **11**, 101 (1990).
10. E. Masuda, S. Kishiro, T. Kitayama and K. Hatada, *Polym. J.,* **23**, 847 (1991).
11. K. Hatada and T. Kitayama, *Macromolecular Design: Concept and Practice* (Macromonomers, Macroinitiators, Macroiniferters, Macroinimers, Macroiniters, and Macroinifers) Pages 85-127, M.K. Mishra, ed., Polymer Frontiers International, Inc., (1994).
12. T. Kitayama, S. Kishiro and K. Hatada, *Polym. Bull.,* **25**, 161 (1991).
13. T. Kitayama, O. Nakagawa, S. Hirotani, T. Nishiura and K. Hatada, *Polym. Prepr. Jpn,* **43**, 1842 (in Japanese), E1126 (in English) (1994).
14. T. Kitayama, O. Nakagawa, S. Hirotani, T. Nishiura and K. Hatada, submitted to *Polym. J..*
15. T. Kitayama, O. Nakagawa, S. Kishiro, T. Nishiura and K. Hatada, *Polym. J.,* **25**, 707 (1993).
16. T. Kitayama, O. Nakagawa and K. Hatada, *Polym. Prepr. Jpn,* **42**, 2241 (in Japanese), E886 (in English) (1993).
17 T. Kitayama, O. Nakagawa and K. Hatada, submitted to *Polym. J..*
18. H. Yuki and K. Hatada, *Adv. Polym. Sci.,* **31**, 1 (1979).
19. K. Hatada, T. Kitayama and K. Ute, *Prog. Polym. Sci.,* **13**, 189 (1988).
20. K. Hatada, K. Ute, K. Tanaka, Y. Okamoto and T. Kitayama, *Polym. J.,* **18**, 1037 (1986).
21. T. Kitayama, T. Shinozaki, T. Sakamoto, M. Yamamoto and K. Hatada, *Makromol. Chem. Suppl.,* **15**, 167 (1989).
22. R. De Vos and E. J. Goethals, *Polym. Bull.,* **15**, 547 (1986).
23. K. Miura, T. Kitayama, K. Hatada and T. Nakata, *Polym. J.,* **22**, 671 (1990).
24. P. J. Flory, *"Principles of Polymer Chemistry",* Chapter XIV, Cornell University Press, (1953).
25. P. J. Flory, *"Principles of Polymer Chemistry",* Chapter X, Cornell University Press, (1953).
26. Y. Tsukahara, K. Tsutsumi, Y. Okamoto and S. Kohjiya, *Polym. Prepr. Jpn,* **42**, 4537 (in Japanese), E1998 (in English) (1993).
27. K. Ute, T. Nishimura and K. Hatada, *Polym. J.,* **21**, 1027 (1989).
28. T. Kitayama, K. Ute, M. Yamamoto, N. Fujimoto and K. Hatada, *Polym. J.,* **22**, 386 (1990).
29. M. Stickler and N. Sutterlin, *"Polymer Handbook",* 3rd Ed., ed. by J. Brandrup and E. H. Immergut, John Wiley & Sons, Inc., VII/183 (1989).
30. A. M. Liquori, G. Anzuino, V. M. Corio, M. D'Alagni, P. de Santis and M. Savino, *Nature* (London), **206**, 358 (1965).
31. E. Schomaker and G. Challa, *Macromolecules,* **22**, 3337 (1989).
32. T. Kitayama, N. Fujimoto, Y. Terawaki and K. Hatada, *Polym. Bull.,* **23**, 249 (1990).
33. T. Kitayama, N. Fujimoto and K. Hatada, *Polym. Bull.,* **26**, 629 (1991).
34. K. Hatada, T. Kitayama, K. Ute, N. Fujimoto and N. Miyatake, *Makromol . Chem., Macromol. Symp.,* in press.

14

ACRYLIC GRAFT COPOLYMERS VIA MACROMONOMERS

Synthesis and Characterisation

Wolfgang Radke, Sebastian Roos, Helga M. Stein, and Axel H. E. Müller[*]

Institut für Physikalische Chemie
Universität Mainz
Welderweg 15
D-55099 Mainz
Germany

ABSTRACT

Comb-shaped poly(methyl methacrylate) (PMMA) and PMMA grafted with poly(n-butyl acrylate) (PnBuA) were prepared by radical copolymerisation of ω-methacryloyl-PMMA with MMA and nBuA, respectively. The comb-shaped PMMA is characterised with respect to radius of gyration by using GPC equipped with a multi-angle laser light scattering detector. The radical copolymerisation of the macromonomer with nBuA in toluene follows complex kinetics. The dependence of the relative reactivity of the macromonomer on absolute concentration and on the ratio of comonomers may be explained by preferential solvation of comonomers by segments of their own kind ("bootstrap effect") or even micelle formation. However, there is no clear evidence for the formation of micelles in toluene. In contrast, NMR studies show micelle formation in the preferential solvent DMSO. The graft copolymers are transparent thermoplastic elastomers. Phase separation is demonstrated by DSC and morphological studies.

INTRODUCTION

Graft copolymers are good substitutes for block copolymers in many applications, e.g. thermoplastic elastomers (TPE's) or compatibilisers for polymer blends. Generally, they are much easier to synthesize than block copolymers. In addition, the branched structure leads to decreases melt viscosities which is important for processing.

Typically, TPE's are tri- or multiblock copolymers having segments of high and low T_g. Thus, Kratons made by Shell are ABA block copolymers with polystyrene A and

---

[*] Address correspondence to Dr. Müller.

Macromolecular Engineering, Edited by M.K. Mishra et al.
Plenum Press, New York, 1995

(hydrogenated) diene B blocks. It is of some interest to replace the hydrocarbon by acrylic monomers due to advantageous properties of the polymers formed, e.g. improved weather-ability or optical properties. Many attempts have been made to prepare triblock polymers with PMMA (A blocks, T_g 110 °C) and poly(n-butyl acrylate) (PnBuA, B blocks, T_g -45 °C). However, the polymerisation of primary acrylates is difficult to control and polymeri-sation systems leading to a controlled polymerisation of acrylates are often less suited for the polymerisation of methacrylates. Replacing block copolymers by graft copolymers offers synthetic advantages.

Graft copolymers have been prepared mainly by "grafting onto", "grafting from", and the macromonomer technique. The latter technique allows the control of both the chain length and the spacing of the side-chains. The former is given by the experimental conditions of the synthesis of the macromonomer (mainly by living polymerisation using functionalised initiator or terminator) and the latter is determined by the molar ratio of the comonomers and the reactivity ratio of the low molecular-weight monomer, r_1. In order to prepare graft copolymers of well-defined structure it is thus essential to know the relative reactivity of the macromonomer, as measured by $1/r_1 = k_{12}/k_{11}$.

The reactivity of macromonomers may be influenced by a multitude of parameters in a complex way. The macromonomer chain length[1-7] and tacticity,[8] the molar ratio[3] and total concentration,[9] of the comonomers, as well as the nature of the solvent[10] have been shown to affect the copolymerisation parameters. The reactivity ratios for a multitude of copolymerisation systems were reviewed recently.[11] It is often very difficult to separate and explain the different effects. A very important complication is the incompatibility of the macromonomer and the backbone of the graft copolymer formed which may lead to preferential solvation or even to microphase separation.[12,13]

In order to study other basic effects on the reactivity of macromonomers we earlier examined a system where incompatibility is excluded, i.e. the copolymerisation of ω-methacryloyl-PMMA and methyl methacrylate (MMA)[14,15]. Here, macromonomer and back-bone have the identical chemical nature resulting in a well-defined comb-shaped PMMA homopolymer. From our earlier studies three conclusions could be drawn:

- the MM reactivity strongly decreases with increasing MM weight concentration. This was attributed to the increasing viscosity which limits the mobility of the macromonomer;
- the MM reactivity decreases with decreasing molar ratio $[MMA]_0/[MM]_0$, i.e. with decreasing spacing between the side-chains. This was explained by the increasing segment density in the vicinity of the propagating centre which hinders the approach of the MM and confirmed by model calculations;
- Under undisturbed conditions, i.e. at low polymer concentrations and high molar ratios ($[MMA]_0/[MM]_0 \wedge 100$) the MM reactivity does *not* depend on the MM chain length.

The dependence of the MM reactivity on the distance between the grafts has an important implication: it means that r_1 depends on the distance of the graft which was last incorporated from the propagating chain end. This indicates that the copolymerisation of macromonomers is a Markov process of high order instead of a Bernoullian process (i.e. terminal model).

In this paper we first present data characterisation by GPC/multi-angle laser light scattering (MALLS) combination of comb-shaped PMMA formed. Then we report kinetic studies on the radical copolymerisation of PMMA macromonomers with n-butyl acrylate and the characterisation of the thermoplastic elastomer formed.

EXPERIMENTAL PART

Methacryloyl-terminated PMMA macromonomers of low polydispersity were pre-pared by group transfer polymerisation using a trimethylsiloxy-functionalised initiator. After polymerisation the trimethylsiloxy protecting group was removed and the resulting OH-ter-minated PMMA was reacted with methacryloyl fluoride or methacryloyl chloride.

Comb-shaped PMMA was prepared by copolymerisation of ω-methacryloyl PMMA with MMA in toluene at 60 °C using AIBN as an initiator. Light scattering experiments were performed in THF by coupling a GPC apparatus to a Wyatt DAWN-F multi-angle laser light scattering (MALLS) detector.

The copolymerisation of methacryloyl-terminated PMMA with n-butyl acrylate (nBuA) was investigated in toluene at 60 °C using tert-butylperoxyneodecanoate as an initiator. Ca. every sixth macromonomer ($DP_n = 68$; $M_w/M_n = 1.05$) was labelled with naphthylcarboxyethyl methacrylate in order to facilitate the determination of the MM conversion by GPC analysis of the copolymer. The conversion of nBuA was determined by GC. The reactivity parameter r_1 was determined by using Jaacks's method.[16]

Glass transition temperatures were determined using a Perkin-Elmer DSC 7 appara-tus at a heating rate of 10 K/min.

RESULTS AND DISCUSSION

Characterisation of Comb-Shaped Polymer Formed by Copolymerisation of PMMA Macromonomers with MMA

The selective detection of side-chains makes the polymers formed good model polymers for the investigation of properties of comb-shaped polymers. By using a multi-an-gle laser light-scattering (MALLS) detector for GPC, the molecular weight and radius of gyration can be determined simultaneously as a function of elution volume and the branching ratio $g = \langle R_g^2 \rangle_{comb} / \langle R_g^2 \rangle_{linear}$ can be compared with theoretical predictions.

First results (Figs. 1 and 2) show that the g-value is smaller and depends stronger on M than expected from the theory of Cassasa and Berry[17] which may be due to additional long-chain branching from transfer reactions. Further experiments are presently being conducted by living copolymerisation.

Copolymerisation of PMMA Macromonomers with n-Butyl Acrylate

The copolymerisation of methacryloyl-terminated PMMA-MM (index 2) with n-bu-tyl acrylate (nBuA; index 1) was investigated in toluene at 60 °C using tert-butylper-oxyneodecanoate as an initiator. The dependence of the relative reactivity of the PMMA-MM on the molar ratio of monomers, $[MM]_0/[nBuA]_0$, and on the total monomer concentration, $[nBuA]_0+[MM]_0$, was studied. In all these experiments the value of r_1 is higher than that for the copolymerisation with MMA ($r_1 = 0.37\pm0.1$),[18] i.e. in the reactivity of the macro-monomer, $1/r_1$, is always lower than that of the low molecular weight analogue.

Fig. 3 shows that the MM reactivity initially increases with increasing total concen-tration of monomers (and thus total concentration of polymer) and then reaches a constant level. This is in strong contrast to the copolymerisation with MMA, where a decrease of MM reactivity was observed due to the viscosity increase. The initial increase may be understood in a classical way by the kinetic excluded volume effect. The constant reactivity at higher concentrations, however, may be due to preferential solvation ("bootstrap effect") as pro-

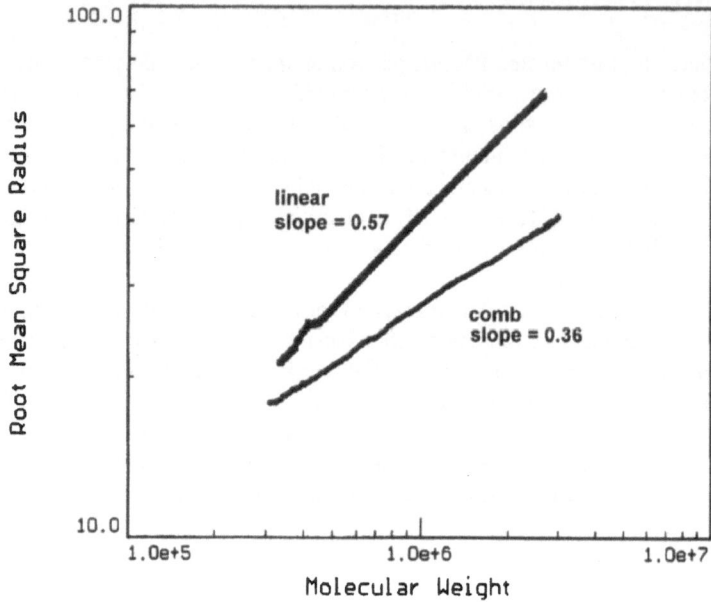

Figure 1. Dependence of mean root square radius of gyration on molecular weight for linear and comb-shaped PMMA.

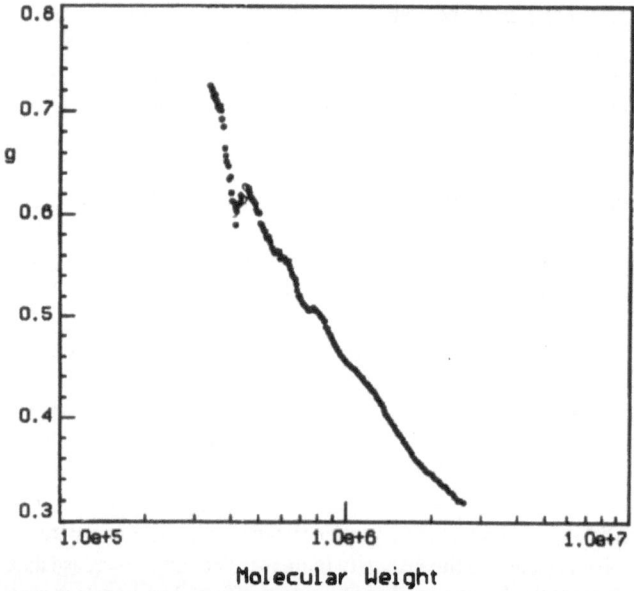

Figure 2. Dependence of branching ratio, *g*, on molecular weight for comb-shaped PMMA.

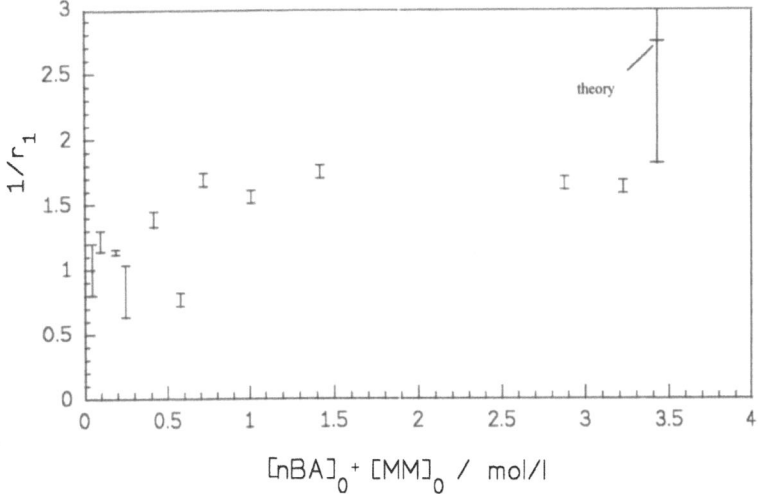

Figure 3. Dependence of PMMA macromonomer reactivity on the total concentration of monomers in the radical copolymerisation with n-butyl acrylate in toluene ($[MM]_0/[nBuA]_0 = 8 \cdot 10^{-3}$).

posed by Harwood.[12] In Harwood's "bootstrap" model it is assumed that the chain end is preferentially surrounded by monomers of its own kind, leading to a increased concentration as compared to stoichiometry (see Fig. 4).

In the extreme case this might even lead to microphase separation, i.e. micelle formation (Fig. 4). A similar effect was found by Wang and Percec for the copolymerisation of styryl-terminated poly(phenylene oxide) macromonomers with MMA.[9,19,13]

Fig 5 shows that the MM reactivity first increases with the increasing molar ratio $[MM]_0/[nBuA]_0$ up to a maximum and then slowly decreases. The initial increase may be explained again by preferential solvation ("bootstrap effect").

The slow decrease of reactivity after the maximum can have many reasons:

- the decreasing spacing leading to a higher segment density around the chain end,

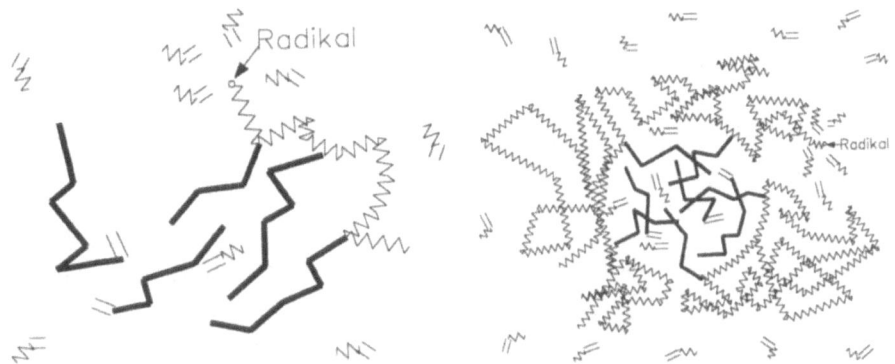

Figure 4. Preferential solvation ("bootstrap effect", left) and micelle formation (right) in the copolymerisation PMMA macromonomers (thick lines) with nBuA (PnBuA backbone = zigzag lines).

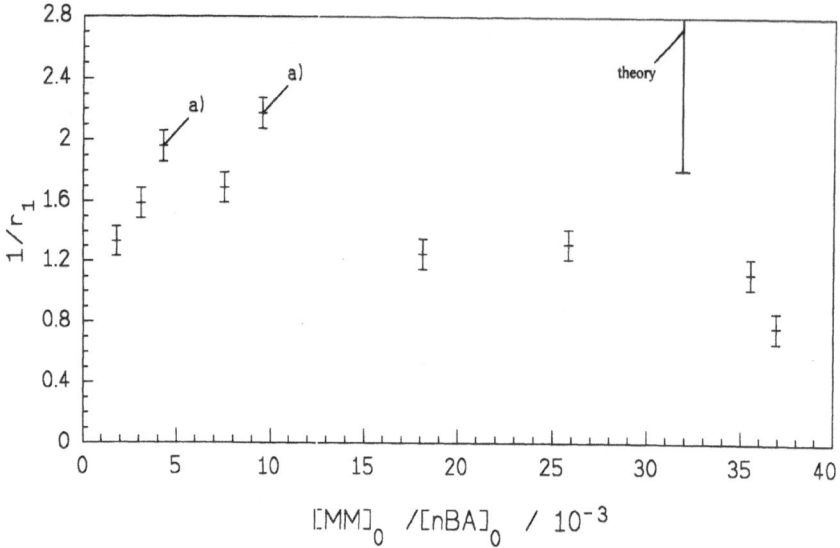

Figure 5. Dependence of PMMA macromonomer reactivity on the ratio of molar concentrations of monomers ($[MM]_0+[nBuA]_0 = 0.6$ mol/l; [a)] $[MM]_0+[nBuA]_0 = 1.1$ mol/l).

- the increase of viscosity, since the increase of the molar ratio is accompanied by an increase in total MM concentration, and by the onset of microphase separation at a certain critical concentration.

Microphase Separation in Solution

The kinetic data in the copolymerisation indicate incompatibility of the graft copolymer backbone with the side chains and the incoming MM molecules. This may even lead to micelle formation. However, this assumption is rather unexpected since both homopolymers have good solubility in toluene. Thus, NMR spectroscopy was used in order to test micelle formation. In case of micelles, the polymer segments in the interior have a decreased mobility. In the limiting case it will approach that of the bulk polymer, making it virtually "invisible" in the NMR experiment. Comparison of the NMR signal intensities, I_{exp}, of the two comonomers with the intensities expected from the known composition of the copolymer, I_{th}, will result a deviation if micelle formation occurs. ^{1}H NMR spectra of various copolymers in toluene-d_8 show ratios of the measured and expected the signals $A = I_{exp}/I_{th}$ between $0.85 < A < 1.2$, indicating no significant effect. Similar results are found in chloroform-d_1 and THF-d_{10}. However, in DMSO-d_6 which is a good solvent for PMMA and a bad solvent for PnBuA, a very drastic effect of 0.1 $A < 0.5$ occurs. The value of A is time and temperature dependent. Obviously, the micelles are not in equilibrium after dissolution in this solvent and it takes long times for equilibration.

Light scattering experiments in toluene sometimes indicate a strong angular dependence of the scattered intensity and a strong concentration dependence of the apparent molecular weights. However, this does not occur with all graft copolymers and may be due to the very low refractive index increments in this solvent which can lead to strong effects of dust particles. Similar, dynamic light scattering experiments sometimes give evidence for a small number of particles with very large diameter, sometimes they do not show large particles at all.

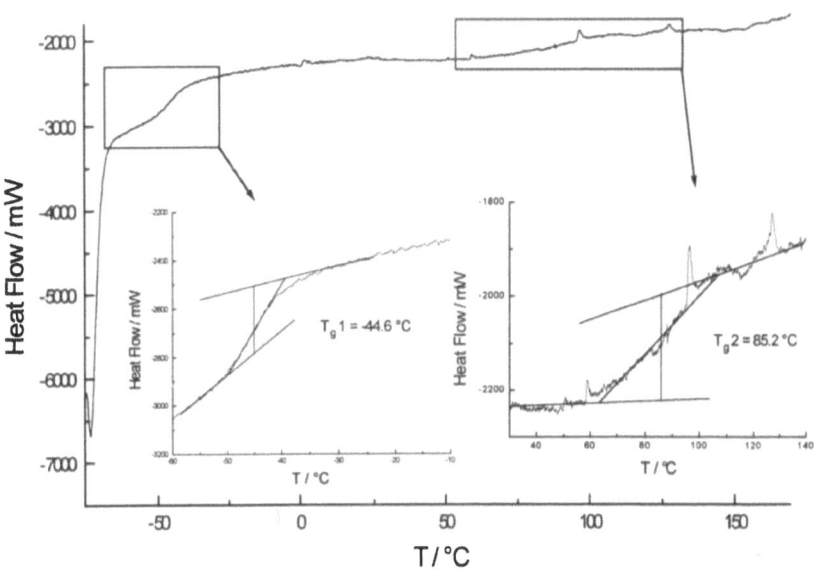

Figure 6. DSC diagram of PnBuA-g-PMMA.

Thus, at present we must conclude that the formation of micelles in toluene is rather improbable.

Characterisation of Graft Copolymers Formed by Copolymerisation of PMMA Macromonomers with nBuA

The graft copolymers formed are transparent and show typical behaviour of thermoplastic elastomers. Typical elongations are 300-600% at 6-10 MPa.

Morphological studies indicate microphase separation with very small domain sizes (radius of spherical PMMA domains < 3 nm).

Microphase separation is also shown by the DSC diagrams (Fig. 6) which exhibit two glass transition temperatures, one at -45 °C for the PnBuA main chains and one at 95 °C for the PMMA side chains. The value for PMMA is lower than that of the homopolymers (or macromonomers) which is 105 °C. This may indicate that the PnBuA segments penetrate into the PMMA microphase acting as a softener.

ACKNOWLEDGMENTS

This work was supported by *the German Minister for Research and Technology* and *Röhm GmbH*, Darmstadt. The authors wish to thank Dr. Hemnalini Kumar, *Wyatt Technology Corp.*, Santa Barbara, CA, for the MALLS measurements.

REFERENCES

1. J.P. Kennedy, M. Hiza, J. Polym. Sci., Polym. Chem. Ed. 21, 1033 (1983)
2. M. Akashi, T. Yanagi, E. Yashima, N. Miyauchi, J. Polym. Sci., Polym. Chem. Ed. 27, 3521 (1989)

3. P. Rempp, P. Lutz, P. Masson, P. Chaumont, E. Franta, Makromol. Chem.;Suppl. 13, 47 (1985)
4. G. G. Cameron, M. S. Chisholm, Polymer 26, 437 (1985)
5. Y. Gnanou, P. Lutz, Makromol. Chem. 190, 577 (1989)
6. K. Ito, Y. Masuda, T. Shintani, T. Kitano, Y. Yamashita, Polym. J. 15, 443 (1983)
7. K. Mühlbach, V. Percec, J. Polym. Sci., Polym. Chem. Ed. 25, 2605 (1987)
8. K. Hatada, T. Kitayama, K. Ute, E. Masuda, T. Shinozaki, M. Yamamoto, Polymer Bull. (Berlin) 21, 165 (1989)
9. V. Percec, U. Epple, J.H. Wang, H.A.. Schneider, Polymer Bull. 23, 19 (1990)
10. Y. Tsukahara, N. Hayashi, X. Jiang, Y. Yamashita, Polym. J. 21, 377 (1989)
11. G. F. Meijs, E. Rizzardo, J. Macromol. Sci., Rev. Macromol. Chem. C30, 305 (1990)
12. J. Harwood, Makromol. Chem., Macromol. Symp. 10/11, 331-354 (1987)
13. V. Percec, J.H. Wang, Makromol. Chem., Macromol. Symp. 54/55, 561 (1992)
14. W. Radke, A.H.E. Müller, Makromol. Chem., Macromol. Symp. 54/55, 583 (1992)
15. W. Radke, A.H.E. Müller, Polym. Prepr. (Am. Chem. Soc., Div. Polym. Chem.) 32(1), 567 (1991)
16. V. Jaacks, Makromol. Chem. 161, 161 (1972)
17. E. F. Casassa, G. C. Berry, J. Polym. Sci., Part A-2 4, 881 (1966)
18. N. Grassie, B.J.D. Torrance, J.D. Fortune, J.D. Gemmell, Polymer, 653 (1965)
19. V. Percec, J.H. Wang, J. Polym. Sci., Polym. Chem. Ed. 28, 1059 (1990)

15

ANIONIC SYNTHESIS OF MACROMONOMERS AND GRAFT COPOLYMERS WITH WELL-DEFINED STRUCTURES

Roderic P. Quirk,[*] Qizhuo Zhuo, Yuhsin Tsai, Taejun Yoo, and Yuechuan Wang

Maurice Morton Institute of Polymer Science
University of Akron
Akron, Ohio 44325

INTRODUCTION

Macromonomers are linear macromolecules carrying some polymerizable functional groups at their chain ends; the polymerizable functional groups can be at one chain end or at both chain ends[1-12]. Macromonomers are macromolecular monomers, often referred to as "Macromers®"[1]. The important feature of macromonomers is that they can undergo copolymerization with other monomers by a variety of mechanisms to form comb-type, graft copolymers[13,14], as shown in Scheme 1. This aspect of macromonomers distinguishes them from telechelic (α,ω-difunctional) polymers[5,8]; telechelic polymers undergo step-growth type chain extension reactions with other monomers to form linear macromolecules, not branched structures. The polymerizable functional group at the chain end of a macromonomer is often a vinyl group, 1, but it can also be a heterocyclic ring such as an oxirane (epoxide) functionality, 2. These functional groups participate in chain reaction polymerizations with other vinyl or heterocyclic monomers, respectively. A condensation-type macromonomer has two functional groups at one chain end (3) which can participate in step-growth (condensation) polymerization with other difunctional monomers; for example, the functional group, X, could be hydroxyl, amino, carboxyl or isocyanate[5,9].

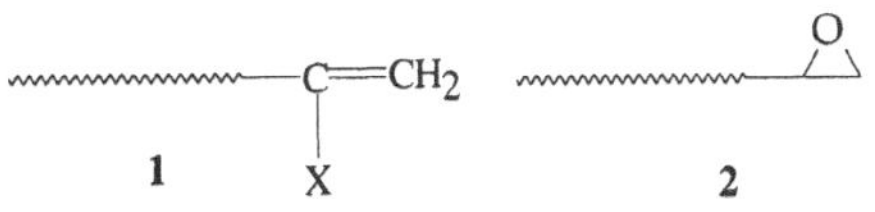

[*] Address correspondence to Dr. Quirk.

Macromolecular Engineering, Edited by M.K. Mishra et al.
Plenum Press, New York, 1995

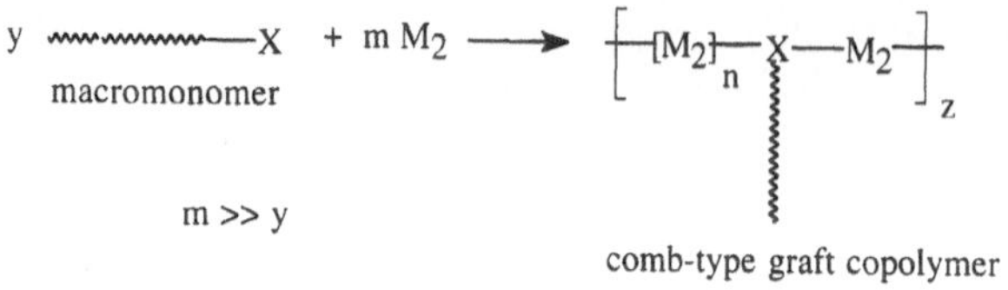

$$y \; \text{\scriptsize wwwwwww} \!-\! X \; + \; m \, M_2 \longrightarrow \left[\!-\! [M_2]_n \!-\! X \!-\! M_2 \!-\! \right]_z$$

macromonomer

$$m \gg y$$

comb-type graft copolymer

Scheme 1

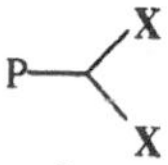

$$P \!-\! C \overset{\displaystyle X}{\underset{\displaystyle X}{\Big<}}$$

3

Macromonomers provide a unique method of preparing graft copolymers with control of the branch structure. Graft copolymers formed from macromonomers (see Scheme 1) are often classified as comb-type, graft copolymers to distinguish them from the normal type, which have a distribution of branch lengths[14]. If a well-defined macromonomer can be prepared with a low degree of compositional heterogeneity, then copolymerization of this macromonomer with other monomers will form a graft copolymer in which the structure of the graft branches is also well-defined. However, the incorporation of the macromonomer will be governed by the statistics of the copolymerization process, i.e., there will be compositional heterogeneity associated with the number and distribution of the graft branches in the graft copolymers. One of our research goals is to develop methods for the synthesis of well-defined graft copolymers in which the number of branches per chain and their distribution along the backbone polymer chain can be controlled.

The molecular weights of macromonomers are generally in the range of 5×10^2 to 2×10^4 g/mole[5]. The choice of molecular weight is often a compromise between copolymerizability and the effect of branch length on physical properties. Obviously, the concentration (number / gram) of functional groups decreases with increasing molecular weight of the macromonomer; thus, the copolymerization efficiency will decrease with increasing molecular weight. The graft copolymers formed from macromonomers are generally heterophase materials whose morphology and physical properties are dependent on their composition[15,16].

ANIONIC SYNTHESIS OF MACROMONOMERS

Living polymerizations provide reliable methods for the synthesis of macromonomers with predictable, well-defined structures. New living polymerization systems have been developed which proceed via a variety of mechanistic types including anionic, cationic, radical, Ziegler-Natta, ring-opening metathesis, coordination and group-transfer polymerization[17]. However, anionic polymerization is one of the most reliable and useful methods for the routine synthesis of macromonomers, especially those based on styrene and alkyl methacrylate backbones, which are of interest for the preparation of comb-type graft copolymers which exhibit the properties of thermoplastic elastomers.

Living anionic polymerization provides an excellent methodology for the synthesis of macromonomers with low degrees of compositional heterogeneity[2,18-20]. The absence of chain termination and chain transfer reactions provides polymers whose molecular weights are precisely predicted and controlled by the stoichiometry of the polymerization as shown in eq. 1[19,20].

$$M_n = (\text{g of monomer})/(\text{moles of initiator}) \qquad (1)$$

In general, these polymers can also be prepared with narrow molecular weight distributions; however, this depends on the relative rates of initiation (R_i) and propagation (R_p) [20]. In addition, the polydispersity (compositional heterogeneity in molecular weight) will decrease with increasing molecular weight in accord with the relationship developed by Flory [21,22] (eq. 2), where the second approximation is valid only for higher degrees of polymerization.

$$X_w/X_n = 1 + [X_n/(X_n + 1)^2] \cong 1 + [1/X_n] \qquad (2)$$

From eq. 2, it is apparent that the polydispersity will be highest for the lower molecular weight macromonomers. Thus, it is generally advisable to use a very reactive initiator such as sec-butyllithium versus n-butyllithium [23]. Since each initiator molecule initiates one polymer chain [eq. 1] by addition of monomer units and the initiator residue is located at the initiating (alpha) chain end of the polymers, a macromonomer can be prepared by utilizing an initiator which contains a polymerizable functional group as shown schematically in eq. 3, where —X is a polymerizable functional group which must also be stable with respect to the anionic polymerization of monomer (M_1).

$$X-R-Li + n\, M_1 \longrightarrow X-R-[M_1]_{n}-Li \xrightarrow{H_2O} X-R-[M_1]_{n}-H \qquad (3)$$

This functionalized initiator procedure is most effect for heterocyclic ring-opening polymerization where the propagating anion is generally not reactive with respect to addition to vinyl functionality such as styryl or methacryloyl groups [24].

Another attribute of living anionic polymerization which is important for the synthesis of macromonomers is that these polymers retain the carbanionic chain end when all of the monomer has been consumed [25]. Therefore, these carbanionic chain ends (e.g. P^-Li^+) can react with a variety of electrophilic reagents (e.g. X-Y) to generate polymers with functional end groups, —Y, as shown in eq. 4 [26].

$$P^-Li^+ + X-Y \longrightarrow P-Y + LiX \qquad (4)$$

Condensation Macromonomers

The reaction of polymeric organolithium compounds with substituted 1,1-diphenylethylene derivatives is an excellent system for development of a general functionalization reaction (eq. 5) because (a) these addition reactions are simple and quantitative; (b) only monoaddition, i.e. no oligomerization, has been reported at stoichiometric concentrations; (c) the rate and efficiency of the crossover reaction can be monitored by ultraviolet-visible spectroscopy; (d) copolymerization of substituted 1,1-diphenylethylene derivatives with other monomers will result in polymers with multiple functional groups along the polymer chain; (e) these addition reactions take place readily in hydrocarbon solution at room temperature and above; and (f) a variety of substituted 1,1-diphenylethylenes with functional groups on the aromatic ring can be prepared readily [26,27].

$$(5)$$

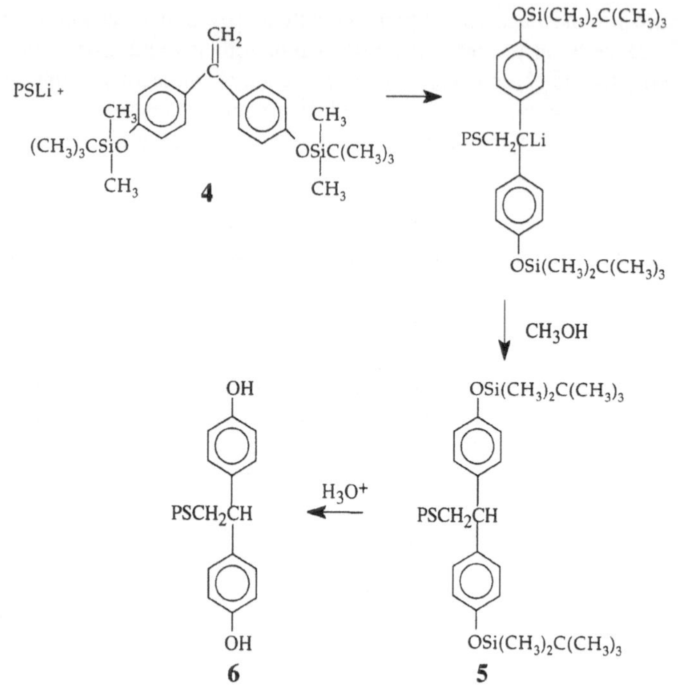

Scheme 2

The 1,1-diphenylethylene functionalization methodology can be used to prepare condensation macromonomers, **3**. ω,ω-Diphenolpolystyrenes, **6**, can be synthesized in quantitative yields by reacting poly(styryl)lithium with 1,1-bis(4-t-butyldimethyl-siloxyphenyl)ethylene, **4**, followed by methanol termination and hydrolysis with dilute acid as shown in Scheme 2[28]. No unfunctionalized polystyrene was detected by TLC analysis of the functionalized polymers (**5,6**) which were also characterized by UV-visible, ^{1}H and ^{13}C NMR spectroscopy and end-group titration. A chain-extension reaction of an ω,ω-diphenol-functionalized polystyrene (M_n=2.6x10^{3}g/mol) was carried out with triphosgene [bis(trichloromethyl)carbonate] in methylene chloride at 51°C (see Scheme 3). The resulting comb-type polymer (**7**) with M_n =1.2 x 10^3 g/mol was formed in high yield with no residual polystyrene macromonomer.

Epoxide-Functionalized Macromonomers

The direct reaction of poly(styryl)lithium (PSLi) (M_n = 2.0 x 10^3 g/mol; M_w/M_n<1.1) with epichlorohydrin (EPC) (normal addition) in benzene produces the corresponding ω-epoxide-functionalized polystyrene, **8**, in only 9% yield (eq. 6); the main product corresponds to dimeric species (70%) and ring-opened products.

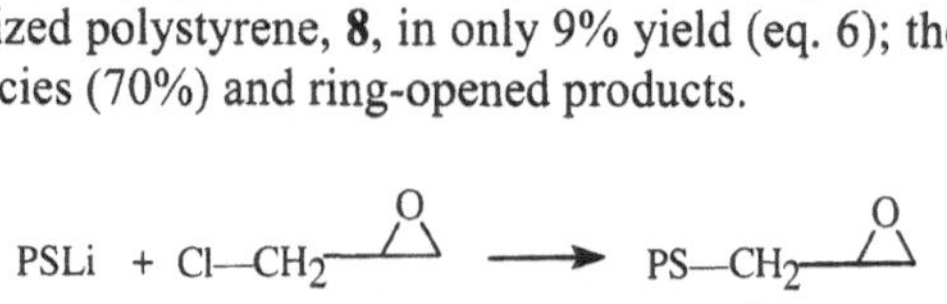

$$\text{PSLi} + \text{Cl—CH}_2\!-\!\!\overset{\displaystyle O}{\triangle} \longrightarrow \text{PS—CH}_2\!-\!\!\overset{\displaystyle O}{\triangle}$$
$$\underset{\textbf{8}}{}\tag{6}$$

The amount of dimer formation was reduced to 22% when poly(styryl)lithium in benzene was added to epichlorohydrin (inverse addition); however, the yield of epoxidized polysty-

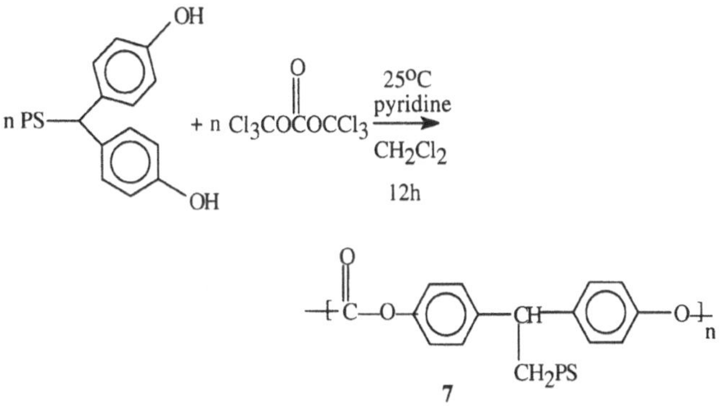

Scheme 3

rene was still low (28%). The other major product formed is the ring-opened product, **9**, formed by direct reaction between poly(styryl)lithium and the epoxide ring as shown in eq. 7; evidence for the ring-opened product was obtained by ^{13}C NMR spectroscopic analysis of the product mixture ($\delta =68.38$ ppm for the methine carbon attached to the hydroxyl group in **9**).

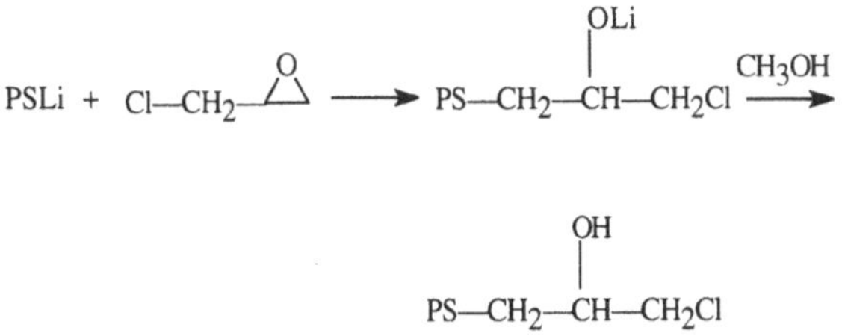

$$(7)$$

In the presence of THF (10 vol% THF for PSLi; only THF for EPC), the yield of epoxide-functionalized macromonomer is 74% with only 10% dimer. The most efficient synthesis of an epoxide-functionalized macromonomer (97% yield, no dimer) was obtained by first end-capping poly(styryl)lithium with 1,1-diphenylethylene as shown in Scheme 4.

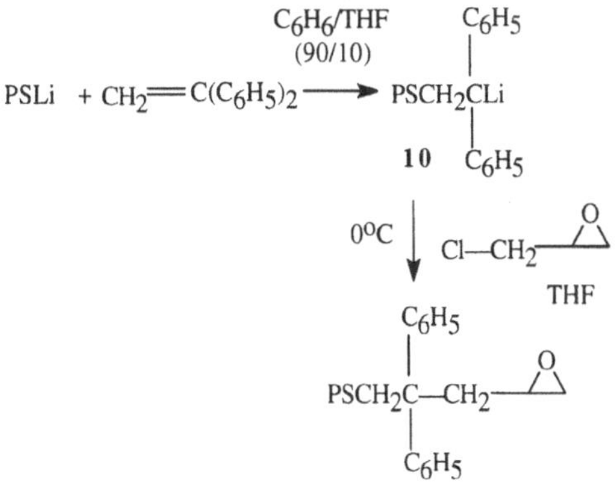

Scheme 4

The decreased reactivity and increased steric requirements of the diphenylalkyllithium chain end in **10** promote selectivity and decrease the formation of dimer in the epoxide functionalization reaction.

Model Graft Copolymer Synthesis

The macromonomer-based copolymerization method for the synthesis of comb-type graft copolymers solves some of the classic problems in traditional grafting reactions: (a) the lack of control of the grafted branch molecular weight and molecular weight distribution; and (b) the contamination of the graft copolymer with the homopolymers of the backbone and the grafting monomer[14]. However, this method does not control the number of graft branches per molecule or the distribution of graft branches along the polymer backbone. In principle, living anionic polymerization with a non-homopolymerizable macromonomer can provide a method of preparing branched and graft polymers with control of all of the structural parameters and low degrees of compositional heterogeneity as illustrated in Scheme 5. Thus, living polymerization will form the first (A) backbone segment, **11**, which will react with a non-homopolymerizable macromonomer,**12**, to only add one macromonomer unit to the chain end to form **13** and maintain the living nature of the polymerization. Addition of more backbone-forming monomer (nM^1) will then generate a new (B) backbone segment whose length will be defined by the ratio of the grams of monomer added to moles of active chain end [see eq. 1]. The living polymer **14** can then react with another equivalent of non-homopolymerizable macromonomer, **12**, to place another graft branch at the precise point on the backbone which was defined by the B segment length (see **15**). Addition of more backbone-forming monomer (mM^1) to the adduct **15** will then generate a new (C) backbone segment whose length will be defined by the ratio of the grams of

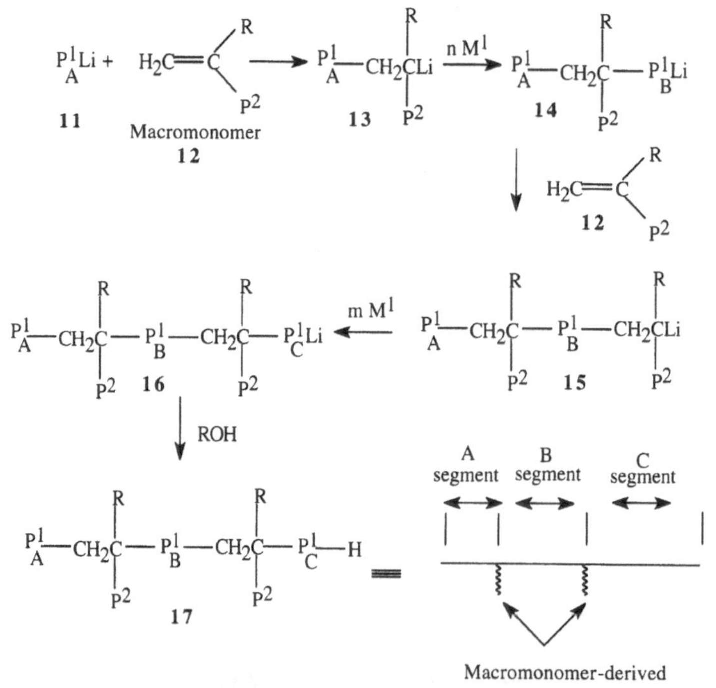

Scheme 5

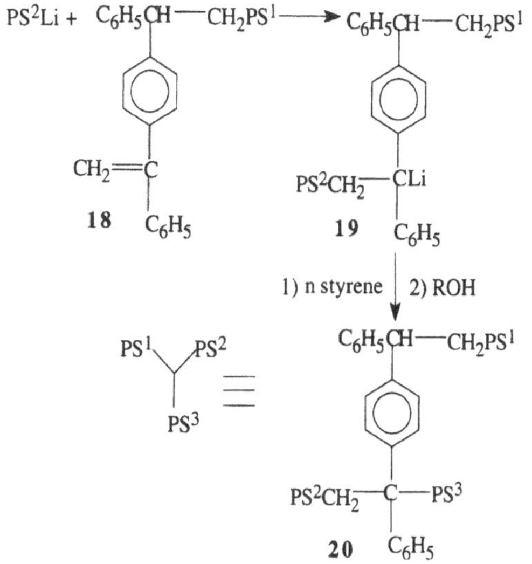

Scheme 6

monomer added to moles of active chain end [see eq. 1]. At this point the whole sequence of macromonomer/comonomer addition could be repeated to generate more graft branches at specific locations along the backbone or the living polymer can be terminated to form the precisely defined graft copolymer with two graft branches located at segment distances A and A + B along the polymer backbone (**17**).

A demonstration of the feasibility of this approach has been performed using a 1,1-diphenylethylene-functionalized polystyrene macromonomer, **18**, to form a hetero three-armed, star-branched copolymer, **20**, as shown in Scheme 6[52]. For the synthesis of hetero 3-armed, star-branched polymers, the first step involved the addition of a polymeric organo-lithium compound, poly(styryl)lithium, with the 1,1-diphenylethylene-functionalized poly-styrene macromonomer, **18**, to form the corresponding coupled product, **19**, a diphenylalkyllithium. For stoichiometric amounts of poly(styryl)lithium and macro-monomer, it was found that the efficiency of this coupling reaction is >96%. This result also shows that the vinyl functionality of the macromonomer is ≥ 96%. Finally, the third arm was formed by addition of monomer, e.g. styrene, in the presence of THF to promote the crossover reaction. SEC analyses of the heteroarm star polymer product (**20**) showed that each of these steps proceeded efficiently to give the expected products; only small amounts of non-star product were observed which corresponded in retention volume to the small amount of unreacted macromonomer and a small amount of polystyrene corresponding to the second arm. A narrow molecular weight distribution star product ($M_w/M_n = 1.02$) was easily obtained by one fractionation step. Work is in progress to extend this methodology to other backbone forming monomers, especially polydienes, to form model graft thermoplastic elastomers.

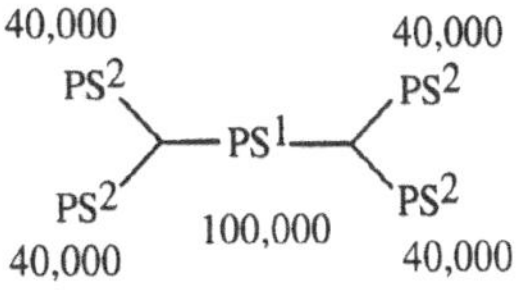

This methodology has been used to synthesize an H-shaped polymer, **21**, as shown in Scheme 7. The reaction of α,ω-dilithiumpolystyrene (**22**) with two moles of 1,1-

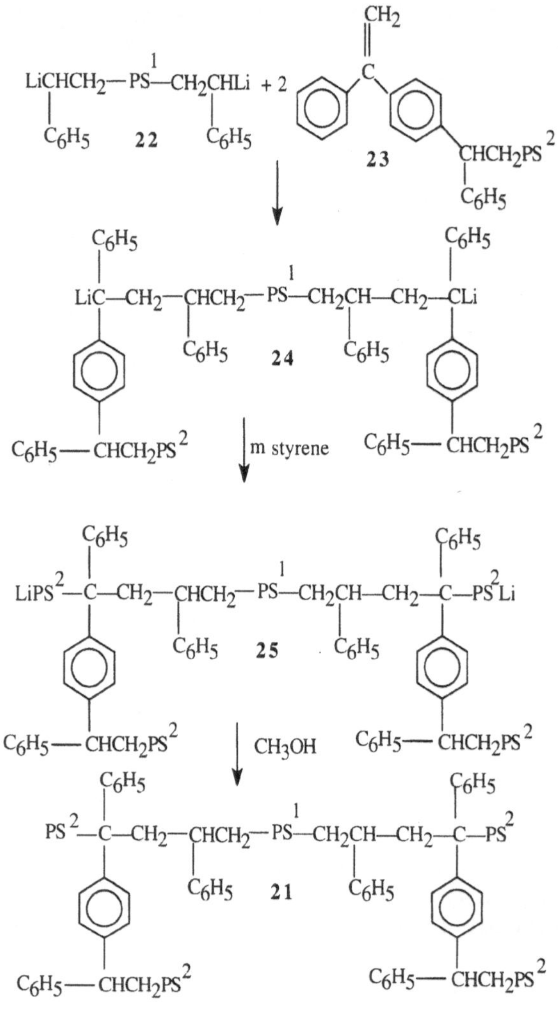

Scheme 7

diphenylethylene-functionalized polystyrene macromonomer (**23**) was followed by addition of styrene monomer to the resulting bis(diphenylalkyllithium)-functionalized polymer (**24**) to generate the H-shaped architecture (**25**). Each step in this synthesis was monitored by a combination of SEC and UV-visible absorption spectroscopy. The diphenylalkyllithium adduct (**24**) absorb at 480nm and the poly(styryl)lithium chain ends (see **22,25**) absorb at 330nm; thus, it was possible to follow the course of each crossover and addition reaction. The resulting unfractionated H-shaped polystyrene (M_w/M_n = 1.18) exhibited a g' value of 0.74 which is agreement with previous literature characterization data for H-shaped polymers[30].

CONCLUSIONS

Macromonomers offer a versatile methodology for the synthesis of comb-type graft copolymers with a wide variety of backbone and branch structures. Macromonomers are readily prepared by living polymerizations by either using a functionlized initiator or by post-polymerization, chain-end functionalization reactions of the living chain ends. Both the macromonomer synthesis and the subsequent copolymerization process can be effected using any polymerization methodology. Non-homopolymerizable macromonomers combined with living polymerization methods provide pathways for the synthesis of model graft copolymers in which the number of branches, the distribution of the branches along the polymer backbone and the branch structure can be controlled.

REFERENCES

1. "Macromer®" is a trademark of CPC International Inc. R.Milkovich and M.T. Chiang, U.S. Patent 3, 786, 116(1974).
2. R. Milkovich in *Anionic Polymerization. Kinetics and Mechanism*, J.E.McGrath, Ed., ACS Symposium Ser. No. 166, American Chemical Society, Washington, D.C., 1981, p 41.
3. G.O. Schulz and R. Milkovich, *J. Appl. Polym. Sci.*, **27**, 4773(1982).
4. G.O. Schulz and R. Milkovich, *J. Polym. Sci., Polym. Chem. Ed.*, **22**, 1633(1984).
5. P.F. Rempp and E. Franta, *Adv. Polym. Sci.*, **58**, 1(1984).
6. P. Rempp, P. Lutz, P. Masson and E. Franta, *Makromol. Chem., Suppl.*, **8**, 3 (1984).
7. P. Rempp, P. Lutz, P. Masson, P. Chaumont and E. Franta, *Makromol. Chem., Suppl.* ,**13**, 47(1985).
8. P. Rempp, E. Franta, P. Masson and P. Lutz, *Progr. Colloid & Polym. Sci.*, **72**, 112(1986).
9. Y. Kawakami in *Encyclopedia of Polymer Science and Engineering*, J.I. Kroschwitz, Ed., Wiley-Interscience, New York, 1987, Vol. 9, p. 195.
10. V. Percec, C. Pugh, O. Nuyken and S.D. Pask in *Comprehensive Polymer Science*, Vol. 6, *Polymer Reactions*, G.C. Eastmond, A. Ledwith , S. Russo and P. Sigwalt, Eds., Pergamon Press, Elmsford, New York, 1989, p. 281.
11. Y. Gnanou, *Ind. J. Technol.*, **31**, 317(1993).
12. *Macromolecular Design: Concept and Practice*, M. K. Mishra, Ed., Polymer Frontiers International, Inc., Hopewell Jct., New York, 1994.
13. "Basic Definitions of Terms Relating to Polymers", *Pure Appl. Chem.*, **40**, 482(1974).
14. P. Dreyfuss and R. P. Quirk in *Encyclopedia of Polymer Science and* Engineering, J. I. Kroschwitz, Ed., Wiley, New York, Vol. 7, 1986, p. 551.
15. G. Riess and G. Hurtrez in *Encyclopedia of Polymer Science and Engineering*, J. I. Kroschwitz, Ed., Wiley, New York, Vol. 2, 1985, p. 324.
16. G. Holden, E. T. Bishop and N. R. Legge, *J. Polym. Sci.*, C **26**, 37(1969).
17. R. P. Quirk and J. Kim, *Rubber Chem. Technol.*, **64**, 450(1991).
18. M. Morton, *Anionic Polymerization: Principles and Practice*, Academic Press, New York, 1982.
19. S. Bywater in *Encyclopedia of Polymer Science and Engineering*, J. I. Kroschwitz, Ed., Wiley, New York, Vol. 2, 1985, p. 1.
20. P. Rempp and E. Franta in *Recent Advances in Anionic Polymerization*, T. E. Hogen-Esch and J. Smid, Ed., Elsevier, New York, 1987, p. 353.
21. P. J. Flory, *Principles of Polymer Chemistry*, Cornell University Press, Ithaca, New York, 1953, p. 338.
22. J.F. Henderson and M. Szwarc, *J. Polym. Sci., Macromol. Rev.*, **3**, 317(1968).
23. S. Bywater and D. J. Worsfold, *J. Organometal. Chem.*, **10**, 1(1967).
24. R. P. Quirk and J. Kim in *Ring-Opening Polymerization*, D. J. Brunelle, Ed., Hanser, New York, 1993, p. 263.
25. R. N. Young, R. P. Quirk and L. J. Fetters, *Adv. Polym. Sci.*, **56**, 1(1984).
26. R. P. Quirk in *Comprehensive Polymer Science, First Supplement*, S. L. Aggarwal and S. Russo, Eds., Pergamon Press, Oxford, UK, 1992, p. 83.
27. R. P. Quirk and J. Kuang, Macromol. Symp., **85**, 267(1994).
28. R. P. Quirk and Y. Wang, *Polym. Int.*, 31, 51(1993).
29. R. P. Quirk and T. Yoo, *Polym. Bull.*, **31**, 29(1993).
30. J. Roovers and P. M. Toporowski, *Macromolecules*, **14**, 1174(1981).

16

NEW SUPERSTRUCTURES FROM BLOCK AND GRAFT COPOLYMERS WITH PRECISELY CONTROLLED CHAIN ARCHITECTURE

C. D. Eisenbach, A. Göldel, H. Hayen, T. Heinemann, U. S. Schubert, and M. Terskan-Reinold

Universität Bayreuth
Makromolekulare Chemie II
D-95440 Bayreuth
Germany

INTRODUCTION

Polymers have a high potential as structural and functional materials. However, synthetic polymer systems are still far away from the elegant schemes of biological systems in self-organization phenomena and the formation of complex structures with specific functions. In biopolymeric systems, the supramolecular structure results from a special sequence of amino acids along the polypeptide chain and/or periodical sequences of other building units along the chain, resulting in highly organized structures which in most cases are stabilized by hydrogen bond formation or ionic interactions. Unique materials with extraordinary ultimate properties result, and it is a challenging question, if and/or to which extent this can be transfered over to synthetic macromolecules.

A primary objective of basic materials science is to develop a detailed knowledge of the structure-property relationships in macromolecular systems in order to understand how the molecular architecture of the individual polymer chain influences the morphology and superstructure of the polymeric materials, e.g., a key problem in view of new polymeric suprastructures and in analogy to the building principles of biopolymeric systems with specific functions is to understand how the overall morphology of microphase separated systems, i.e., the formation of supramolecular structures and the conformation of the single chain are controlled by the individual properties of covalently linked segments. This is illustrated in Scheme 1 with the example of segmented block copolymers consisting of a periodical sequence of two or more segments differing in their constitution. In a hierarchical view, the simplest case of self-organization would be the segretion to a phase separated system, e.g., with lamellar morpholoy. The next higher level is represented by phase morphologies, where the segments have adopted their thermodynamically preferred conformation and packing order as represented in Scheme 1 by the micordomains formed from

Macromolecular Engineering, Edited by M.K. Mishra et al.
Plenum Press, New York, 1995

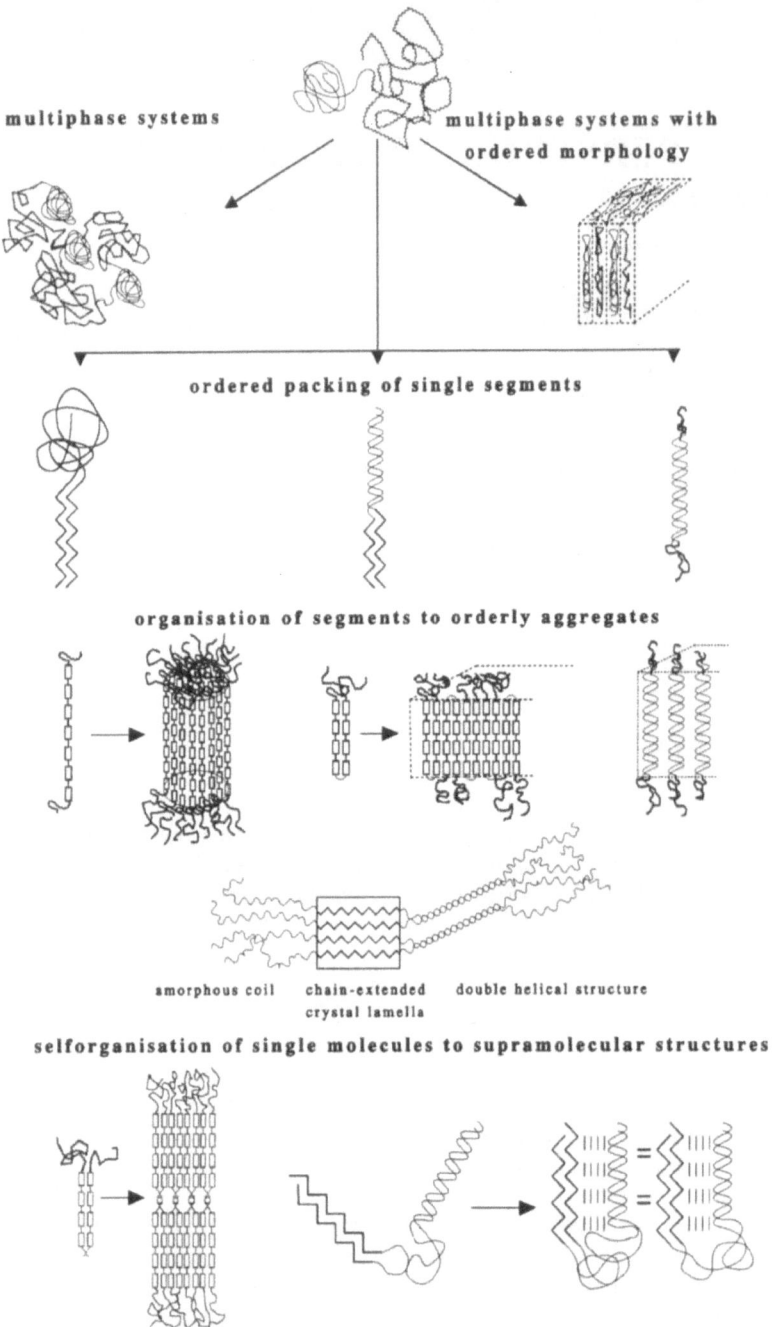

multiphase systems

multiphase systems with
ordered morphology

ordered packing of single segments

organisation of segments to orderly aggregates

amorphous coil   chain-extended   double helical structure
                    crystal lamella

selforganisation of single molecules to supramolecular structures

Scheme 1. Schematic of a hierarchical order in the self-organization and structure formation of segmented block copolymers composed of incompatible and/or self-recognizing segments.

chain-extended or chain-folded crystallized segments, or double helices. The most sophisticated systems are represented further below, where the individual chains have adopted a tertiary structure and have organized into well-ordered quaternary structures by recognition-directed self assemblies.

Block copolymers of the AB, ABA, BAB and $(AB)_x$ type consisting of highly flexible A segments and molecularly uniform B segments of well-defined architecture capable of either the formation of chain-extended and chain-folded crystallites or double helical metal complexes have been chosen as model system (Scheme 2). This selection not only allows to study how the shape of macromolecules and the packing in nano to mesoscopic dimensions is determined by the constituion of the whole macromolecule, specific constitutional units and the interaction between chain segments, but also to mimick typical conformations found in biopolymers such as a coil, a double helix, and chain-extended or chain-folded (β-sheet) structures.

Here we report on our studies of the phase behaviour and self-organization of such segmented block copolymers and graft copolymers with polydisperse polyether or polydimethylsiloxane soft segments and molecularly uniform piperzine (PIP)-based poly(N-alkylurethane) hard segments or grafts, respectively, or segments containing the 6,6'-disubstituted bipyridine (bpy) constitutional units. The syntheses of the respective

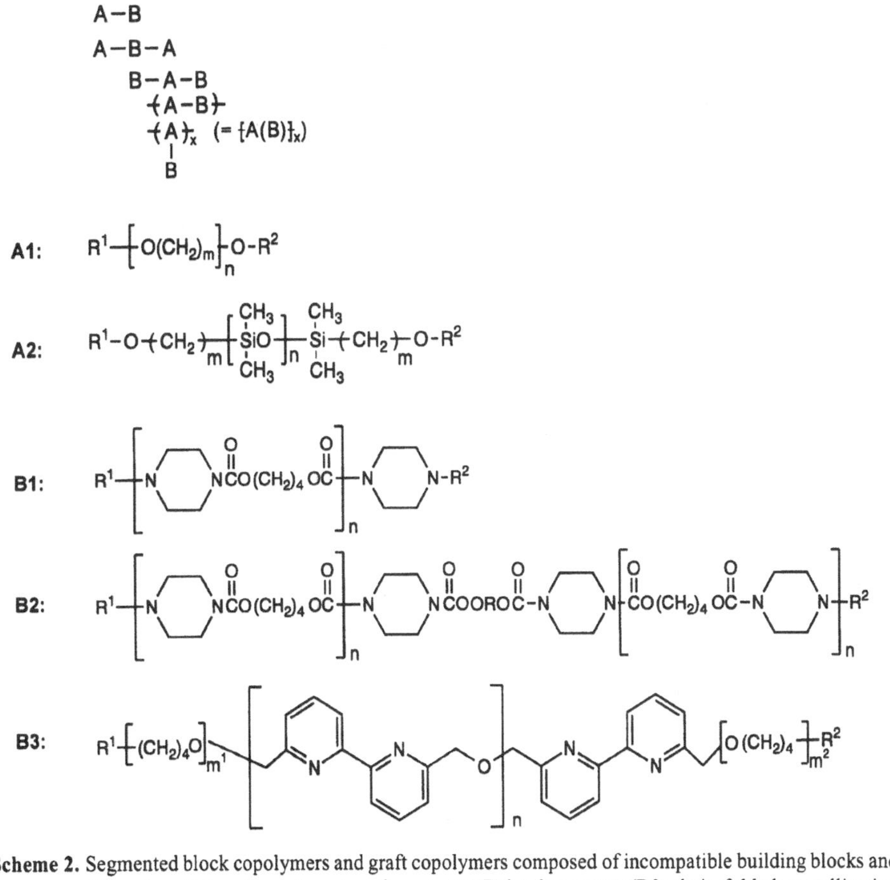

Scheme 2. Segmented block copolymers and graft copolymers composed of incompatible building blocks and segments with preferential conformation; A: soft segments; B: hard segments (B2: chain-folded crystallization for, e.g., $R = (CH_2)_8$; B3: double helix formed with Cu(I).

molecularly uniform building blocks have already been described elsewhere (Ref. 1-8): The approach for block and graft copolymers with regularity in the primary structure of the macromolecule and containing such molecularly uniform segments or grafts is given in condensation polymers. The linkage of building units in a well-defined way can be achieved by using bifunctional starting materials, where one of the two functional groups is temporarily blocked (cf., e.g. Ref. 2,3). Selective removal of the protecting group regenerates the functionality and allows a subsequent addition of another building unit. This ultimately leads to a macromolecule with the desired specific chain structure (cf., e.g., Ref. 3,5,6,9). Furthermore, the step-by-step synthesis of the starting blocks A or B from constitutional unit precursors not only allows the introduction of any irregularity along the chain at a selected site, but also to place a spacer between the blocks or to vary the end groups $R^{1,2}$ (Ref. 6-8,10). In this context it has to be emphasized that the chosen poly(N-alkylurethane) building block is distinguished by an excellent thermal stability as compared to classical urethanes containing the -NHCOO- linkage and thus can be repeatedly melted up to about 300°C without any decomposition (Ref. 2,5); this qualifies such piperazine based polyurethanes as model systems of choice in condensation polymers because of the N-alkylurethane linkage, and the structure-properties relationships derived from the studies of these systems are considered to be exemplary for segmented block copolymers.

RESULTS AND DISCUSSION

Segmented Polyether-urethane and Polysiloxane-urethane Block Copolymers

Segmented block copolymers with polyurethane (PU) hard segments and polyether or other soft segments form a microphase separated system due to the inherent incompatibility of the two segments (Ref. 11,12) It is generally accepted that phase separation proceeds via hard segment segregation, and that the resulting polyurethane hard domains act as multifunctional, thermoreversible netpoints and as fillers in these thermoplastic elastomers. A problem which has not been tackled so far is which morphology results in blends of segmented polyurethanes where the blend components differ by, e.g., the hard-soft segment linkage, or even by usually incompatible soft segments. A detailed picture of the segregation phenomena, morphology and molecular dynamics of $(A-B)_x$ multiblock copolymers in general has been established by the study of various model PU systems with molecularly uniform PU oligomers and/or uniform hard segment length (Ref. 2,10,13-23).

The mutual influence of the hard-soft segment linkage on segment mobility and packing in the interphase is illustrated from the study of polydimethylsiloxane (PDMS)-urethane multiblock copolymers $(A2-B1)_x$ with different spacers between the hard and soft segment (Ref. 10). The molecularly uniform hard segments are composed of 4 repeating units and are linked to the PDMS soft segment by a tri- or hexamethylene spacer. The hard domain melting (Fig. 1a) is significantly lower for the sample with the trimethylene spacer ($T_m = 433$ K) than for the sample with the longer hexamethylene spacer ($T_m = 446$ K). The hard domain melting temperature of the later sample is close to $T_m = 451$ K of the pure tetramer model compound, indicating similar packing order in both cases. The lower melting temperature of the hard domain in the PDMS-urethane multiblock copolymer with the shorter spacer is a result of the close-by dimethyl substituted siloxane moiety, which sterically hinders the hard segment packing at the interphase and thus increasing the surface defect enthalpy. Blends of the two multiblock copolymers are completely miscible, i.e., co-segregation of the hard segments results in hard domains which are built from both types

Figure 1. a) DSC curves (heating rate 20 K/min) of PDMS-urethane multiblock copolymers $(A2-B1)_x$ ($A2$: $R^{1,2}$ = nihil. $B1$: $R^{1,2}$ = carbonyl; $n = 4$) with trimethylene ($A2$ with $m = 3$, curve 1) and hexamethylene ($A2$ with $m = 6$, curve 2) spacer between soft and hard segment, and 1:1 (w/w) blend (curve 3). b) DSC curves of PEO-urethane $(A1-B1)_x$ ($A1$: $R^{1,2}$ = nihil; $m = 2$, $\bar{n} = 9$. $B1$: $R^{1,2}$ = carbonyl; $n = 4$) (curve 1) and PDMS-urethane $(A2-B1)_x$ (see a), curve 2) multiblock copolymers, and 1:1 (w/w) blend (curve 3); number of repeating units $n = 4$ for PU segments in all samples.

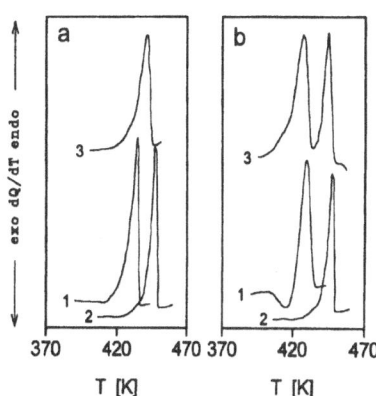

of multiblock copolymers; this is evident from the single melting endotherm of binary mixtures (see curve 3, Fig. 1a).

The DSC traces of a 1:1 (w/w) mixture of a PDMS-urethane $(A2-B1)_x$ and a poly(ethyleneoxide) (PEO)-urethane $(A1-B2)_x$ multiblock copolymer are dipicted in Fig. 1b together with the DSC traces of the individual blend components; the differences in the hard domain melting temperature of the two elastomers ($T_m = 446$ K vs. 428 K) can be associated with differences in hard-soft segment compatibilites. In contrast to the PDMS-urethane blend discussed above, where only one hard domain melting endotherm appeared (Fig. 1a, curve 3), two melting endotherms at 427 K and 444 K appear. This means that co-crystallization of the hard segments of the two different multiblock copolymers does not occur and the DSC trace reflects the melting of the domains composed only of hard segments of the one or the other blend component. Obviously, the phase separation is first controlled by the incompatibility of the soft segments of the two blend components, and not by the segregation of the hard segments. The resulting multiphase polymeric system can be described as a two-phase system of the PDMS and PEO soft segments in which hard segment micro-domains are embedded. In this context it has to be added that the same features were observed for $(A1-B3)_x/(A2-B3)_x$ multiblock copolymer blends even after the formation of the bipyridine-Cu(I) complex.

Polyether-Urethane Graft Copolymers

The study of a homologous series of molecularly uniform polyurethanes B1 and of segmented polyether-urethanes $(A1-B1)_x$ ($m = 4$) with corresponding monodisperse PU hard segments has shown that the polyurethanes crystallize without chain folding in these model systems (Ref. 2,22,24). It was found that chain folding is induced upon replacing only a single tetramethylene unit by a more flexible unit such as hexamethylene or octamethylene because this defect (in B2) in the otherwise regular structure (in B1) acts as a joint (Ref. 3-5). The change from chain-extended to chain-folded hard segments results in a change from cylindrical microdomains of $(A1-B1)_x$ to lamellar microdomains in $(A1-B2)_x$ of half the height of the hard domain cylinders (see orderly aggregates in Scheme 1), which caused a distinct lowering of the melting temperature and different dynamic mechanical properties (Ref. 3,5,25-27).

The similarities between the schematic appearance of single polyether-urethane structural units -A1-B2- with the hard segment in the chain-folded conformation (see Scheme 1) and a branched polymer stimulated the synthesis and investigation of graft copolymers

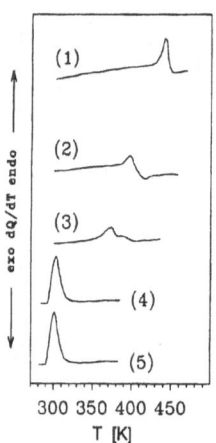

Figure 2. DSC curves of graft copolymers [A1(B1)]$_x$ with polyether A1 (R^{1,2} = carbonyl; m = 4, $\bar{n}$ = 28) main chain and molecularly uniform oligourethane B1 (R^2 = methylcarboxylate; backbone linkage R^1 = 1,4-piperazinediyl-2-yl-methyl carboxylate) side chains with 7 (curve 1), 4 (curve 2), 3 (curve 3) and 2 (curve 4) piperazine/butanediolbischloroformate based repeating units per side chain, and of the unsubstituted polether main chain (POTM 55.000, curve 5); all samples were films cast from CHCl$_3$ solution. Heating rate 20 K/min.

[A(B)]$_x$ with poly(oxytetramethylen) main chain and molecularly uniform oligourethane side chains (Ref. 6,7,25,26).

Above a certain length of the side chains, [A1(B1)]$_x$ graft copolymers with methyl carboxylate graft end groups and 1,4-piperazinediyl-2-yl-methyl carboxylate graft-backbone linkage exhibit the typical properties of thermoplastic elastomers, i.e., the grafts segregate to form semi-crystalline hard domains dispersed in a continuous soft phase formed by the poly(oxytetramethylene) backbone. The investigation of the thermal and mechanical properties in comparison with analogously built segmented polyether-urethane elastomers (A1-B1)$_x$ reveals (backed by thermal and x-ray analysis of the corresponding PU model compounds) that the crystallization of the grafts is limited by the nature of the linkage to the main chain (Ref. 6,26).

Depending on their length, the side chains of these graft copolymers segregate to semi-crystalline domains. This is shown by the DSC-traces in Fig. 2. The materials with grafts with three, four and seven 1,4'-piperazinediylcarbonyloxytetramethyleneoxycarbonyl repeating units are characterized by melting endotherms of which the temperature increases with increasing length of the side chains. The reason for those endotherms has to be seen in the melting of semi-crystalline domains formed by the segregation of the oligourethane grafts. For the material with only two repeating units per side chain, one endotherm is observed at the same temperature where crystalline areas of the unsubstituted backbone polymer polyoxytetramethylene (Mn = 55.000) (compare curves 4,5 in Fig. 2) melt: In this case the short oligourethane grafts do not seem to form semi-crystalline domains, or their melting is not observed in the DSC experiment because of overlaying endotherms.

The occurrence of microphase separation, i.e., the existence of hard domains formed by the oligourethane side chains dispersed in a flexible amorphous phase formed by the main chain in the case of these graft copolymers without spacer between the backbone and side chain is proved by the dynamic mechanical experiment (Fig. 3). The storage modulus G' of [A1(B1)]$_x$ with n ≥ 3 repeating units in the graft shows a temperature dependency characteristic for thermoplastic elastomers. The final decrease of the modulus, reflecting the melting of the oligourethane hard domains, is determined by the graft length: With increasing degree of oligomerisation (n in B1) the extension of the plateau region increases, which is in agreement with the results of the DSC experiment (Fig. 2).

The graft copolymer with only 2 repeating units in the side chain shows not elastomeric properties as could already be expected from the DSC experiments; above 200 K a continuous decrease of G' and no plateau are observed. In contrast, the materials with the same number (n = 2 in B1) of repeating units but an alkylene(methyl)aminooxymethylene

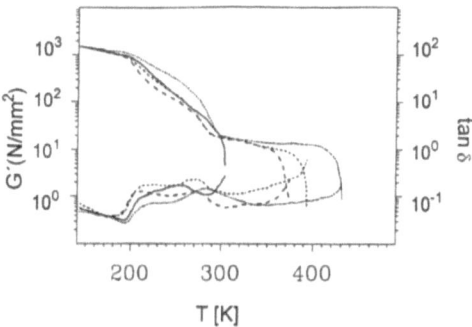

Figure 3. Dynamic mechanical properties of graft copolymers $[A1(B1)]_x$ (see Fig. 2) with poly(oxytetramethylene) main chain and uniform oligourethane side chains consisting of 2 (—), 3 (———), 4 (- - -) and 7 (····) 1,4-piperazinediylcarbonyloxytetramethyleneoxycarbonyl repeating units per side chain. G': storage modulus; tan δ: logarithmic decrement.

spacer instead of the methylene group (e.g., alkylene = ethylene) exhibit a distinct endotherm characteristic for hard domain melting in the DSC trace and showed the typical feature of a thermoplastic elastomer (Fig. 4): A characteristic plateau region not seen in the sample without spacer occurs at temperatures higher than 300 K; this was even more pronounced for alkylene = hexamethylene (not shown here).

Besides by varying the distance between the side chains and backbone, the formation of ordered structures in graft copolymers could also be influenced by the end group: Only by replacing the methyl carboxylate with a methylcarbaminate end group in the graft copolymers without a spacer, leads to the formation of a semi-crystalline hard phase as revealed from DSC analysis and results in an elastomer (Fig. 4).

Block Copolymers with Bipyridine Containing Segments

These studies were directed to the synthesis and the investigation of the phase behaviour and self-organization of segmented triblock and multiblock copolymers with the potential of the formation of double helical structures (Ref. 28,29). As it is known from organic supramoelcular chemistry, bipyridine oligomers exhibit spontaneous self-organization in the presence of Cu(I) ions to form a double-stranded helical complex, which possesses

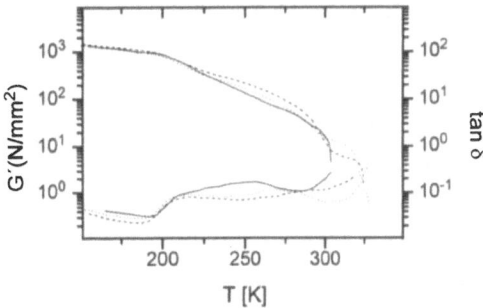

Figure 4. Dynamic mechanical properties of the graft copolymers $[A(B1)]_x$ with poly(oxytetramethylene) main chain (see Fig. 3) and uniform oligourethane side chains B1 (see Fig. 3) with two 1,4-piperazinediylcarbonyloxytetramethyleneoxycarbonyl repeating units (n = 2 in B1) but different constitution: (—): methylcarboxylate end group R^2 and no spacer (as in Fig. 3, curve 1); (- - -): methylcarboxylate end group R^2 and ethylene(methyl)aminooxymethylene spacer (backbone linkage R^1: 1,4-piperazinedyl-2-yl-methyleneoxycarbonylamino(methyl)ethylene carboxylate); (····): N-methylcarbaminate end group R^2 and no spacer.

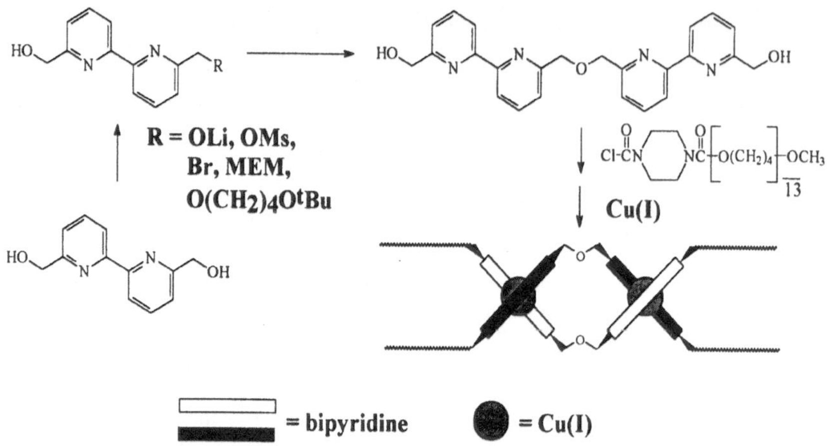

= bipyridine   = Cu(I)

Scheme 3. Synthesis of unsymmetrical functionalized 2,2'-bipyridine (bpy) building blocks B3 ($R^1 \neq R^2$; n = 0), bifunctionalized oligo(bpy) molecule B3 ($R^{1,2}$ = OH; $m^{1,2}$ = 0,1; n = 1), triblock copolymer A1-B3-A1 (A1 with R^1 = nihil and R^2 = CH_3: m = 4; $\bar{n}$ = 13. B3 with $R^{1,2}$ = oxycarbonyl-1,4-piperazinediylcarbonyl: $m^{1,2}$ = 0; n = 1), and schematic of the dinuclear helical complex.

characteristic features reminesent of the DNA double helix (Ref. 28,29); one aspect of this work is also to bridge or to apply these principles to macromolecular systems (cf. Ref. 30).

Precursor building blocks containing 6,6'-disubstituted 2,2'-bipyridine (bpy) consti- tutional units B3 (Scheme 2) had to be synthesized first. A suitable synthetic route for bifunctional oligo(bpy) in preparative scale had to be developed first in order to obtain test materials in sufficient quantities. Monofunctionalized and symmetrical bifunctional bpy (B3; n = 0) starting materials are obtained in large scale quantities by the route via bipyridine N-oxide (Ref. 8,31). The key intermediate in the synthesis of unsymmetrical substituted bpy building blocks is the monolithium salt of the bishydroxy-bpy (B3, $m^{1,2}$ = n = 0, $R^{1,2}$ = OH) which can be further functionalized and also employed in the preparative scale synthesis of bisfunctionalized oligo(bpy) (Scheme 3). The synthesis of generations of functionlized oligo(bpy) units via sequential Williamson condensation (Scheme 4) is facilitated by groups which would increase the poor solubility of oligo(bpy) ligands and could also be used in subsequent reactions; here it should be emphasized that the unsymmetrically functionalized molecules open new avenues for the use of oligobipyridines. The conversion of functional- ized B3 with functionalized A1 or A2 gives various segmented block copolymers.

The Cu(I) complex of bpy units is immediately formed after addition of, e.g., a Cu(I) trifluoromethanesulfonate solution to the block copolymer solution in chloroform/acetroni- trile as indicated by the deep orange-red colouring of the solution. The UV spectra of the complexed block copolymers (Fig. 1) are identical with those of oligo(bpy)/Cu(I) complexes if sterical interactions at the linkage bpy-polyether are avoided by the use of a spacer (m = 4 in B3) (Ref. 8,9); without spacer (m = 0 in B3), the charge-transfer-transition band (t_{2g} - π^*) is blue shifted to about 400 nm (Ref. 31).

Whereas the complexation of the three-block copolymers B3-A1-B3 resulted in a chain extension reaction (Ref. 31), the addition of Cu(I) ions to the multiblock copolymers $(A-B3)_x$ results in a crosslinking by formation of tetrafunctional Cu(I) complexes (Ref. 9). The dynamic mechanical behaviour of the polymer-ion complexes of the three-block copolymers B3-A1-B3 is typical for two-phase thermoplastic elastomers (Fig. 6, curves 1 and 2). Elastomeric behaviour is also found for the multiblock copolymers $(A-B3)_x$ if excess

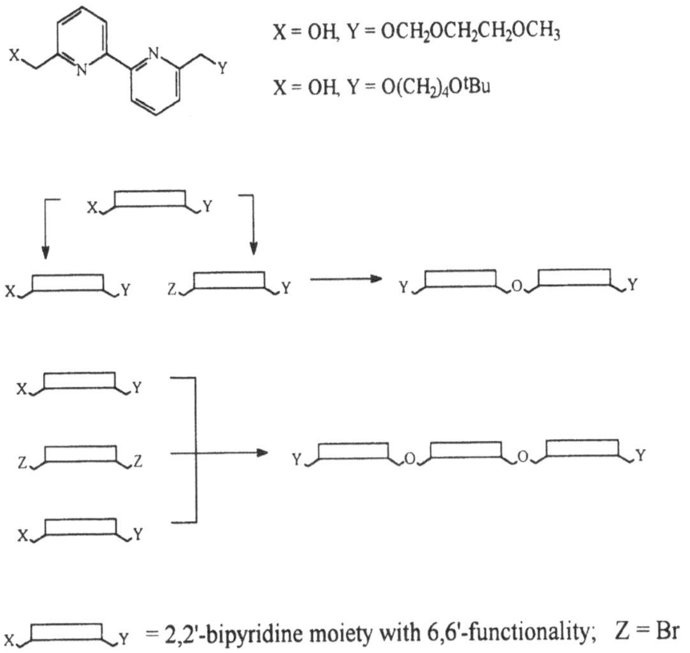

X = OH, Y = OCH2OCH2CH2OCH3
X = OH, Y = O(CH2)4OtBu
X□Y = 2,2'-bipyridine moiety with 6,6'-functionality; Z = Br

Scheme 4. Schematic representation of the Williamson condensation leading to functionalized oligo(bpy) strands B3.

Cu(I) with regard to the stoichiometrical [Cu(I) $(bpy)_2$] complexes is used (Fig.6, curves 3 and 4).

In the case of the three-block copolymers, the softening which was associated with the distruction of hard segment Cu(I) complex domains starts at about 310K followed by yielding for the B3 end block with n = 0 and at about 435K for the system with the B3 end group consisting of two bpy constitutional units (B3 with n = 1). This means that the strength of the domains of Cu(I) complexed bis(bpy) segments is as high as in a segmented

absorption
λ [nm]
450 nm
1,0
0,8
0,6
0,4
0,2
0,0
250
300
350
400
450
500
550
600

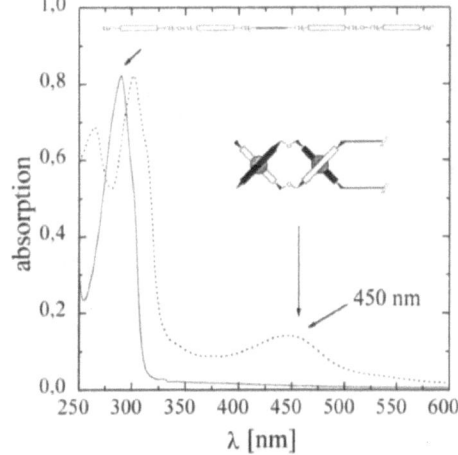

Figure 5. UV/Vis spectra of the uncomplexed (—) and Cu(I)-complexed (····) triblock copolymer B3-A1-B3 in $CHCl_3$; (A1 with $R^{1,2}$ = nihil: m = 4; $\bar{n}$ = 28. B3: n = 1; R^1 = H with m^1 = 0 and R^2 = oxycarbonyl-1,4-piperazinediylcarbonyl with m^2 = 1).

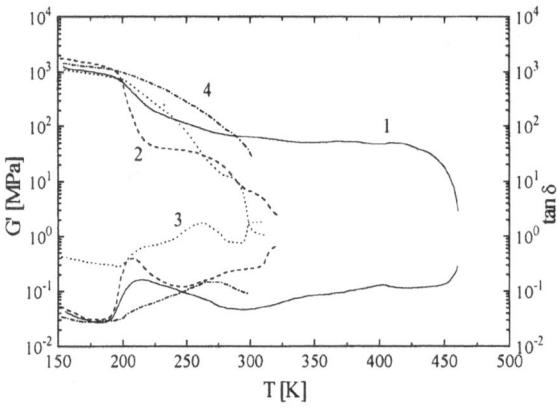

Figure 6. Dynamic mechanical properties of Cu(I) complexed block copolymers; (—; 1): B3-A1-B3 (see caption Fig. 5); (- - -; 2): B3-A1-B3 (see caption Fig. 5, but B3: $m^{1,2} = 0$; $n = 0$); (···; 3) and (-··-; 4): (A1-B3)$_x$ (see caption Fig. 5, but B3 with $R^{1,2}$ = oxycarbonyl-1,4-piperazinediylcarbonyl with $m^{1,2} = 1$ and $n = 0$). Molar ratio bpy/Cu(I) = 2 except for (3) with 2/1.25.

polyether-urethane (A1-B1)$_x$ (A1: $m = 4$, $\bar{n} = 28$) with molecularly uniform poly(N-alkylurethane) hard segments consisting of 7 piperazine/butandiolbischloroformate based repeating units (B1: $n = 7$) (cf. Ref. 2). The occurrence of elastomer properties for the complexed multiblock copolymer (A-B3)$_x$ in the presence of excess Cu(I) can only be explained by the existence of Cu(I)-bpy clusters composed of both the tetrahydral [Cu(I)(bpy)$_2$] complexes and [Cu(I)(bpy)L$_2$] (together with [Cu(I)L$_4$] with L being the ligand of the employed Cu(I) for bipyridine complex formation.

The occurrence of microphase separated systems in bulk with nano to mesoscopic superstructures consisting of Cu-bpy complex aggregates in a polymer matrix was confirmed for both thermoplastic elastomer systems by transmission electron microscopy (TEM) (Ref. 9). Bright field images of both solution cast and ultramicrotomed films show domains which contain copper and nitrogen as was established by complementary element spectroscopic imaging (ESI) TEM. The TEM images can be tentatively explained by the existence of almost hexagonal structures of columnar stacks of [Cu(I)(bpy)$_2$] complexes in the polyether phase; identical copper and nitrogen net element distributions were obtained. High resolution TEM images revealed a fine structure of the copper-nitrogren rich domains which are in accordance with stacks of [Cu(I)(bpy)$_2$] complexes of B3-A1-B3 (Fig. 7) (Ref. 32). The distance of about 8Å between the interference lines seen in the electron micrograph correlates with the interatomic Cu-Cu distance in copper-bpy and copper-oligo(bpy) single crystals (Ref. 29,33). No evidence for a microphase separated system was seen for the multiblock copolymer with a 2:1 bipyridine/Cu(I) ratio. However, all copolymer systems show features typical for the controlled self-organization of bpy units in the presence of Cu(I) ions, and the triblock copolymers give highly ordered suprastructures.

CONCLUSION

Macromolecules with a regular chain architecture exhibited very specific self-organization properties, as has been demonstrated in these studies with the example of segmented block copolymers consisting of hydrophobic/hydrophilic segments and/or segments containing bipyridine or bis(bipyridine) units capable of spontaneous recognition-directed assembly to give tetra-coordinated Cu(I) complexes, i.e., a double helical structure in the later case. Furthermore, the example of specially designed graft copolymers showed that the very same morphology as obtained from segmented polyether-urethanes with tailored chain-folding hard segments can be created by macromolecules of quite different constitution. In all cases

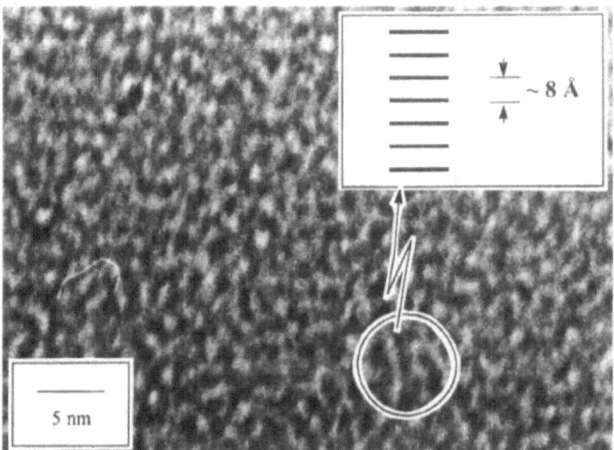

Figure 7. Elastic bright field image of Cu(I) complexed triblock copolymer B3-A1-B3; same sample as in Fig. 6, curve 2 (A1 with $R^{1,2}$ = nihil: m = 4; $\bar{n}$ = 28. B3: n = 0; R^1 = H with m^1 = 0 and R^2 = oxycarbonyl-1,4-piperazinediylcarbonyl with m^2 = 0); magnification 250.000; ultrathin film cast from a $CHCl_3$ solution. The schematic close-up illustrates a stack of bpy-Cu(I) ion complexes.

it was evident that the direct covalent linkage of different building blocks containing crystallizable units affected their segregation and molecular packing into domains and that thermodynamically preferred morphologies could be obtained when separating these building blocks by a suitable spacer.

The synthetic approach for such macromolecules with specific constitution and supermolecular structures was given in condensation polymers when precursor building blocks with different and adjustable functionalities are employed. In view of the variation of material properties with the special chain architecture it has to be explored as to which extent of irregularities as compared to the material with the precisely controlled chain architecture is tolerated in view of the balance between the creation of new materials with special properties and the synthetic efforts. However, the study of mutal effects of different building blocks on a nano to a mesoscopic scale in segregation and self-organization processes is considered to significantly contribute a broadening of the knowledge towards functional supermolecular systems, and to have a strong impact on future polymer materials science and development.

ACKNOWLEDGMENT

Financial support of these studies by the German Ministry of Research and Technology (Grants no. 03M40436 and 03C2013/4), Bayer AG (Leverkusen) and the German Science Foundation (SFB 213, Universität Bayreuth) is gratefully acknowledged. T.H. wishes to thank the Fonds der Chemischen Industrie for a doctoral fellowship.

REFERENCES

1. Harrell, L.L., Jr., Macromolecules **2**, 607 (1969)
2. Eisenbach, C.D., Nefzger, H., in "Contemporary Topics in Polymer Science", ed. by W.M. Culbertson, Plenum Publ. Corp., New York 1990, Vol. 6, p. 339

3. Eisenbach, C.D., Hayen, H., Nefzger, H., Macromol. Chem., Rapid Commun. **10**, 463 (1989)
4. Eisenbach, C.D., Stadler, E., Enkelmann, V., Macromol. Chem. Phys. **196**, 833 (1995)
5. Eisenbach, C.D., Stadler, E., Macromol. Chem. Phys. **196**, (1995), in press
6. Eisenbach, C.D., Heinemann, T., Macromolecules **28**, (1995), in press
7. Eisenbach, C.D., Heinemann, T., Macromol. Chem. Phys. **196**, (1995), in press
8. Eisenbach, C.D., Schubert, U.S., Baker, G.R., Newkome, G.R., J. Chem. Soc., Chem. Commun., **1995**, 69)
9. Eisenbach, C.D., Göldel, A., Terskan-Reinold, M., Schubert, U.S., Macromol. Chem. Phys. **196**, (1995), in press
10. Eisenbach, C.D., Hayen, H., Popp, G., Polym. Prepr. (Amer. Chem. Soc., Div. Polym. Chem.) **35/1**, 583 (1994)
11. Oertel, H., Bayer Farbenrevue **11**, 1 (1965); Chemiker Zeitung **98**, 344 (1974)
12. Estes, G.M., Seymour, R.W., Cooper, S.L., Macromolecules **4**, 452 (1971)
13. Blackwell, J., Gardner, K.H., Polymer **20**, 13 (1979)
14. Camberlin, Y., Pascault, J.P., Letoffé, J.M., Claudy, P., J. Polym. Sci., Polym. Phys. **24**, 1401 (1986)
15. Yang, W.P., Macosco, C.W., Wellinghoff, S.T., Polymer **27**, 1235 (1986)
16. Fu, B., Feger, C., MacKnight, W.J., Schneider, N.S., Polymer **26**, 889 (1985)
17. Miller, J.A., Lin, S.B., Hwang, K.K.S., Wu, K.S., Gibson, P.E., Cooper, S.L., Macromolecules **18**, 32 (1985)
18. Christenson, C.P., Harthcock, M.A., Meadows, M.D., Spell, H.L., Howard, W.L., Creswick, M.W., Guerra, R.E., Turner, R.B., J. Polym. Sci., Polym. Phys. **24**, 1401 (1986)
19. Quay, J.R., Blackwell, J., Lee, C.D., Hespe, H., Born, L., J. Macromol. Sci., Phys. B **24**, 61 (1985)
20. Eisenbach, C.D., Baumgartner, M., Günter, Cl., in "Advances in Elastomers and Rubber Elasticity", Lal, J., Mark, J.E., Eds. Plenum Publ. Corp., New York 1987, p. 51
21. Fischer, E.W., Makromol. Chem., Macromol. Symp. **12**, 123 (1987)
22. Kornfield, J.A., Spiess, H.W., Nefzger, H., Hayen, H., Eisenbach, C.D., Macromolecules **24**, 4787 (1991)
23. Boese, D., Eisenbach, C.D., Fischer, E.W., Hayen, H., Nefzger, H., Planer-Kühne, G., Reynolds, N., Spiess, H.W., Makromol. Chem., Macromol. Symp. **50**, 191 (1991)
24. Eisenbach, C.D., Nefzger, H., Enkelmann, V., Macromol. Chem. Phys. **195**, 3325 (1994)
25. Eisenbach, C.D., Heinemann, T., Ribbe, A., Stadler, E., Angew. Makromol. Chem. **202/203**, 221 (1992)
26. Eisenbach, C.D., Heinemann, T., Ribbe, A., Stadler, E., Macromol. Symp. **77**, 125 (1994)
27. Eisenbach, C.D., Stadler, E., Colloid Polym. Sci. **273**, (1995), in press
28. Lehn, J.-M., Sauvage, J.-P., Simon, J., Ziessel, R., Piccini-Leopardi, C.P., Germain, G., Declercq, J.P., van Meerssche, M., Nouv. J. Chim **7**, 413 (1983)
29. Lehn, J.-M., Rigault, A., Siegel, J., Harrowfield, J., Chevrier, B., Moras, D., Proc. Natl. Acad. Sci. U.S.A., **84**, 2565 (1987)
30. Lehn, J.-M., Makromol. Chem., Macromol. Symp., **69**, 1 (1993)
31. Eisenbach, C.D., Schubert, U.S., Macromolecules **26**, 7372 (1993)
32. Eisenbach, C.D., Göldel, A., Terskan-Reinold, M. Schubert, U.S., in preparation
33. Burke, P.J., McMillin, D.R., Robinson, W.R., Inorg. Chem. **19**, 1211 (1980)

17

MOLECULAR ORGANIZATION OF POLYSTYRENE AND POLYMETHYLMETHACRYLATE WITH FLUOROCARBON SIDE CHAINS

Sergei Sheiko,[1] Alexei Turetskii,[2] Jens Höpken,[3] and Martin Möller[1*]

[1] Organische Chemie III/Makromolekulare Chemie
Universität Ulm
D-89069 Germany
[2] Karpov Institute of Physical Chemistry
ul. Obukha 10
103064 Moscow, Russia
[3] CIBA-GEIGY Ltd
CH-4002 Basel
Switzerland

ABSTRACT

Polystyrene and polymethylmethacrylate with perfluoroalkyl side chains, $F(CF_2)_n(CH_2)_m$ have been prepared and studied with respect to structure formation and utilization for oil and water repellent coatings. As the length of the side chains exceeds $n=4$ and $m=2$, the polymers crystallize in bilayers which itself form a multilayered structure. Solution casting yielded spontaneous assembling of the bilayers parallel to the substrate plane and gave laminate coatings. Very low critical surface tension values were determined by dynamic contact angle measurements. IR absorption and X-ray photoelectron spectroscopy revealed a highly ordered layer of fluorocarbon segments at the air/polymer interface. This is consistent with scanning force microscopy measurements showing a surface formed by regularly packed CF_3-groups. Anisotropic variations of the linear spacing in directions parallel and perpendicular to the layers were detected by X-ray diffraction measurements at elevated temperatures and upon a glass transition.

[*] Address correspondence to Dr. Möller.

Macromolecular Engineering, Edited by M.K. Mishra et al.
Plenum Press, New York, 1995

INTRODUCTION

Fluorinated polymers are widely employed as coatings for the modification of metal as well as hydrocarbon polymer surfaces, imparting soil resistance, water- and oil-repellent properties, and low surface friction [1]. Because fluorocarbon polymers do not adhere to most nonfluorinated surfaces, special techniques have to be adopted in order to supply sufficient adhesion between the substrate and the coating. The adhesion problem can be eliminated when hydrocarbon polymers or copolymers with monomers carrying highly fluorinated side chains are used as surface modifiers for hydrocarbon polymers [2-8]. The polymer backbone can be tailored to provide good adhesion to the substrate, while the fluorocarbon segments are not compatible with the hydrocarbon matrix and concentrate at the coating/air interface. The effectivity of the surface protection depends (i) on the coverage of the surface with fluorocarbon segments, and (ii) on the degree of ordering in the surface layer. Studies on perfluoroalkylalkanes demonstrated that strong incompatibility of per-fluorinated molecule segments with hydrocarbons can cause peculiar self-organization [9,10]. Recently, surface active polymers were obtained in which fluorocarbon substituents are attached to the main chain via flexible hydrocarbon spacer [4-8]. Critical and dispersion surface tensions were attained as low as 6 mN/m and 9 mN/m respectively. This communication reports on molecular ordering of perfluorinated side chains of polystyrene and polymethylmethacrylate in bulk and in the air/coating interface.

EXPERIMENTAL

Synthesis of p-1H,1H,2H,2H-perfluoroalkyloxymethylstyrene, $F(CF_2)_n(CH_2)_m$-OCH_2-C_6H_4-$CH=CH_2$ with n=4, 8; m=2 and perfluoroalkyl-alkylenemethacrylate $F(CF_2)_n(CH_2)_m OOC$-$(CH_3)C=CH_2$ with n= 8, 10, 12 and m= 2, 6 has been described before [6,13]. Radical polymerization was carried out with a small amount of solvent added in order to avoid incomplete conversion due to vitrification during polymerization.

A Perkin-Elmer DSC-7 equipped with a PE-7700 computer and TAS-7 software was used to monitor the thermal transitions at scan rates of 10 K/min. Sample weights were typically chosen around 10 mg. Cyclohexane and indium were used as calibration standard. The onset of the recorded endotherm upon heating was taken as the transition temperature.

X-ray diffraction patterns have been recorded with films cast from 1 wt.% trichloro-trifluoroethane solution on a polyimide film (Kapton, Du Pont) using Ni-filtered CuK_α radiation and a flat camera with a sample to film distance of 90 mm. Powder diffractograms at elevated temperatures were recorded by means of a Guinier-Simon camera with mono-chromatic CuK_α radiation with specimen filled into a quartz capillary. Samples were heated by means of a constant flow of nitrogen gas and the temperature was controlled by a thermocouple introduced into the nozzle within the direct vicinity of the specimen. The temperature-variation in the sample compartment was checked with an additional thermo-couple and was found to be ±2K. Prior to measurements, the sample was heated at a rate of 5K/min to the required temperature, followed by annealing for 30 min. Exposure times were approximately 3 h. Optical density data were collected from the photographically obtained patterns using a linear microdensitometer LS20 (Delft Instruments) controlled by SCANPI software.

Surface tension, AFM and XPS experiments were performed on thin films spin cast at 2000 rpm on glass slides from 1 wt.% trichlorotrifluoroethane solution. The average roughness of the surface was controlled by AFM and found to be 3-4 Å.

Dynamic contact angles Q were measured against different liquid n-alkanes at 25°C according to the Wilhelmy method [14] by means of a Krüss K12 tensiometer. The average advancing and receding contact angles were calculated by regressive extrapolation of the force to zero immersion depth. A Platinum plate with a perimeter of 59.5 mm was used for the surface tension measurements of the liquids.

Angle dependent XPS experiments were done with a hemispherical analyser and a MgK_α X-ray source. The fraction of the fluorine atoms at the interface was determined from the C 1s, O 1s and F 1s spectra with incidence angles 0° and 60° corresponding to efficient sampling depths of 45 Å and 22 Å respectively.

IR absorption spectra of solution cast films were measured at room temperature with a FTS-60 (Digilab, Bio-Rad) spectrometer at 2 cm^{-1} resolution. A germanium plate and gold coated glass slides were used as substrates for the measurements of transmission (electric field vector **E** is parallel to the substrate) and grazing incidence reflection (**E** is perpendicular to the substrate) spectra respectively.

A Nanoscope 2 (Digital Instruments, CA, USA) was employed to visualize the surface structure of the spin cast films at ambient conditions with a Si_3N_4 tip mounted on a cantilever with a spring constant 0.06 N/m. The applied force (repulsive) was minimized to 5 nN to reduce surface deformation.

RESULTS AND DISCUSSION

1. Polymer Powders

Data from the differential scanning calorimetry experiments are listed in Table 1. They can be summarized as follows: (i) All polymers displayed an endothermic isotropization transition upon heating; (ii) the perfluoroalkyl-alkylenemethacrylates revealed in most cases a solid state transition prior to the isotropization; (iii) as far it could be observed, a glass transition was found to be lower than that of pure PS and PMMA.

The melting transition vanished and a glass transition became easily detectable when the number of CF_2-groups was less than n=6 and also when the fluorinated side chains got diluted by copolymerization with styrene or methacrylate units [7]. From these observations, ordering of the side chains can be deduced for samples precipitated from solution and cooled from the melt as well.

Figure 1 shows X-ray powder patterns obtained at ambient temperature for F8H2-PMA, F8H2-PS and F12H6-PMA. In all cases, a series of small angle reflections were observed with a halfwidth of 0.4 degrees 2θ and whose intensity decreased with 2θ. These

Table 1. DSC data, heating rate 10 K/min

Sample name	T_g °C	ΔCp J/g K	T_d °C	ΔH J/g	T_i °C	ΔH J/g
F8H2PS[b]	55	2.4	—	—	82	2.1
F8H2PMA[a]	—	—	79	2.5	90	2.7
F8H2PMA[b]	8	0.07	—	—	88	4.3
F10H6PMA[a]	—	—	—	—	104	14.1
F10H6PMA[b]	—	—	97	8.5	107	2.1
F12H6PMA[a]	—	—	115	0.5	139	22.6
F12H6PMA[b]	84	0.1	—	—	137	23.8

g-glass transition, d-disordering, i-isotropization, a)- 1st heating, b) -
2d heating

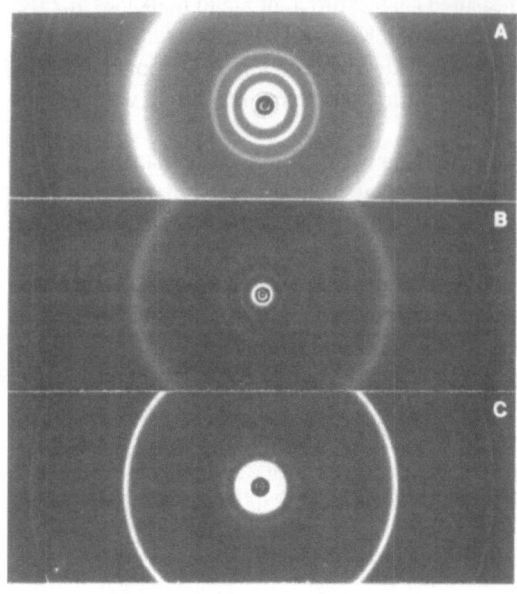

Figure 1. X-ray diffraction pattern of the polymers containing different semifluorinated side chains: F8H2-PS (a); F8H2-PMA (b); F12H6-PMA (c).

reflections correspond to the first three orders for a spacing L of 3.1 nm for F8H2-PMA, 3.7 nm for F8H2-PS and 5.6 nm for F12H6-PMA. All studied polymers show in addition a wide angle reflection with a halfwidth of 2 degrees 2θ which is centred at 17.6 degrees 2θ and corresponds to a spacing of $d = 0.50$ nm.

The L-values are about twice the length of the respective side chains in their extended conformation suggesting formation of bilayers with the axis of the side chains oriented perpendicularly to the basic layer plane. The reflection with the spacing of $d = 0.50$ nm can be assigned to the lateral packing of the fluorinated side chains. A similar value of about 0.49 nm is characteristic for the hexagonal dense packing of n-perfluoroalkanes and corresponds to the scattering at the (100) planes [15].

Powder diffractograms of F8H2-PS have been measured at different temperatures in order to elucidate structural changes upon heating. Fig. 2 shows that heating results in a shift of the strong small angle reflections towards wider angles in combination with a gradual decrease in intensity for the higher order reflections. The latter effect indicates a distortion of the spatial long range correlation of the layers.

Above the melting point at 85°C, the diffractogram in Fig. 2 retained a small angle reflection, which indicates the bilayer ordering in the melt state. The poor degree of the ordering is seen from the low intensity and the large halfwidth of the reflection at 95°C. The reflection indicate that the bilayers assemble even in the molten state.

In contrast to the small angle scattering, the reflection with $d = 0.50$ nm shifts to smaller angles and increases in its halfwidth. The broadening might be caused by a peak splitting which is clearly observed at higher temperatures.

Anisotropic temperature coefficients in the linear dimensions might be expected because of the alternating hydrocarbon and fluorocarbon layers with their particular packing behaviour. Although the data indicate shrinkage normal to the bilayers, the volume temperature coefficient may well be positive. Dilatometry experiments have to be performed to answer this question [16].

For F8H2-PS, the temperature dependence of the L-values is shown in Fig.3. Like in the case of the other polymers, the heating results in a decrease of the L spacing. Unusually,

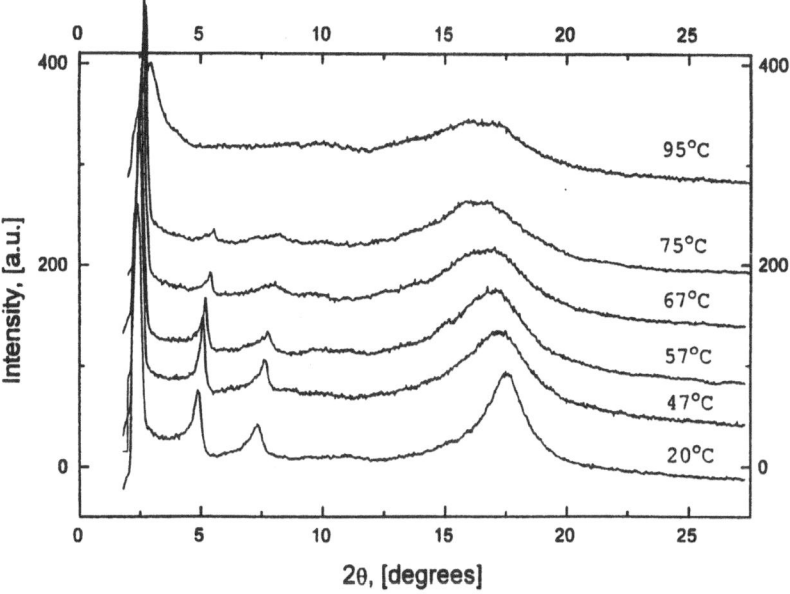

Figure 2. X-ray diffractograms of F8H2-PS powder measured at elevated temperatures.

the slope changes rather sharply at the glass transition, i.e., the negative linear expansion coefficient of L decreases further above the glass transition temperature.

The temperature dependent transformations described here are not fully reversible. Diffractograms obtained after cooling the melt to ambient temperature did not coincide completely with those for the original samples: The L-values became smaller as 36 Å (the

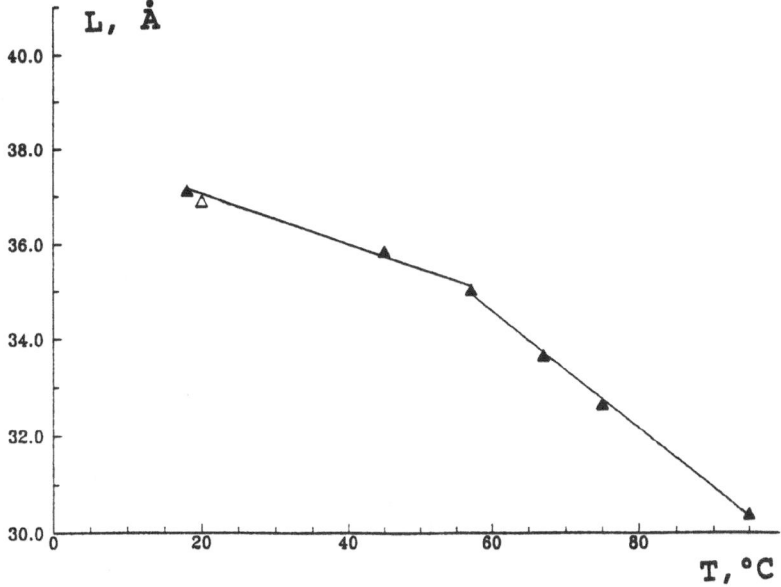

Figure 3. Temperature dependence of the layer spacing (L).

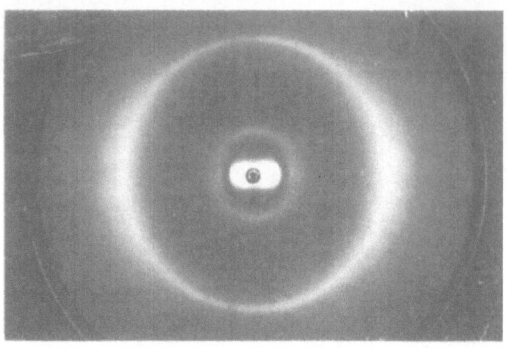

Figure 4. X-ray diffraction patterns of the F12H6-PMA film cast from solution on a polyimide substrate.

original value was 37 Å), while the maximum of the wide angle reflection was positioned at 0.505.

2. Solution Cast Films

The structure of solution cast films was studied employing a polyimide film as a substrate. Fig.4 shows the X-ray diffraction patterns obtained for a F12H6-PMA film, which is representative for the patterns displayed by the other polymers. In this experiment the primary beam was parallel to the film surface and substrate plane. The substrate normal is defined to be the equatorial direction of the patterns.

It can be seen that the intensity of the small angle reflection with its higher orders concentrates on the meridian. These reflections correspond to a well developed layered structure with a thickness $L = 5.6$ nm. The orientation of the reflection corresponding to the d-spacing of 0.50 nm is less pronounced and can be described as a ring with an intensity maximum at the equator.

Fig. 5 shows results on the wetting behavior from dynamic contact angle experiments. The inverse square root of the liquid surface tension is plotted versus the cosine of the advancing contact angle Θ to allow extrapolation of the dispersive surface tension [17].

An extremely low value of $\gamma^d = 8.7$ mN/m was extrapolated for F10H2-PMA. Correspondingly, the critical surface energy according to Zisman [18] resulted to 6 mN/m. Such low surface energies have been reported only for mono-layers of perfluorocarbon acids or thiols [19], where the fluorocarbon chains are packed to form a regular surface from CF_3-groups.

Consistent with these results, XPS measurements demonstrated an excess of fluorine in the surface layer. The fractions of fluorine atoms in dependence of the depth are presented in Table 2. The doubling of atomic ratios F/C and F/O found at reducing the sampling depth is consistent with a model where the extended side chains cover the surface completely and are oriented perpendicular to the surface.

IR spectroscopy was employed to monitor the orientation of the side chains (Figure 6) based on the bands which have been assigned to CF_2 stretching modes, i.e., 1150, 1205 and 1240 cm^{-1} for a helical of a $(CF_2)_n$ chain [20]. The conformation of short n-perfluoroalkanes was shown to be all-trans zig-zag slightly distorted from planarity [21].

Yet, the polarization of the CF_2 stretching modes is not determined unambiguously for the side chain fluorinated polymers [5,22]. It was proposed, that there are two sets of modes polarized perpendicular to each other and to the axis of the planar chain. The 1150 and 1240 cm^{-1} modes were assigned to one set, and the 1210 cm^{-1} to the other one. The origin of the modes between 1300 and 1400 cm^{-1} had not been defined so far [22]. However, these

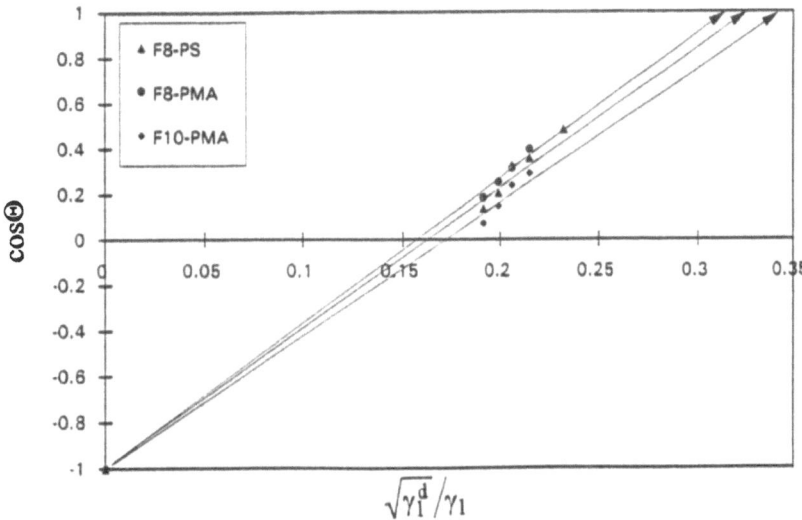

Figure 5. Girifalco-Good-Fowkes-Young evaluation [15] of the dispersion surface energy of F8H2-PS, F8H2-PMA and F10H6-PMA.

bands can be easily assigned to CF_3 modes by comparison with the spectra of perfluorooctyl- and decyl iodide as model compounds.

Different relative intensities in transmission and reflection spectra in Fig. 6 indicate a rather strong orientation of the fluorinated side chains in the cast polymer films studied here. Qualitatively, the orientation of the side chains can be determined to be almost perpendicular to the surface plane [22].

The question to which extent the oriented side chains in the surface layer are laterally ordered can be answered by SFM experiments. Figure 7 shows a scanning force micrograph of F8H2-PS with a regular pattern on the molecular scale. Bright spots in the micrograph correspond to elevations. These elevations can be assigned consistent with the spectroscopy observations to -CF_3 end groups which are packed with a 7-9 Å periodicity. The strongly distorted structure does not allow to identify a unit cell and lattice parameters rigorously. The 7-9 Å spacing are not in agreement with a 5.7 Å interchain distance expected for hexagonally packed perfluoroalkyl chains [15,23]. However, the discrepancy might be caused by deformation of the surface during the scanning process.

Table 2. Atomic density ratios of F8H2-PMA at different sampling depths

a [deg]#	d [Å]+	F/C	F/O
60	22	2.4	18
0	45	1.25	9.5
side chains*		2.13	17
polymer*		1.21	8.5

- incidence angle, + - a sampling depth

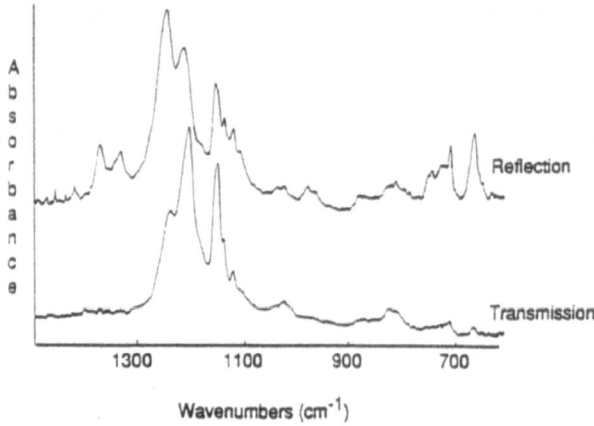

Figure 6. IR absorption spectra of the solution cast F8H2-PS measured in transmission (**E** parallel to the substrate plane) and in reflection (**E** is perpendicular to the substrate plane).

CONCLUSION

In summary, polystyrene and polymethylmethacrylate modified with fluorocarbon side chains lead to a layered molecular organisation in bulk and at the air/polymer interface. Semifluorinated side chains form bilayers which are organized in multilayers upon crystallisation from solution and from the melt.

In particular, solution cast films reveal a laminate structure where bilayers, which might be already formed in solution, organize themselves parallel to the substrate. The CF_3 end groups assemble at the coating/air interface and result in an extremely low critical surface tension. The spontaneous assembling process does not require neither special amphiphilic molecules being employed in LB procedure nor specially organised substrates which are necessary in adsorption and epitaxy deposition techniques[24].

X-ray measurements at elevated temperatures revealed unusual anisotropic change of the spacing in directions parallel and perpendicular to the layers. Heating results in a decrease of the layer spacing L while the distance between the side chains increases. The spacing L shows a negative change of the linear expansion coefficient upon the glass transition. The anisotropic behavior upon heating is a consequence of the layered structure

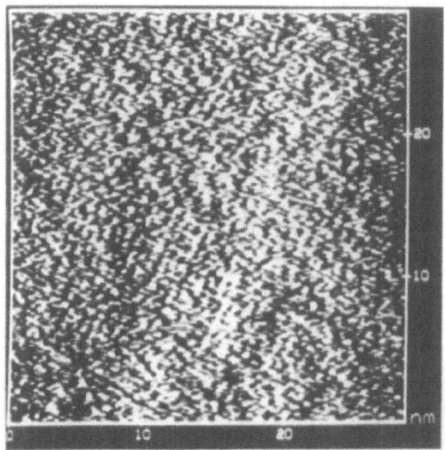

Figure 7. Atomic force micrograph of the surface of a poly(p-1H,1H,2H,2H-perfluorodecyloxy-methylstyrene) spin-coated on a glass slide.

where alternating hydrocarbon and fluorocarbon sublayers exhibit different temperature and phase behavior. However, dilatometry measurements are required to elaborate temperature variations of the volume.

ACKNOWLEDGMENTS

To S. Magonov for performing AFM measurements with atomic resolution.

REFERENCES

1. Kissa, E. "Fluorinated Surfactants, Synthesis - Properties - Applications", Marcel Dekker, New York, 1984
2. Fox, H.W.; Zisman W.A. *J. Colloid Sci.* **1952**, *7*, 109.
3. Kobayashi, H.; Owen, M.J. *Macromolecules* **1990**, *23*, 4929.
4. Jonson, R.E.,Jr.; Dettre, R.H. *Polym. Prepr. (Am. Chem. Soc., Div. Polym. Chem.)* **1987**, *28*, 48.
5. Sneider, J.; Erdelen, C.; Ringsdorf, H.; Rabolt, J.F. *Macromolecules* **1989**, *22*, 3475.
6. Höpken, J.; Möller, M. *Macromolecules* **1992**, *25*, 1461.
7. Höpken, J.; Sheiko, S.; Czech, J.; Möller, M. *Polym. Prepr. (Am. Chem. Soc., Div. Polym. Chem.)* , 937.
8. Katano, Y.; Tomono, H.; Nakajima, T. Macromolecules **1994**, *27*, 2342.
9. Höpken, J.; Möller, M. *Macromolecules* **1992**, *25*, 2482.
10. Russel, T.P.; Rabolt, J.F.; Twieg, R.J.; Siemens, R.L.; Farmer, B.L. *Macromolecules* **1986**, *19*, 1135.
11. Viney, C.; Russel, T.P.; Depero, L.E.; Twieg, R.J. *Mol. Cryst. Liq. Cryst.* **1989**, *168*, 63.
12. Mahler, W.; Guillon, D.; Skoulios, A. *Mol. Cryst. Liq. Cryst.* **1985**, *2*, 11.
13. Höpken, J.; Faulstich S., Möller, M. *Mol. Cryst. Liq. Cryst.* **1992**, *210*, 59.
14. Smith, L.; Doule, C.; Gregoris, D.E.; Andrade, J.D. *J. Appl. Polym. Sci.* **1982**, *26*, 1269.
15. Bunn, C. W.; Howells, E. R. *Nature*, 1954, *18*, 549; Schwichert, H.; Kimmig, M.; Strobl, G.; *J. Chem. Phys.* **1991**, *95*, 2800.
16. Lermann, E.; Sheiko, S.; Möller, M. Structure and Phase behaviour of Perfluorododecyleicosane (in preparation).
17. according to $\cos\Theta = -1 + 2(\gamma_s^d)^{1/2}\gamma_l^{-1/2}$, with γ_s^d = dispersive surface energy of the solid, and γ_l = surface tension of the wetting liquid; Fowkes, F. M. *J. Phys. Chem.*, **1962**, *66*, 382.
18. Shafrin, E.G.; Zisman, W.A. *J. Phys. Chem.* **1960**, *64*, 519.
19. Dann, J.R. *J. Colloid Interface Sci.*, **1970**, *32*, 302.
20. Rabolt, J.F.; Fanconi, B. *Macromolecules*, 1978, *11*, 140.
21. Campos-Vallete, M.; Rey-Lafon, M.; *J. Mol. Struct.* **1983**, *101*, 23.
22. Chidsey, C.E.D.; Loiacono, D.N. *Langmuir* **1990**, *6*, 682.
23. Magonov, S.N.; Kempf, S.; Kimmig, M.; Cantow, H.-J. *Polym. Bulletin* **1991**, *26*, 715.
24. Wittmann, J.C.; Lotz, B. Prog. Polym. Sci., **1990**, *15*, 90.

18

SYNTHESIS OF 1-10 MICRON POLYMER PARTICLES BY THE NEW GRAFTING-PRECIPITATION METHOD (GPM)

W. De Winter, D. Timmerman, and R. Declercq

Central Research Dept.
Agfa-Gevaert
B-2640 Mortsel
Belgium

INTRODUCTION

In the technology of polymer synthesis in heterogeneous (preferentially aqueous) media, the techniques of emulsion- and suspension polymerisation are well known (10). Emulsion polymerisation results in very small polymer particles (size 0.06 - 0.2 μm) and suspension polymerisation yields much larger polymer particles (size $\geq$ 20 μm).

The preparation of polymer beads of intermediate size (0.2 - 20 μm) requires the implementation of alternative or modified polymerisation technologies as, e.g., seeding polymerisation (5, 12, 13, 32), precipitation polymerisation (16), modified suspension polymerisation (19, 25), or, most frequently described, dispersion polymerisation (1, 2, 4, 11, 20, 21, 33) (see Table 1).

All these techniques yield useful results, although their industrial application may still show some drawbacks. Seeding polymerisation, for example, involves a complicated

Table 1. Polymerisation techniques in aqueous medium for the preparation of polymer particles

Emulsion polymerisation	Other methods	Suspension polymerisation
Latex	- seeding polymerisation - homodisperse particles - processing complicated/several steps - precipitation polymerisation - heterodisperse particles - reproducibility technologically complex - dispersion polymerisation - heterodisperse particles - originally - many improvements by different ways	coarse dispersion
Particle size : 0.05-0.2 μm $\Rightarrow$	0.2-20 μm $\Rightarrow$	$\geq$ 20 μm

Macromolecular Engineering, Edited by M.K. Mishra et al.
Plenum Press, New York, 1995

Table 2. Parameters for dispersion polymerisation

| Parameter | Action | Results on | |
		Particle size	Homodispersity
Polarity of reaction medium	↑	↓	—
Monomer concentration	↑	↑	↓
Degree of conversion	↑	↑	—
Initiator concentration	↓	↑ - ↓	↑↓
Polymeric stabiliser concentration	↓	↑	↓
Reaction temperature	↑	↑↓	↑
Stirring intensity	↑	↑	↓

technology and is quite expensive, and therefore should only be used for specific small volume applications. Precipitation polymerisation based on crosslinking, on the other hand, is mainly described in organic media, and therefore not so well suited for production under present-day environmental regulations. The modified suspension-polymerisation methods yield more homodisperse particles than those obtained via the regular suspension-polymerisation technique; however, the obtained particle-size distribution is still too broad for most specialty applications envisaged.

Most promising, and also most widely described in literature is the dispersion polymerisation technique. In dispersion polymerisation the reaction mixture starts out as a homogeneous solution, and the resulting polymer precipitates as spherical particles stabilised by a steric and/or electrostatic barrier of dissolved polymer. Specific graft- or block copolymers seem to be particularly useful for this purpose. The influence of several reaction parameters on the obtained particle size and distribution has been determined in various systems. Although this influence may depend strongly on the system studied, some relationships seem to be universally valid, even when some contradictions still remain (see Table 2).

For the explanation of certain observed phenomena, e.g. why some particular reaction conditions yield an extremely narrow particle-size distribution, no satisfactory hypothesis could so far be formulated in the literature (18, 22).

In the perspective of further improvements and understanding, we developed in our research laboratories a proprietary GPM (Graft Precipitation Method), related to the described dispersion polymerisation technique, for the preparation of particularly homodisperse, spherical particles in the range of 1 to 10 μm in diameter, and usable in our photographic materials. This GPM technique, which had to be extrapolatable to an industrial preparative scale, will be further described in this publication (28, 29).

THE AGFA-GEVAERT MANUFACTURING PROCESS FOR MICROPEARLS

Description of the GPM (Graft Precipitation Method)

The polymerisation begins as a regular solution polymerisation in a mixed aqueous organic medium, in which the polymeric chains start to precipitate once they reach a given size. In order to ensure that the formed polymer particles have a controllable spherical shape, rather than an amorphous powder form, and that they do not show any tendency to

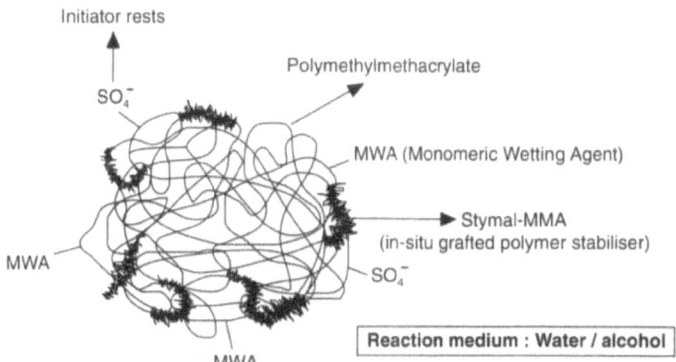

Figure 1. Schematic drawing of a polymer particle.

agglomerate, an extra sterical and electrostatical stabilizer is required. The GPM technique relies on the use of an ionisable polymeric stabilizer-emulsifier that becomes grafted onto the growing polymer grain during its preparation. Furthermore, a monomeric wetting agent (MWA), that becomes incorporated at the outer surface of the growing polymeric particle, may be used as an additional stabilizer.

In this publication the GPM technique will be illustrated on the basis of the system composed of methyl methacrylate (MMA) as a monomer, copoly(styrene/Na-maleate) (= stymal) and 4-(12-(methacryloylamino)-dodecanoylamino)-benzenesulphonic acid (= MWA) (8) as stabilizers, potassium persulphate as an initiator, and a water-alcohol mixture as the reaction medium (see fig. 1). These four ingredients are also the most important reaction parameters.

An accurate mutual adjustment of these four parameters is required in order to control the particle size and the homogeneous degree of dispersion. The influence of all these parameters on the conditions of the synthetic process will be discussed in the next chapter. Furthermore it should be stated that, next to these parameters, also the reaction temperature and -profile, and the mode and intensity of agitation play an important role in determining the optimal conditions of the reaction (15). The influence of these parameters will be discussed partly in this publication, and partly in a forthcoming one.

Standard Experimental Procedure

A double-walled 20-L reaction vessel, equipped with a stirrer and a thermometer is charged at room temperature under a nitrogen blanket with a solution of 156.6 g of copoly(styrene/maleic anhydride) (adjusted with NaOH to pH 7) and 48.6 g of potassium persulphate in 6 liter of water, and stirred continuously at 140 rpm.

A solution of 3 kg of methyl methacrylate in 5.4 L of ethanol is added. The reaction mixture becomes turbid and is stirred for 120 minutes at room temperature (prereaction). Then the reaction vessel is heated gradually with water of 65°C up to 60-65°C over 30 minutes. Above 30°C the reaction mixture becomes transparent, and above 55-60°C it turns turbid again, showing the onset of polymerisation.

The exothermic polymerisation reaction increases the temperature to 80°C in about 45 minutes. A milky white dispersion is obtained. Then the temperature starts dropping over about 30 minutes back to 65°C, where it is kept for 16 hours, and then cooled to 30°C, all under continuous stirring.

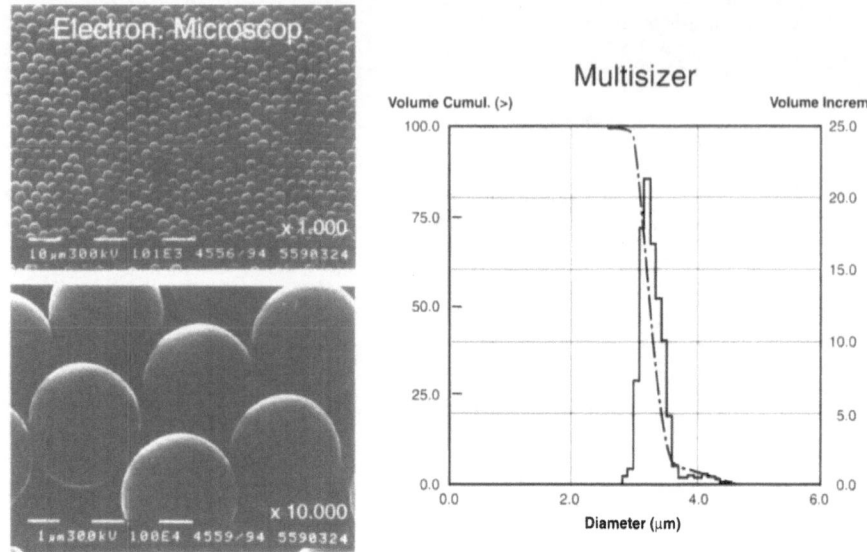

Figure 2. Measurement of particle size and distribution.

After filtration through a Nylon cloth with meshes sizing 75 x 75 μm, 13.2 kg of a polymer-particles dispersion is obtained (yield 98.4 %).

Measurement of Particle Size and Distribution (See Fig. 2).

Three methods have been used :

Nanosizer. Coulter Nanosizer measures the average size of particles in suspensions between 0.04 and ca. 5 μm diameter. The measuring principles used are those of Brownian motion and autocorrelation spectroscopy of scattered laser light.

An indication of the size distribution width is also given as a single digit between 0 and 9. A monosized particle suspension has 0 or 1, whereas 8 or 9 indicates that the ratio of largest to smallest particles is large.

Multisizer. The Coulter Multisizer II is a single particle counter that employs the Coulter electrical impedance method to provide a particle distribution analysis within the overall range of 0.4 to 1200 μm. Each result is displayed graphically as a percentage of channel content, which can be selected to represent either volume (weight), number (population) or surface area.

- dma = number median
- dmv = volume median
- polydispersity = dmv/dma

Electron Microscopy. The scanning electronic microscope (Philips Type 505) makes a topographical image of a surface formed by a gold-covered monolayer of the microparticles to be measured.

Table 3. Influence of the water/ethanol ratio

| Sample no. | Ethanol/water (% by vol.) | Particle properties | | | | | Remarks |
| | | Nanosizer | | Coulter | | | |
		Size (μm)	Dispersitydegree	dm_a	dm_v		
Ref.	0/100	0.06	—	—	—		regular emulsion polymerisation
BB	30/70	1.1	3-4	0.81	2.36		—
BA	40/60	1.4	1	0.85	0.98		—
AV	47/53	2.3	0	1.85	2.23		—
BD	50/50	2.3	0	2.30	2.56		—
CL	53/47	2.8	0	2.35	3.10		—
CI	54/46	3.2	0	2.18	3.38		see fig. 2
BF	55/45	2.4	0	1.03	3.26		large amount of precipitate
AY	60/40						amorphous powder

Reaction conditions: [MMA] - 2.07 mole/L; [$K_2S_2O_8$] 12.5 x 10^{-3} mole/L; copoly(styrene/Na-maleate) 5.2 g/100 g MMA.

EXPERIMENTAL RESULTS

Influence of Different Reaction Parameters

Polarity of Reaction Medium. In a first series of experiments the most suitable organic solvent to be added to the aqueous medium had to be selected. It was concluded that ethanol outperformed other alcohols (methanol, iso- and n-propanol) and other candidates as, e.g., dioxane, tetrahydrofuran, etc. with regard to applicability in the GPM process for obtaining the desired polymer beads.

The results of the experiments with varying ratio water/ethanol are represented in table 3.

Upon using only water as a reaction medium a regular emulsion polymerisation is performed, yielding small (nanometer-size) latex particles. Adding more and more ethanol to the reaction medium gives rise to increasing particle size; the best particle shape (spherical) and homodispersity are obtained (in the particular system described here) with a water/ethanol ratio of 53/47, for obtaining beads of 2.3 μm diameter, without precipitate formation, and with a very narrow size distribution.

Further increase of the alcohol content results in an ever increasing degree of precipitate formation, ending up in a completely formless mass of polymer.

Polymeric Stabilisation. Homodisperse spherical micro-size particles could not be obtained without the use of a polymeric stabiliser. In a first series of experiments (see table 4) a large series of possible polymeric stabilisers were tested.

Different types of polymers, e.g., polyvinyl alcohol derivatives, polyethylene oxides, polyvinylpyrrolidones, copolymers with acrylic acid, could be used in the GPM process for the synthesis of the desired beads. The best results, however, were obtained with stymal as a polymeric stabiliser.

In a second series of experiments, only with this stymal as a stabiliser, the optimal amount of it was determined (see table 5).

Table 4. Influence of the type of polymeric stabiliser

Sample no.	Tested polymer	Particle properties*			% of amorphous precipitate
		Size μm	Degree of dispersity	Shape	
A	Copoly(vinyl alcohol/vinyl acetate) 88-12	1.52	3	spherical	10
B	" 75-25	2.53	2-3	"	0.3
C	polyethyleneglycol	1.50	0	"	0.5
K	polyvinylpyrrolidone	0.35	3	"	< 0.3
AV	co(styrene/Na-maleate) (= stymal)	2.3	0	"	0
H	polystyrene-Na-sulphonate	—	—	—	100
E	copoly(styrene/Na-acrylate)	1.9	0	spherical	0.3
X	copoly(methyl methacrylate/Na-methacrylate)	—	—	—	100

Small amounts of stabilising polymer (≤ 2.5 g per 100 g of monomer) lead to large amorphous particles, while too large quantities of stabilising polymer ($\geq$ ca. 7 g per 100 g of monomer) do not satisfy either. The optimal quantity, under the conditions studied here, seems to be situated between 3 and 7 g of polymeric stabiliser per 100 g of monomer.

In a third series of experiments a small part of the methyl methacrylate monomer (0-5 mole%) was substituted by the monomeric wetting agent (MWA) 4-(12-(methacryloylamino)-dodecanoyl-amino)-benzenesulphonic acid. It was observed that the use of MWA instead of the polymeric stabiliser did not lead to the production of the desired micron-size spherical homodisperse particles. When using MWA in combination with the polymeric stabiliser, however, an additional stabilising effect of the MWA was observed. So, the total amount of polymeric stabiliser necessary for obtaining the suitable beads could be significantly decreased by incorporating small amounts of MWA, viz. ca. 2 mole % versus the total monomer content.

Prereaction Conditions. A prereaction is performed by heating the polymeric stabiliser in the presence of the free radical initiator before the so-called graft-precipitation polymerisation.

Table 5. Influence of the amount of stabilising polymer

Sample no.	g of stabilising polymer per 100 g of MMA	Particle Properties**			% of amorphous precipitate
		Size μm	Degree of dispersity	Shape	
HA	0	—	—	—	100 sticky mass
AC	1.3	10-50	—	irregular	not filtratable
V	2.6	2,1	0	spherical	3.2
AV	5.2	2,3	0	"	< 0.3
AB	7.8	1,6	0	"	2
BL	10	—	high	"	very heterogeneous

- *Reaction conditions*: [MMA] - 2.07 mole/L; [$K_2S_2O_8$] 12.5 x 10^{-3} mole/L; ethanol/water 47/53.
* Copoly(styrene/Na-maleate) = stymal
** Nanosizer measurements.

Table 6. Influence of the preactivation of the stabiliser precursor

Sample no.	Preactivation time (hr)	Particle properties (Coulter)		
		dm_v	dm_a	dm_v/dm_a
CL	1	3.17	2.53	1.25
CM	2	2.87	2.65	1.08
CN	3	2.82	2.60	1.08
CO	4	2.81	2.65	1.06

Reaction conditions :[MMA] = 2.07 mole/L; EtOH/H_2O = 53/47 % by vol.; [stymal] = 7.0 % of MMA; [$K_2S_2O_8$] = 12.5 x 10^{-3} mole/L. Reaction temperature = 70 °C

Upon varying the duration and the temperature of this prereaction from 1 to 4 hours, and from room temperature to 70°C, different particle sizes and distributions were obtained as presented in Table 6.

It was observed that a preactivation time of at least 2 hours was necessary in order to obtain homodisperse polymer particles. A further increase in prereaction time hardly had any additional effect on the particle-size distribution.

Initiator Concentration. Table 7 represents the results of the experiments with various initiator concentrations.

Potassium persulphate was selected over ammonium persulphate as most suitable free radical initiator, because of better stability at room temperature. Other free radical initiators, e.g., azobis-isobutyronitrile, benzoyl peroxide, hydrogen peroxide, and di-t.-butyl peroxide did not yield better results. The optimal amount of potassium persulphate to be used in the particular system described here lies in the range from ca. 6 to 20 x 10^{-3} mole per L. Lower as well as higher concentrations lead to the production of more heterodisperse particles, the formation of agglomerates, and non-spherical shapes.

Monomer Concentration. Table 8 represents the results of experiments of the system MMA, stymal, and potassium persulphate in water-ethanol with changing concentration of the monomer (MMA).

Table 7. Influence of the initiator concentration

Sample no.	[$K_2S_2O_8$] mole . 10^{-3}			Particle properties				Remarks
	per L	per 100 g MMA	pH	Nanosizer		Coulter		
				Size μm	Dispersity degree	dm_a	dm_v	
BT	1	0.48	—	—	—	—	—	shapeless mass
BU	5	2.41	7	2.36	0	1.58	2.55	some precipitate
BX	12.5	6.0	5.3	2.38	0	1.74	2.35	spherical
BV	20	9.64	4.9	2.16	0	1.80	2.25	"
BW	40	19.28	3.3	1.79	1	1.0	2.20	agglomerates

- *Reaction conditions*: [MMA] - 2.07 mole/L; copoly(styrene/Na-maleate) : 6 g per 100 g MMA; water/ethanol 53/47.

Table 8. Influence of the MMA-monomer concentration

| Sample no. | [MMA]mole/L | Nanosizer | | Coulter | | Remarks |
		Size (μm)	Dispersity degree	dm_a	dm_v	
AW	1.0	2.15	0	1.91	2.03	spherical, no precipitation
AV	2.07	2.3	0	1.85	2.23	"
AX	2.5	2.1	3-4	0.89	4.0	small amount of precipitate

Reaction conditions: Copoly(styrene/Na-maleate) 5,2 g/100 g MMA; [$K_2S_2O_8$] 12.5 x 10^{-3} mole/L; ethanol/water 47/53.

Higher MMA concentrations show an evolution towards a higher degree of heterodispersity, and worse filtrability due to more agglomerate formation and to the presence of shapeless amorphous polymer pieces. Concentrations ranging from 1 to 2.2 mole/L seem to yield the best results with regard to the desired properties.

DISCUSSION OF THE RESULTS

Theory of Particle Nucleation and Growth Processes

All the reaction parameters described in the experimental part affect the particle-nucleation process and hence the particle size and size distribution of the final polymer beads. A good knowledge of the general nucleation process is indispensable to understand and eventually predict the influence of the different reaction parameters on the particle size. In addition, by studying the particle size with respect to a chosen set of reaction parameters, a more accurate nucleation mechanism can be proposed for the system studied.

The coagulative nucleation mechanism, recently proposed by Morrison, Gilbert and Napper and described by Croucher and Winnik (7) combines the theory of homogeneous nucleation, proposed by Fitch and Tsai (9) for the emulsion polymerisation, with the theory of coagulation of the primary particles formed in the beginning of the nucleation event. In the coagulative nucleation theory it is postulated that the initiation and initial polymerisation both take place in solution. During this stage of the polymerisation, often referred to as the induction period, in our system soluble poly-MMA-oligomers are formed. The reaction medium stays transparent during this induction period.

The poly-MMA-oligomers grow to the point where they reach their solubility limit. At this point a colloidally unstable species is formed. The authors call this entity a precursor. Once such a precursor has been formed it can grow by propagation of the polymer chain, and/or undergo homocoagulation with another precursor dependent on the ratio of reaction rates. Alternatively, stabilisation can take place when the precursor collides with an amphiphilic stabiliser, which can then be adsorbed on the surface of the particle.

Homocoagulation of the precursors and stabiliser adsorption lead to the formation of stable nuclei, which can grow further by swelling with the monomer. The nucleation phase is terminated when the newly formed soluble oligomers are captured by a growing particle

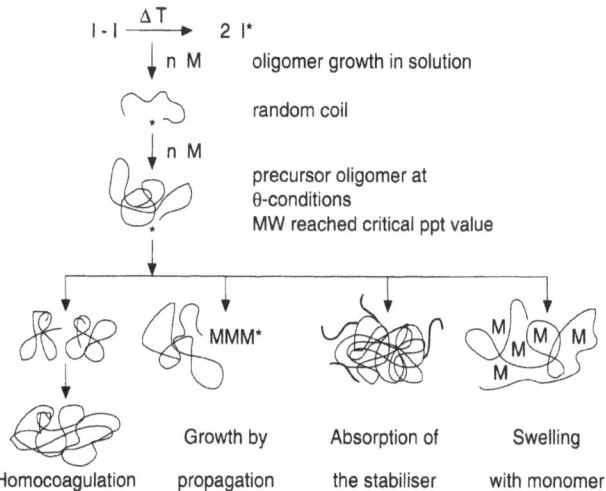

Figure 3. Possible reaction schemes.

before they reach the molecular weight necessary to precipitate in the reaction medium or when the precursor disappears by heterocoagulation with post-precursor particles before they are stabilised. These different possibilities are presented in the scheme below.

Once a stable nucleus is formed, swelling with the monomer is thought to be the main mechanism for particle growth. Further polymerisation takes place in the monomer-swollen particles. Consequently, from a kinetic viewpoint the dispersion polymerisation obeys the laws of bulk-polymerisation kinetics.

In the following discussion the impact of the different reaction parameters, varied in this study, on the nucleation process as compared to the process described by Morrison,

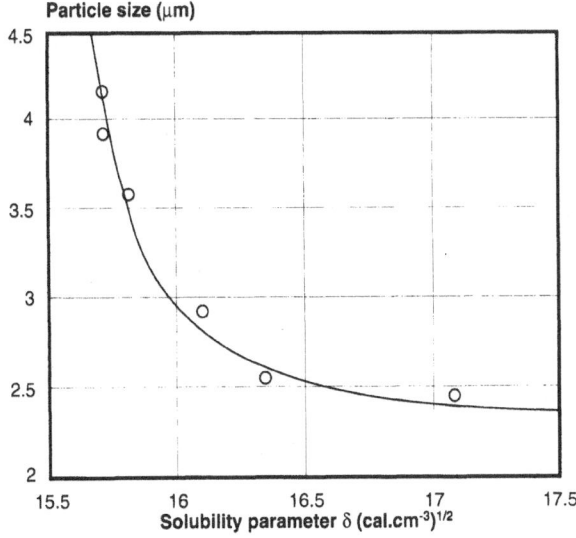

Figure 4. Influence of the solubility parameter on the particle size.

Gilbert and Napper and their influence on the final particle size and size distribution are described.

Solvency of the Reaction Medium

For the system (MMA, stymal, $K_2S_2O_8$) and for the polymerisation conditions (volume, temperature profile, stirring conditions) studied within the scope of this publication, it must be concluded that the water/ethanol ratio has probably the most important influence on the size, the shape and the dispersity of the fabricated polymer beads. Figure 4 gives a schematic representation of the relationship between the particle size of the polymer beads and the solubility parameter of the reaction medium at the onset of the polymerisation (reaction conditions as given in table 3).

The solubility parameter was calculated as the root-mean-square volume-weighted average of the solubility parameters of water, ethanol and methyl methacrylate. Due to the low concentrations of initiator and stabiliser precursor, their contributions to the solubility parameter were neglected. Although the monomer concentration plays an important role in the determination of the final particle size, all experiments used in figure 4 were carried out with he same monomer concentration. Consequently, the particle size was governed solely by the ethanol/water ratio used. Similar observations were made by Lok and Ober (18, 22, 24), and by Almog and Levy (1, 2).

By increasing the ethanol/water ratio the critical degree of polymerisation at which the poly- MMA-oligomers formed in the early stages of the polymerisation nucleate is increased.

This results in a diminished particle formation and an increased probability of oligomer capture by the latex particle before the oligomer reaches the critical degree of polymerisation. Therefore, fewer and larger particles will be formed.

As pointed out earlier by Paine (26), the polarity of the medium also changes the interaction of the graft-copolymer stabiliser and the formed particles. In the system described in this publication, a graft-copolymer is formed in situ by the grafting of MMA-chains or units onto the stymal polymeric stabiliser, as will be further described in the next paragraph. A higher ethanol/water ratio improves the solubility of the poly-MMA-moiety of the graft-copolymer stabiliser and reduces the tendency of the stabiliser to associate with the surface of the precursors. At the same time a higher alcohol content also reduces the solubility of the stymal and decreases the stabilising efficiency of the adsorbed stabiliser. Both effects promote the homocoagulation between two precursors and the heterocoagulation between precursors and postprecursor entities. As a result less and larger particles will be formed. Due to the important contribution of hetero- and homocoagulation to the formation of the polymer beads at higher ethanol content, a more heterodisperse dispersion is obtained.

In order to control the particle size the ratio ethanol/water can be varied between well-defined limits. The criteria to whom the reaction medium should respond could be determined as follows:

- at the onset of the reaction, the reaction medium should be a solvent for all the ingredients;
- during the polymerisation the medium should also be a solvent for the in-situ formed graft copolymer, but a precipitation agent for the formed polymethyl methacrylate;
- finally it should allow the formed polymer beads to be maximally stabilised in dispersed form, i.e. allow neither agglomeration, nor coagulation.

As water is a non-solvent for the monomeric MMA and its homopolymer, but a solvent for the stymal-stabiliser, and since ethanol is a solvent for MMA, but a non-solvent for poly(MMA) and stymal, the mixing of both yields many possible combinations. Varying the water/ethanol ratio, therefore, allows perfect tailoring of the size, shape and distribution of the obtained beads.

High alcohol content ($\geq$ 60 %) yields mainly amorphous precipitate, due to insufficient solubility of the polymeric stabiliser. The behaviour is similar to the one observed when no polymeric stabiliser is used at all.

Increasing the water content to 45 % improves the solubility of the polymeric stabiliser and leads to the preparation of properly shaped particles that have the desired size, but a high degree of heterodispersity; this is probably due to still insufficient solubility of the polymeric stabiliser.

Further increase of the water content, from 45 to 60 %, yields an improved stabilising ability, both sterically and ionically, resulting in smaller but very homodisperse polymer beads. 46-47% water yields pearls of ca. 3 μm diameter, 53 % water leads to particles of ca. 2 μm, and 60 % water finally gives 1-μm beads, all under analogous experimental conditions, with good homodispersity, as presented in table 3. When 70 % water is used, the MMA monomer is insoluble in the reaction medium, which leads to micel formation, and eventually to the production of very heterodisperse polymer particles.

Polymer Stabilisation

The role of the polymeric stabiliser consists in the stabilisation of the growing particles, through an ionic double layer as well as through a sterical barrier mechanism. For the GPM system described herein, it is absolutely necessary that the stabilising polymer be not only physically adsorbed to the growing polymer particle, but also chemically bound to it by a grafting reaction.

The graft-copolymer stabiliser is formed in situ by H-abstraction of stymal followed by grafting with MMA. In the literature such grafting on homopolymers has already been described (14, 17, 18, 19, 22, 27).

Of utmost importance for the formation of the graft copolymer is the induction period during which the poly-MMA-oligomers are formed and the grafting of the precursor should take place. When the stabiliser precursor is grafted with poly-MMA side-chains an amphiphilic graft- copolymer containing a number of hydrophobic poly-MMA side-chains and a hydrophilic ionised poly(styrene-co-monosodium maleate) is formed. During the nucleation period the graft copolymer is adsorbed to the final polymer beads, as schematically represented in figure 1.

In order to achieve homodisperse particles of the desired size and shape, two criteria have to be met. The hydrophobic side chains of the stabilising graft copolymer and the bulk of the monomer to be polymerised have to be compatible, in order to obtain a perfect intermingling of the polymer chains. The hydrophilic chains of the stabilising copolymer are preferentially soluble in the ethanol/water reaction medium. As these hydrophilic chains remain connected to the growing particles, they give the perfect stabilisation behaviour, i.e. the ionic nature of the dissociated sodium maleate groups yields the double-layer-type stabilisation, and the adsorbed stymal chains yield the sterical stabilisation.

Higher amounts of stabiliser precursor lead to the formation of smaller particles, which implies that the number of nuclei formed increases with increasing stabiliser concentration. This is normal, since more stabiliser can stabilise a larger surface area and decreases surface energy. According to the coagulative nucleation theory the oligomer capture by the already formed particles and hetero- and homocoagulation is hindered by the increased stabilisation. This allows more oligomers to become mature stable particles and therefore more and smaller particles are formed.

To achieve a higher degree of grafting the stabiliser precursor can be preactivated before being added to the reaction mixture by heating it in the presence of the initiator in aqueous medium (14). As a result the graft copolymer formed contains a higher number of hydrophobic poly-MMA side-chains and has an increased tendency to combine with the newly formed particle precursors. A faster adsorption of the stabiliser to the precursors hampers the coagulation and smaller particles will evolve. This is demonstrated by the results of table 6. By increasing the duration of the prereaction during which the initiator and stabiliser precursor are heated, smaller polymer beads with an improved homodispersity are obtained.

Initiator Concentration

The influence of the initiator concentration on the particle size has been described by several authors, who came to different conclusions for different systems. So, an increase in initiator concentration is claimed to lead to larger as well as to smaller particles (18, 22, 23, 30, 31, 34), or even to have no influence at all (6). In the system studied in this publication it was found that with increasing initiator concentration, smaller poly-MMA-beads were obtained. These findings are in favour of the homogeneous nucleation mechanism proposed by Fitch and Tsai (9) for the emulsion polymerisation, and they support the view that the coagulative nucleation is not the size determining process. Presumably, even at high initiator concentration, the precursor particles are sufficiently stabilised by the amphiphilic graft copolymer and by the dissociated groups on the particle surface, in order to largely prevent the coagulation.

Monomer Concentration

It has been described in literature that the particle size increases with increasing monomer concentration. However, as this size growth is much greater than the one expected solely because of the increase in particle volume due to the higher concentration, an additional effect of the nucleation process is generally accepted (3). Two possible explanations of this phenomenon have been put forward.

A first possible explanation could be called the solvency effect. Since methyl methacrylate is a good solvent for its polymer, an increase of the initial monomer concentration allows the oligomethacrylate chains to grow to a higher degree of polymerisation before they form stabilising precursors. This implies an increased probability for an oligomer to be captured by a latex particle before it is able to form a precursor particle; hence fewer and larger particles are formed.

A second possible explanation is that the adsorption of the amphiphilic graft copolymer is counteracted by the better solvency of the reaction medium for the polymethacrylate side chains. As the solubility of the hydrophilic main chains is simultaneously decreased, their sterically hindering activity is weakened. Both these effects favor the coalescence of newly formed particles and can explain the formation of larger particles and the increased polydispersity at higher monomer concentrations.

Specifically for the system described in this publication, and upon using a variation in monomer concentration from 1 to 2 moles per liter, no significant change in particle size was observed. Above this concentration of 2 mole/L, however, an increase in degree of polydispersity could be observed.

CONCLUSIONS

This publication describes a one-step technique for the reproducible preparation on an industrial scale of very homodisperse polymethyl methacrylate pearls in the range of 2-7 μm by the Graft Precipitation Method (GPM).

The influence of the four most important chemical parameters on the size and on the distribution of the produced polymer particles is described and explained : solvency of the medium, role of the polymeric stabiliser, and concentration of initiator and monomer. In a forthcoming publication the influence of the more technological parameters - reactor shape and size, stirring conditions, reaction temperature and profile - on the size and on the polydispersity of the obtained particles will be further elaborated.

In addition, an attempt is made to achieve a better understanding of the polymerisation mechanism, particularly during the nucleation phase of the synthesis. Most specific for the process described is the use of the graftable amphiphilic polymeric precursor stabiliser as one of the main controlling agents for steering the size and the polydispersity of the resulting PMMA-beads.

REFERENCES

1. Almog, Y., Reich, S. and Levy, M., Brit.Polym.J., *14*, 131, (1982)
2. Almog, Y. and Levy, M., J.Polymer Sci., Pol.Chem.Ed. *19*, 115, 1981.
3. Antl, L., Goodwin, J.W., Hill, R.D., Ottewil, R.H., Owens, S.M., Papworth, S. and Waters, J.A., Colloids and Surfaces, *17*, 67, (1986).
4. Barrett, K., Brit.Polym.J., *5*, 259, (1973).
5. Chung-li, Y., Goodwin, J. and Ottewil, R., Prog.-Coll.&Pol.Sci., *60*, 173, (1976).
6. Corner, T., Colloids and Surfaces, *3*, 119, (1981).
7. Croucher, M.D. and Winnik, M.A., Nato Asi Ser., Ser.C., *303*, 35 (1990).
8. De Winter, W., Mariën A., and Michiels, E., Bull.Soc.Chim.Belge, *99*, 977, (1990).
9. Fitch, R.M. and Tsai, C.H., in Polymer Colloids, Ed. R.M. Fitch, Plenum Press, New York, 73,(1971).
10. Flory, P., Principles of Polymer Chemistry, Ithaca Press, New York, 1953.
11. Fornasari, B., Zunino, A., Besio, M. and Massirio, S., Eur.Pat.Appl. 0 610 522, pr. 8-2-93, to Minnesota Mining & Manuf.
12. Frazza, M., Ho, K., Raney, R., Vogel, M. and Kowalski, A., Eur.Pat.Appl. 0 448 391, pr. 20-3-91, to Rohm & Haas Co.
13. Goodwin, J., Ottewil, R., Pelton, R., Vianello, G. and Yates, D., Brit.Polym.J. *10*, 173, (1978).
14. Graetz, C.W., Thompson, M.W. and Waters, F.A., Eur.Patent 0 013 478 (1982) to Imperial Chemial Industries.
15. Judat, H., Chemie Ing.Techn. *47* (7), 303, (1975).
16. Li, K. and Stöver, H., J.Polymer Sci. part A, *31*, 3257, 1993.
17. Lok, K.P. and Ober, C.K., U.S.Patent 4,247,434 (1981) to Xerox Corporation.
18. Lok, K.P. and Ober, C.K., Can.J.Chem., *63*, 209, (1985).
19. Maadhah, A.G., Amin, M.B. and Usmani, A.M., Polymer Bull. *14*, 433, 1985
20. Napper, D.H., Ind.Eng.Chem., Prod.Res.Dev., *9*, 467, (1970).
21. Napper, D.H., J.Colloid Interface Sci., *58*, 390, (1977).
22. Ober, C.K., A.C.S.Polymer Preprints, *28* (1), 248, (1987).
23. Ober, C.K. and Hair, M.L., J.Polym.Sci., Polym.Chem.Ed. *25*, 1395, (1987).
24. Ober, C.K. and Lok, K.P., Macromolecules, *20*, 268 (1987).
25. Omi, S., Katami, K., Yamamoto, A. and Iso, M., J.Appl.Polymer Sci. *51*, 1, (1994).
26. Paine, A.J., J.Polym.Sci., Part A, Polym.Chem. *28* (9), 2485, (1990).
27. Paine, A.J., Macromolecules, *23* (12), 3109, (1990).
28. Timmerman, D., Priem, J. and Janssens, J., Eur.Pat.Appl. 0 080 225, pr. 10-11-82, to Agfa-Gevaert.
29. Timmerman, D., Thijs, V., De Winter, W., Claes, F.H. and Vandenabeele, H., U.S.Pat. 3,941,727, pr. 23-10-73, to Agfa-Gevaert.
30. Tsing, C.M., Lu, Y.Y., El-Aasser, M.S. and Vanderhoff, J.W., J.Polym.Sci., Part A, Polym.Chem.Ed. *24*, 2995 (1986).
31. Tuncel, A., Kahraman, R. and Piskin, E., J.Appl.Polym.Sci., *50*, 303, (1993).
32. Ugelstad, J., Kaggerud, K.H., Hansen, F.K. and Berge, A., Makromol.Chem. *180*, 737, (1979).
33. Vincent, B., Adv.Colloid Interface Sci., *4*, 193, (1974).
34. Winnik, M.A., Lukas, R., Chen, W.F., Furlong, P. and Croucher, M.D., Makromol.Chem., Macromol.Symp., *10/11*, 483, (1987).

19

DESIGN AND CONTROL OF THE STRUCTURE OF POLYMERS AND MOLECULAR AGGREGATES IN THE SOLID LATTICE: SYNTHETIC AND SELF-ASSEMBLY APPROACH

Synthetic and Self-Assembly Approach

S. Valiyaveettil, U. Scherf, V. Enkelmann, M. Klapper, and K. Müllen[*]

Max-Planck Institute for Polymer Research
Ackermannweg-10
55128 Mainz
Germany

INTRODUCTION

The design and synthesis of oligomers and polymers with a well-defined structure has shown considerable success in recent years [1,2]. However, to fine tune the structure of a macromolecule and to obtain the desired electronic, photonic and thermal properties which are often the direct result of macroscopic order of molecules in the solid lattice, remains to be a challenge [3]. Two approaches are pursued towards designing polymers with well-defined structure. The first method rests on the careful choice of a monomer with multiple functional groups. This is particularly important for the synthesis of defect-free double stranded ladder-type polymers. The second method explores the self-assembly of small molecules and polymeric materials to form supramolecular structures. Here, the weak intermolecular interactions, such as hydrogen bonding and van der Waals forces, are used to tune the structural organization of the self-complementary molecular components. Both of these approaches have been exploited by many research groups [4,5].

Our major research goal is to generate new polymers with high molecular weight and defect-free structures via the above mentioned methods. The following discussion will focus on the design concept and the results obtained for the above mentioned classes of molecular and macromolecular systems.

[*] Address correspondence to Dr. Valiyaveettil or Dr. Mullen.

Macromolecular Engineering, Edited by M.K. Mishra et al.
Plenum Press, New York, 1995

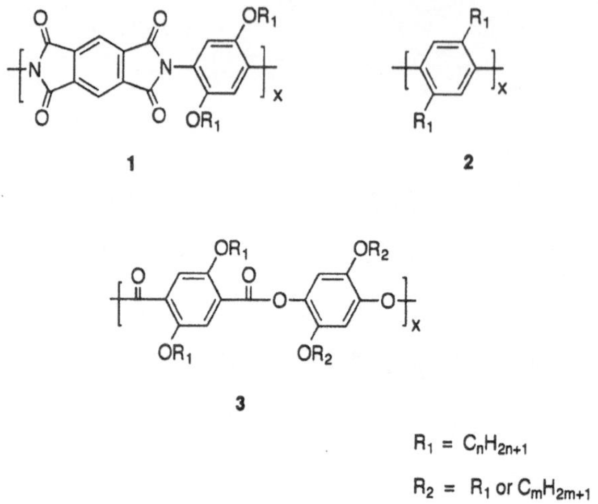

$$R_1 = C_nH_{2n+1}$$

$$R_2 = R_1 \text{ or } C_mH_{2m+1}$$

Figure 1. Some of the rigid rod structures.

RESULTS AND DISCUSSION

Synthetic Approach

The primary structure of a polymer is generally established by using available spectroscopic methods. However, to characterize the secondary or the supramolecular structure of polymer chains in the lattice is often difficult, especially in the case of polymers possessing a flexible backbone. The secondary structure of polymers with linear rigid structure (rigid rods) such as polyimides (**1**) [6], poly(*para*-phenylene)s (PPP)s (**2**) [7] and polyesters (**3**) [8] are straightforward to analyze due to their restricted conformational mobility (Fig. 1). It is also possible to tune the supramolecular organization of these polymers

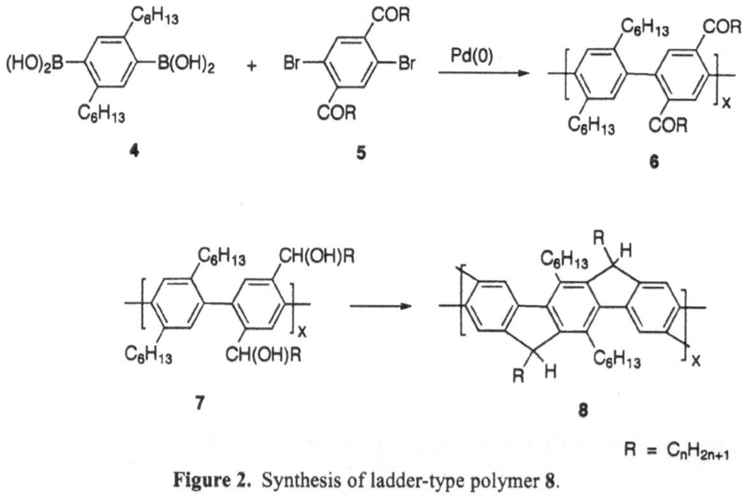

$$R = C_nH_{2n+1}$$

Figure 2. Synthesis of ladder-type polymer **8**.

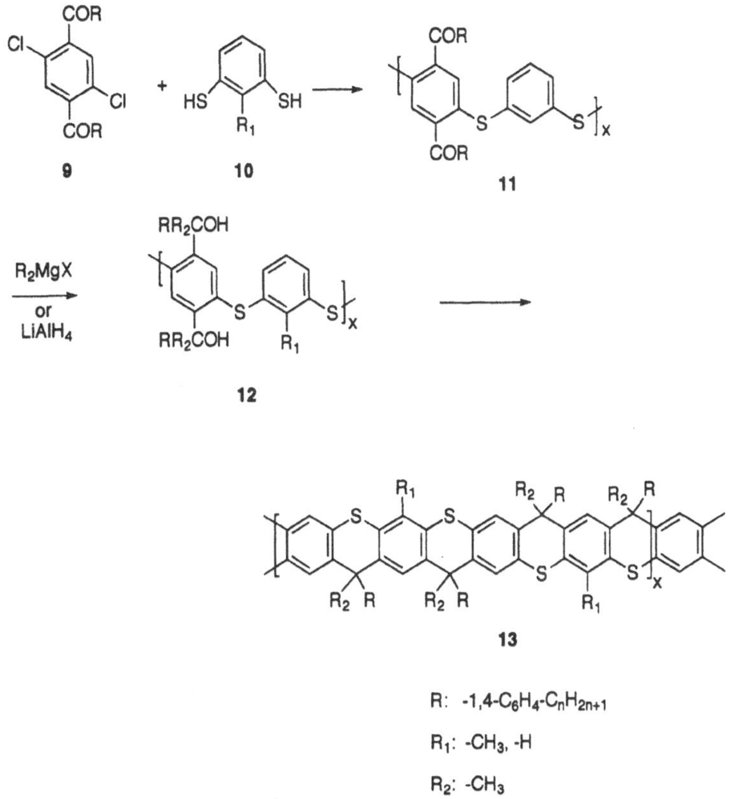

Figure 3. Synthesis of the ladder-type polymer **13**.

by incorporating secondary interactions such as alkyl chain crystallization and hydrogen bonding between the rod-type structures.

Soluble poly(*para*-phenylene)s (PPP)s are particularly interesting because of their promising optical and electronic properties [9]. High molecular weight PPPs are synthesized by Suzuki-type coupling of bifunctional dibromo- and diboronic acid monomers [10]. By using monomers substituted with alkyl chains, highly soluble rigid PPPs are obtained [11]. The alkyl substituted PPPs form well organized domains both in solution and in the melt [12].

The persistence in shape of PPP both in the solid lattice as well as in the solution can be increased by introducing a bridge between the adjacent phenyl rings along the chain. This gives rise to a double strand topology for the PPP. The synthesis of ladder-type poly(*para*-phenylene)s was recently reported by Scherf and Müllen [13]. A clever approach was made to derive a defect-free structure via a two-step synthesis. In the first-step, a PPP with functional groups which can undergo polymer-analogous cyclizations was synthesized and after the quantitative cyclization step, a well characterizable high molecular weight ladder-type polymer **8** was obtained.

The X-ray analyses of these polymers indicate well organized polymer chains in the solid lattice [14].

The above mentioned non-concerted synthesis is a versatile method for synthesizing double stranded ladder-type polymers. The polymer **13** was synthesized by using a variation of the above method [15]. Here the aryl-aryl single bonds were replaced by heteroaromatic bridges (-O- or -S-) to obtain ladder-type polymers with conjugation-decoupled polymer backbones. High molecular weight polymers were obtained by this method. Moreover, they can be considered as precursors for the preparation of fully unsaturated ribbons via dehydrogenation of the methylene bridges.

In all the above mentioned cases, a careful design and synthesis of the monomer is necessary to obtain a defect-free structure. The characterization and processing of these polymers depend on their solubility and melting behavior. Improved solubility of the polymers is achieved by incorporating alkyl substituents into the polymer backbone. One of the major problems in the step-wise synthesis of a ladder-type polymer is the requirement of a quantitative polymer-analogous cyclization reaction to achieve the desired structure. This involves the synthesis of a pre-polymer with appropriate functional groups placed along the chain. Since this approach relies on irreversibly making covalent bonds, any undesired connection would lead to major defects in the topology of the molecule. This can only be corrected by an energy requiring bond breaking and making step which is impossible for a molecule with ladder-type topology. In order to establish a more simple route towards a well-defined topology, self-assembly of small molecules[16] with complementary functional groups is explored and discussed in the following section. Here the molecules are connected through relatively weak non-covalent hydrogen bonds which can be formed reversibly. These bonds are weak enough to be rearranged for correction of an error made during the self-assembly process [17].

Self-Assembly Approach

Self-Assembly of Small Molecules. Recent understanding of intermolecular interactions through weak forces such as hydrogen bonding, electrostatic interactions and hydrophobic effects -the fundamental forces of molecular recognition- make them amenable for manipulation towards synthesizing a new generation of supramolecular polymeric architectures. The structure and net properties of supramolecular architectures formed from simple *molecular* model systems can be controlled by tuning both the primary molecular structure (covalent bonding) and the secondary supramolecular structure (weak bonding such as hydrogen bonding, hydrophobic interaction and *Pi*-stacking). A combination of hydrogen bonding and hydrophobic interaction is used to achieve the necessary control of the self-assembly of the complementary components. The major advantage of this approach is the easy accessibility of the complementary molecular components to achieve the desired topology.

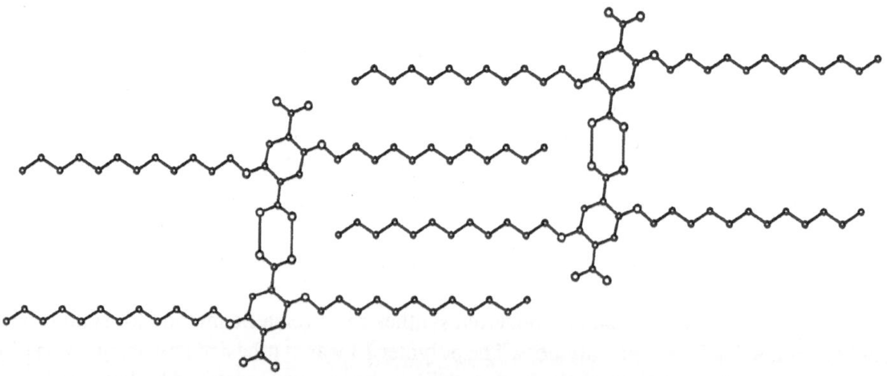

Figure 4. Structure of bis(dodecyloxy)terephthalic acid.

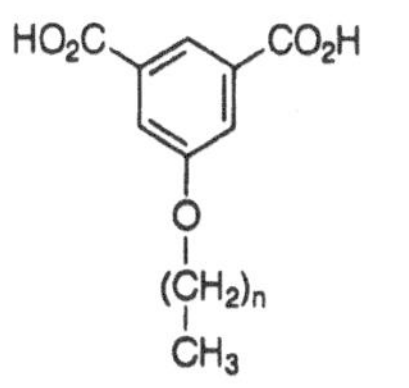

HO$_2$C ⟋⟍ ⟋ CO$_2$H

O

(CH$_2$)$_n$

CH$_3$

14   CnISA,   n = 1 - 20

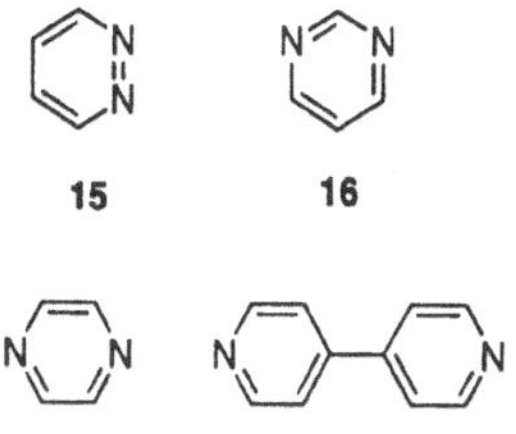

15   **16**

17   **18**

Figure 5. Left: Hydrogen bond donor molecules. Right: Hydrogen bond acceptor molecules.

Linear structures are easy to self-assemble from complementary building blocks. Both non-functionalised terephthalic acid and isophthalic acid show a linear tape-type structure in the solid lattice [18]. The crystal structure of bis(dodecyloxy)terephthalic acid is shown in Fig. 4. The linear structure of the terephthalic acid molecules is maintained along with the alkyl chain crystallization towards extending the organization of the molecules in the solid lattice [19]. This structure has a close relationship with that of alkyl substituted PPPs in the solid state. In the case of alkyl substituted PPPs, the alkyl chains are not organized to the extent as seen for bis(alkoxy)terephthalic acid [12]. Interdigitation, followed by crystallization of the alkyl chains was never observed in the structure of alkyl substituted PPPs.

Another molecular system is based on the monoalkyl derivative of 5-hydroxy isophthalic acid **14** (CnISA) and bifunctional aromatic bases such as pyridazine (**15**), pyrimidine (**16**), pyrazine (**17**) and bipyridine (**18**) (Fig. 5). In the following section we discuss the design principles and the structural organization of 5-alkoxy isophthalic acid (**14**) and its 1:1 mixtures with diamines (15-18) in the crystal lattice [20].

The diacid molecules such as CnISA form hydrogen bonded chains in the crystal lattice when n < 6 or n > 12. In both cases, the dimerization of the carboxyl groups and the formation of a hydrogen bonded tape are the main structural motif in the crystal lattice. From the hydrogen bonded arrays of dicarboxylic acid molecules with long alkyl chains, the alkoxy substituents from the adjacent molecules point towards opposite directions (Fig. 6) [21].

Since the distance between two adjacent alkyl groups of a single hydrogen bonded tape is too large to allow for effective side chain packing, infinite sheets with interdigitating alkyl chains are formed in the crystal lattice.

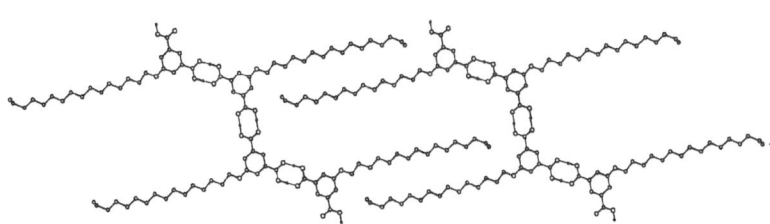

Figure 6. Crystal structure of C16ISA.

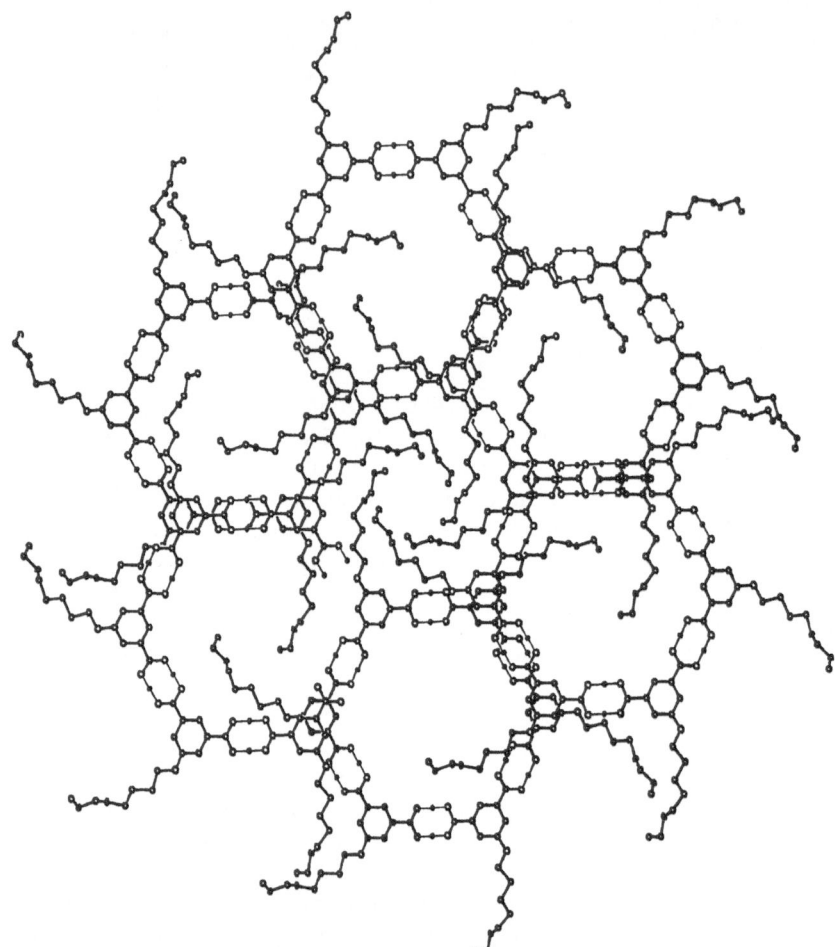

Figure 7. Channel-type structure of C8ISA (along the c axis of the hexagonal cell).

However, 5-alkoxyisophthalic acid molecules with side chain lengths between C6 and C10 show cyclic structures [21]. The crystal structures for C6ISA, C8ISA, C9ISA and C10ISA have been determined. They all crystallize in isomorphous structures with the rhombohedral space group R-3. The structures can be characterized as channel-type, obtained from the stacking of the cyclic hexamers with a ring-type hydrogen bonding motif.

In the projection shown in Fig. 7, all hydrogen bonded rings are placed on top of each other giving rise to a honeycomb structure for the channels. The walls of the channels are made by the isophthalic acid hexamers and the channels are filled with the side chains.

The addition of bifunctional hydrogen bond acceptors induces a change in the hydrogen bond pattern seen in the crystal lattice [20]. Two different cases are considered:

 i. Angular bifunctional acceptors such as pyridazine (**15**) and pyrimidine (**16**), which can be incorporated into the hydrogen bonded chain of the bifunctional diacid molecules. The resulting structure is a tape in which all alkyl substituents are oriented in the same direction (Fig. 8). Thus by interdigitation of the alkyl groups of the two adjacent hydrogen bonded tapes, molecular ribbons are observed with only weak van der Waals interactions between the adjacent ribbons.

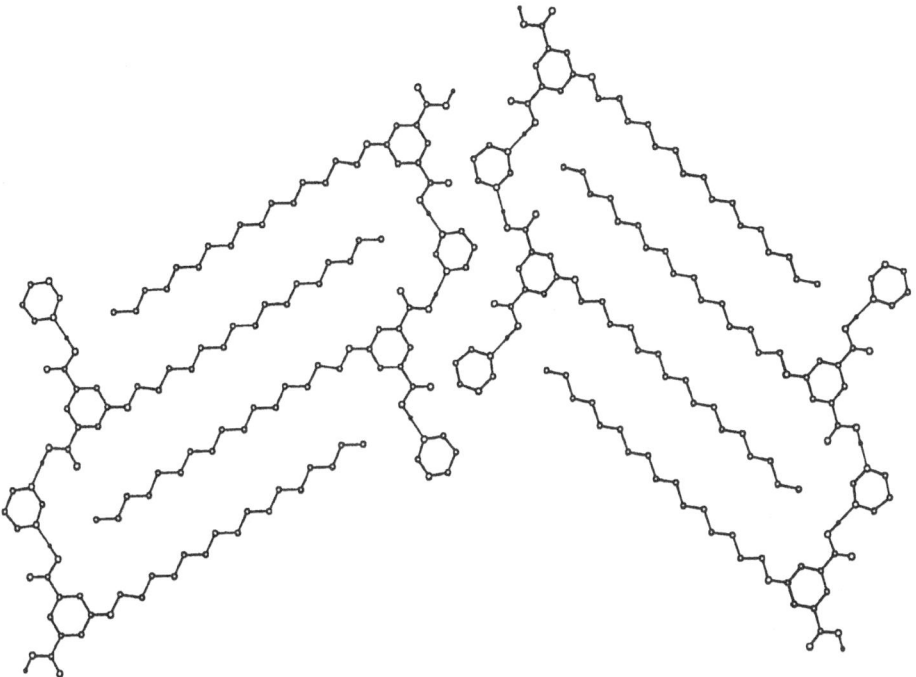

Figure 8. Crystal structure of C16ISA and pyrimidine.

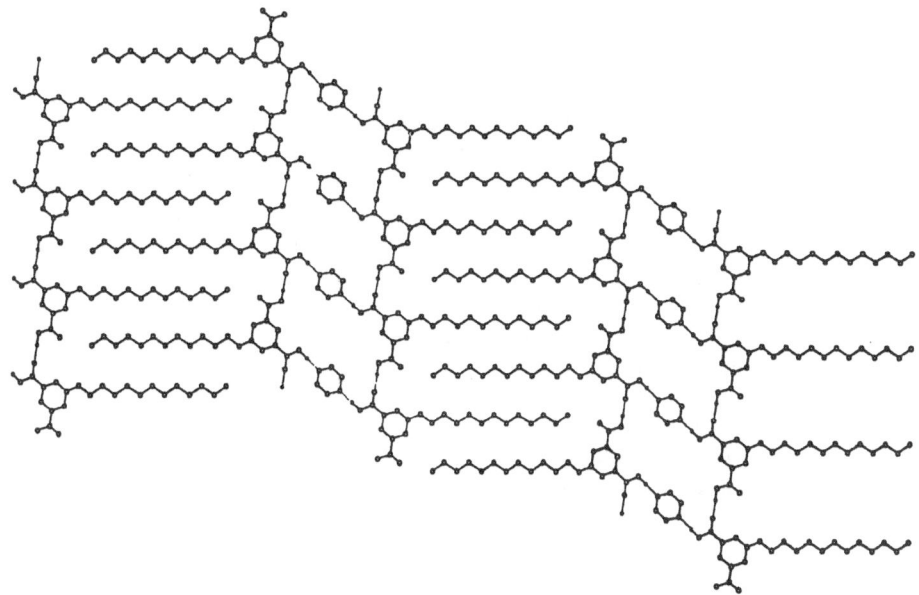

Figure 9. Crystal structure of C12ISA and Pyrazine.

a)   b)

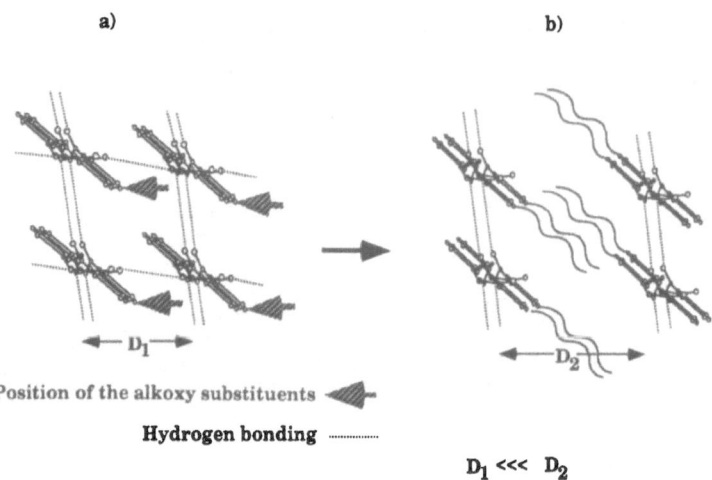

Position of the alkoxy substituents ◄▬

Hydrogen bonding ┄┄┄┄┄

$D_1 <<< D_2$

Figure 10. Representation of the lattice structures of unsubstituted (a) and substituted (b) poly(isophtha-

ii. Linear bifunctional acceptors like pyrazine (**17**) and bipyridine (**18**) crosslink hydrogen bonded tapes of the diacid molecules. Both hydrogen bonding and side chain crystallization from neighboring molecules occur in the closely packed crystal lattice [20]. The resulting structure is a unique double strand, similar to a ladder-type structure, constructed through hydrogen bonding between the complementary components (Fig. 9).

Interdigitation of the alkyl chains from these hydrogen bonded ladder-type structures gives rise to a sheet-type crystal lattice.

In general, it is possible to incorporate molecules with hydrogen bond donor-acceptor sites in the hydrogen bonded tape structure of the diacid molecules. Depending upon the orientation of the hydrogen bond donor-acceptor sites on the complementary components, different structures can be envisaged. As seen above, structures with different topologies are formed either from a single component [21] or a multicomponent molecular system[20]. However, the low solubility of hydrogen bonded aggregates, sensitivity of the weak intermolecular interactions towards certain conditions such as high temperature and the lack of suitable characterization techniques have made this self-assembly approach unpopular.

The second section focuses on the combination of the above two approaches towards structurally well defined, easily accessible and processible polymers. It is a considerable challenge to extend the interplay of hydrogen bonding and hydrophobic interactions seen in small molecular components to polymeric systems and thereby control the polymer chain conformation in the solid lattice. Recently, Stadler et. al. reported the synthesis of polymers with varying amounts of hydrogen bonding sites on the polymer backbone [22]. They have shown that the incorporation of small numbers of hydrogen bonding sites such as acid groups increases the self organization of polymer chains in the solid lattice. Our research aims at the simple synthesis of macromolecules with appropriate functional groups such as hydrogen bonding sites and alkyl chains on the polymer backbone to induce inter-chain interaction in the lattices.

Self-Assembly of Polymer Systems. The first polymeric system studied involves a series of poly(isophthalamide)s, due to its extended hydrogen bonded structure. In the crystal

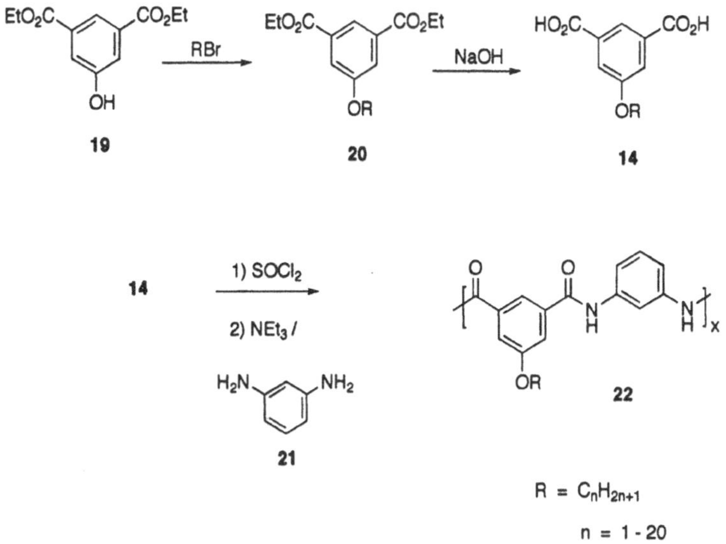

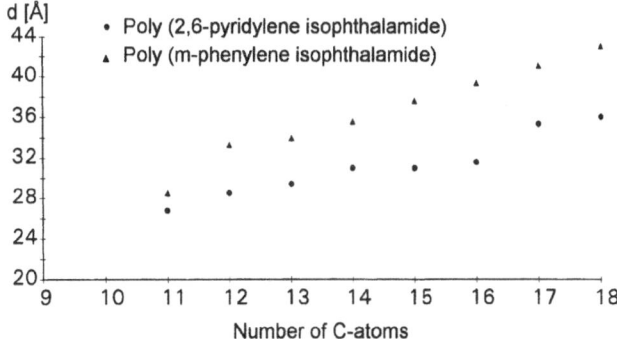

Figure 11. Synthesis of poly(isophthalamide)s (22).

lattice of unsubstituted poly(isophthalamide) each polymer strand is hydrogen bonded to four neighboring strands (Fig. 10) [23].

We chose to systematically alter the hydrogen bonding pattern of the poly(isophtha-lamide) chains in the solid lattice by introducing alkyl substituents at the C-5 position of the isophthalic acid monomer. The incorporation of alkyl chains on the polymer backbone increases the distance between two polymer strands and also forms a separate domain in the polymer lattice due to crystallization.

The 5-alkoxyisophthalic acid (CnISA) is used as one of the monomers in the synthesis of substituted poly(isophthalamide) (22). The hydrogen bonded strand observed in the C16ISA*Pyrimidine co-crystal structure (Fig. 8) may be compared with the poly(isophtha-lamide) strand in the polymer lattice. The hydrogen bonded pyrimidine moiety in the crystal structure is *replaced* by a similar molecule such as 1,3-diaminobenzene (21) in the polyamide molecule. The structural features, especially alkyl chain crystallization, observed in the crystal lattice of 5-alkoxyisophthalic acid molecules can be anticipated in the polymer lattice.

Figure 12. Plot of the *d* spacings vs the number of carbon atoms in the side chain.

The synthesis of the polymers was achieved in good yields through polycondensation of the appropriate diacid chloride and the diamine monomers (Fig. 11) [24].

The structure of the substituted poly(isophthalamide) lattice (Fig. 10, b) was established from the powder diffraction pattern of a homologous series of poly(isophthalamide)s bearing alkyl chains of varying length. The X-ray diffraction studies indicate a lamellar structure for the lattice as predicted from the model shown above [24]. The plot of the number of carbon atoms in the alkyl chains versus the d spacings gives a straight line for the homologous series which indicates an ordered structure for the polymer chains in the lattice (Fig. 12).

The thermal stability of these polymers measured by TGA and DSC is comparable to that of the unsubstituted poly(isopthalamide). This also indicates an orderly polymer lattice due to the maximization of hydrogen bonding and alkyl chain crystallization. Since further functionalization of 5-alkoxyisophthalic acid is synthetically feasible, a functional polyamide lattice is a future target of our research.

CONCLUSION

In this paper both synthetic and self-assembly directed approaches are described as methods for synthesizing well-defined structures. The synthesis of ladder-type polymers (**8**, **13**) is one of the fine examples of a covalent approach, and the polymers described above are now being studied to develop suitable materials for electroluminescence [25]. Self-assembly of molecular and polymeric systems is discussed as an alternative approach towards the design of "supramolecular" polymers. The hydrogen bonded aggregates offer an easy synthetic route to complex structures. Another important feature for the comparison of the two approaches involve the formation of ring and linear structures in polymer synthesis under various conditions. As shown above we can also achieve similar topologies from self-assembly of 5-alkoxyisophthalic acid molecules in the solid lattice. Contrary to the other systems known in the literature,[26] there is no need to pre-organize the complementary components described here. We are currently extending the above mentioned self-assembling systems to design new functional supramolecular architectures for ion transport, LC and NLO active materials.

REFERENCES

1. a) Baumgarten, M., Bunz, U.,. Scherf U., K. Müllen, *NATO ASI Series*, **1994**. b) Bein, T. (Edr.), Supramolecular Architecture, ACS Symposium Series 499, American Chemical Society, Washington, DC, **1992**, c) Tour, J. M., *Trends in Polym. Sci.,* **1994**, *2*, 332, d) Moore, J. S., Zhang, J., Wu, Z., Venkataraman, D., Lee, S., *Macromol. Symp.*, 77, 1994, 295.

2. a) Rothe, M., Chemistry and Physics of Macromolecules, Fischer, E.W., Schulz, R.C., Sillescu, H., Edn, **1991**, 39 b) Baumgarten, M., Müllen, K., *Topics in Curr. Chem.*, **1994**, *169*, 1.

3. a) Fahnenstich, U., Koch, K.-H., Pollmann, M., Scherf, U., Wegner, M., Wegner, S., Müllen, K., *Macromol. Symp.*, **1992**, *54/55*, 465. b) Scherf, U., Müllen, K., *Synthesis*, **1992**, 23. c) Müllen, K., Scherf, U., *Synth. Met.*, **1993**, *55/57*, 739.

4. Percec, V., Tirrell, D. A., (Eds), *Macromol. Symp.*, **1994**.,*77*, 1-421

5. Whitesides, G. M., Simanek, E. E., Mathias, J. P., Seto, C. T., Chin, D. N., Mammen, M., Gordon, D. M., *Acc. Chem. Res.*, **1995**, *28*, 37. and the references sited therein, b) Desiraju, G. R.., Crystal Engineering, The Design of Organic Solids, Elsevier, Amsterdam, **1989**. Organic Solid State Chemistry, Elsevier, Amsterdam, **1987**., c) Lehn, J.-M., *Macromol. Symp.*, **1993**, *69*, 1., *Angew. Chem.,Int. Ed. Engl.*, **1990**, *29*, 1304. d) Kotera, M., Lehn, J.-M., Vigneron, J.-P., *J. Chem. Soc., Chem. Commun.*, **1994**, 197 and references cited therein.

6. Wenzel, M., Ballauff, M., Wegner, G., *Makromol. Chem*, **1987**, *188*, 2865.

7. a) Vahlenkamp, T., Wegner, G., *Macromol. Chem, Phys.*, **1994**, *195*, 1933. b) Schluter, A.-D., Wegner, G., *Acta. Polym.*, **1993**, *44*, 59., c) Witteler, H., Liesser, G., Wegner, G., Schulze, M., *Makromol. Chem., Rapid. Commun.*, **1993**, *14*, 471.

8. Rodriguez-Parada, T. M., Durani, R., Wegner, G., *Macromolecules*, **1989**, *21*, 2507., b) Stern, R., Ballauff, M., Leiser, G., Wegner, G., *Polymer*, **1991**, *32*, 2096. c) Biswas, A., Blackwell, J., Deutscher, K., Wegner, G., *Acta. Polym.*, **1994**, *45*, 182.

9. Tour, J. M., Lamba, J. J. S., *Macromol. Symp.*, **1994**, *77*, 389 and the refernces cited therein.

10. Miyaura, N., Yanags, T., Suzuki, A., *Synth. Commun.*, **1981**, *11*, 513.

11. a) Heitz, W., *Chem-ztg*, **1986**, *110*, 385. b) Rehahn, M., Schlüter, A.-D., Wegner, G., Feast, W. T., *Polymer*, **1989**, *30*, 1060., c) Kallitsis, J. K., Rehahn, M., Wegner, G., *Makromol. Chem.*, **1992**, *193*, 1021., d) Rehahn, M., Schlüter, A.-D., Wegner, G., *Makromol. Chem.*, **1990**, *191*, 1991.

12. McCarthy, T. F., Witteler, H., Pakula, T., Wegner, G., submitted to *Macromolecules*, **1994**.

13. Scherf, U., Müllen, K., *Makromol. Chem. Rapid. Commun.*, **1991**, *12*, 489.

14. Scherf, U.et. al., Unpublished results.

15. Freund, T., Scherf, U., Müllen, K., *Angew. Chem.*, **1994**, *106*, 2547.

16. Lindsey, J. S., *New. J. Chem.*, **1991**, *15*, 153.

17. Philp, D., Stoddart, J. F., *Synlett*, **1991**, 445.

18. a) Derissen, J. L., *Acta Crystallogr. Sect. B*, **1974**, *30*, 2764 b) Bailey, M., Brown, C. J., *Acta Crystallogr. Sect. B*, **1967**, *22*, 387.

19. Valiyaveettil, S. et. al., Unpublished results

20. Valiyaveettil, S., Enkelmann, V., Müllen, K., *J. Chem. Soc., Chem. Commun.*, **1994**, 2097.

21. Valiyaveettil, S., Enkelmann, V., Moessner, G., Müllen, K., submitted to. *J. Chem. Soc. Chem. Commun.*,**1995**.

22. a) Stadler, R., de Lucca Freitas, L. L., *Colloid Polym. Sci.*, **1988**, *266*, 1102., b) Hilger, C., Stadler, R., de Lucca Freitas, L. L., *Polymer*, **1990**, *31*, 818., c) Hilger, C., Stadler, R., *Makromol. Chem.*, 1991, 192, 805., *Polymer*, **1991**, *32*, 17.

23. Kakida, H., Chatani, Y., Tadokora, H., *J. Pol. Sci., Pol. Phys. Ed.*, **1976**, *14*, 427.

24. Valiyaveettil, S., Gans, C., Klapper, M., Gereke, R., Müllen, K., *Polymer Bulletin*, **1995**, *34*, 13 .

25. Grem, G., Paar, C., Stampfl, J., Leising, G., Huber, J., Scherf, U., *Chem. Mater.*, **1995**, *7*, 2., Huber, J., Müllen, K., Salback, J., Schenk, H., Scherf, U., Stehlin, T., Stern, R., *Acta. Polym.*, **1994**, *45*, 244., Gruner, J., Wittmann, H. F., Hamer, P. J., Friend, R. H., Huber, J., Scherf, U., Müllen, K., Moratti, S. C., Holms, A. B., submitted to *Synth. Met.*, **1994**.

26. a) Mathias, J. P., Simanek, E. E., C. T. Seto, Whitesides, G. M., *Macromol. Symp.*, **1994**, *77*, 157. b) Seto, C. T., Mathias, J.-P., G. M. Whitesides, *J. Am. Chem. Soc.*, **1993**, *115*, 1321. c) Persico, G. F., Wuest, J. D., *J. Org. Chem.*, **1993**, *58*, 95. d) Zimmerman, S. C., Duerr, B. F., *J. Org. Chem.*, **1992**, *57*, 2215

20

POLYADDITION OF H₃PO₄ AND ITS DERIVATIVES TO DIEPOXIDES VIA ACTIVATED MONOMER MECHANISM

Polymer Structures and Functionalization

S. Penczek, P. Kubisa, and A. Nyk

Center of Molecular and Macromolecular Studies
Polish Academy of Sciences
90-363 Lodz
Sienkiewicza 112
Poland

The cationic polymerization of cyclic ethers, esters, imines and amides proceeds in the presence of compounds containing an active hydrogen atom (e.g. -OH) by the activated monomer mechanism, first introduced for the polymerization of cyclic ethers in 1984 [1].

Its major feature involves repetitive nucleophilic addition of the -OH ended macro-molecules to the protonated monomer molecules:

$$...\text{-CH}_2\text{OH} + \text{CH}_2\text{---CHR} \longrightarrow ...\text{-CH}_2\text{OCH}_2\underset{R}{\text{CHOH}} + ("\text{HA}")$$

(and the second isomer)   Eq.1

More recently, in our studies of the polymerization of glycidol, we observed that two types of repeating units are formed. One results from a simple polymerization of the three membered ring, and the second type has the four atoms repeating unit. It can only be formed if the activated monomer mechanism does indeed operate. In this way the direct evidence for this mechanism [2] has been provided:

$$...\text{-CH}_2\text{OH} + \text{CH}_2\text{---CH---CH}_2\text{OH} \longrightarrow ...[\text{CH}_2\text{OCH}_2\underset{\text{OH}}{\text{CH}}]\text{CH}_2\text{OH}$$

etc.   Eq.2

The Activated Monomer Mechanism (AMM) competes with the Active Chain End Mechanism (ACEM), involving charged macromolecules, having - in the polymerization of cyclic ethers - tertiary oxonium ions at the chain ends:

Macromolecular Engineering, Edited by M.K. Mishra et al.
Plenum Press, New York, 1995

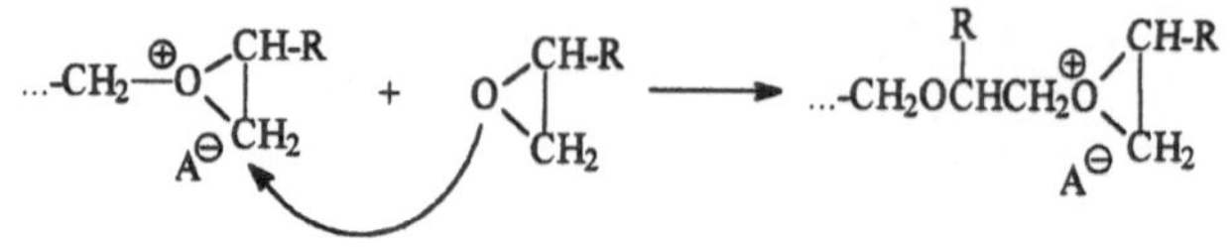

Eq.3

The corresponding contributions of both AMM and ACEM to the chain growth depend on the polymerization conditions. The AMM contribution increases, when the ratio of the instantaneous concentrations of hydroxyl groups to monomer is high. Thus, if the high contribution of the AMM is required, monomer should be added to the polymerizing mixture with the rate comparable to that of the monomer consumption. Then, the steady-state concentration of monomer can be kept at the desired level. In this way living homo- and copolymerization of some oxiranes could be achieved.

It has been shown more recently, that reaction of H_3PO_4 and some derivatives of this acid, with diepoxy compounds leads, depending on the structure of the starting substrates and other reaction conditions, to the linear or branched chains or reticulated ionic gels, highly swelling in water [3,4]. Generally, when e.g., methylphosphoric acid is used this polyaddition proceeds in the following way:

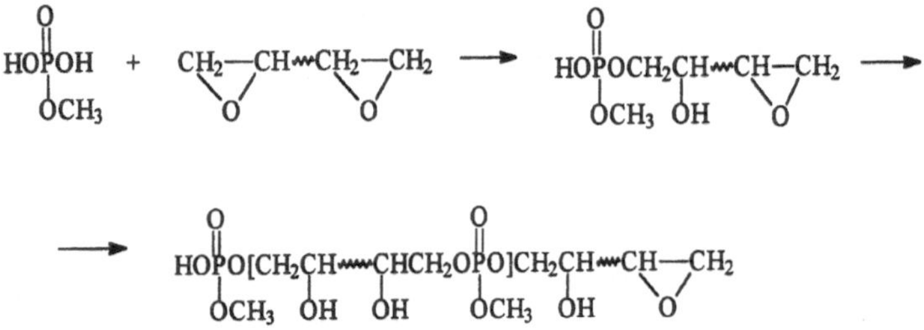

Eq.4

Model studies have shown, that the rate controlling step involves the elementary reaction of the chemically activated oxirane ring with dimeric acid [5,6]. This oxirane activation in addition reaction is related to the AMM of polymerization of oxiranes discussed above [7,8,9].

Initiation

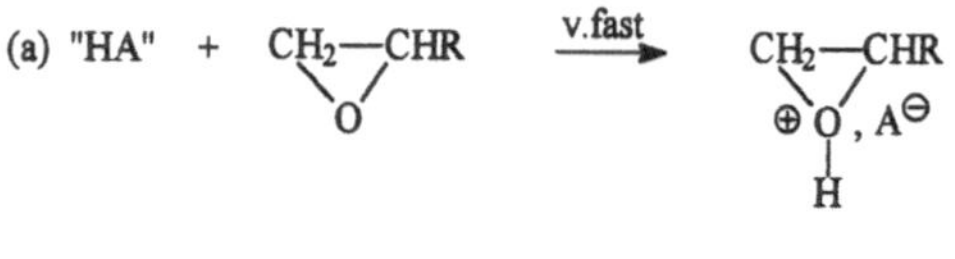

(a) "HA" + CH_2—CHR $\xrightarrow{v.fast}$ CH_2—CHR

(b) R'OH + CH_2—CHR ⟶ $R'OCH_2CHOH$ + ("HA")

Propagation
(e.g.)

...-OCH_2CHOH + n CH_2—CHR ⇌ ...-$(OCH_2CH)_nOCH_2CHOH$ + ("HA")

Eq.5

In the AM mechanism of propagation the protonic acid HA plays (mostly) a role of a catalyst and the alcohol R'OH is an initiator, attached finally to the head-chain end.

Polymerization proceeds as long, as the catalyst HA is available and not overpowered by side reactions. The most important one is trapping of protons by the growing chains:

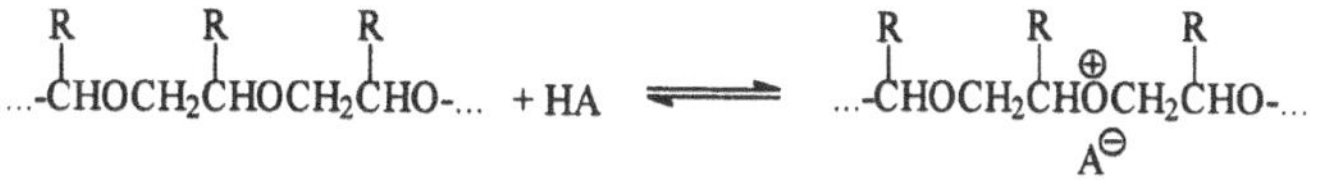

$$...\text{-CHOCH}_2\text{CHOCH}_2\text{CHO-}... + \text{HA} \rightleftharpoons ...\text{-CHOCH}_2\overset{\oplus}{\text{CH}}\text{OCH}_2\text{CHO-}...$$

(with R groups on each CH, and A$^{\ominus}$ below)   Eq.6

If protonic acids, able to form covalent bonds are used, then the acid itself may play simultaneously several functions: of an initiator, catalyst, and chain terminating agent. This is shown below for the reaction ("polymerization") of ethylene oxide with dimethylphosphoric acid:

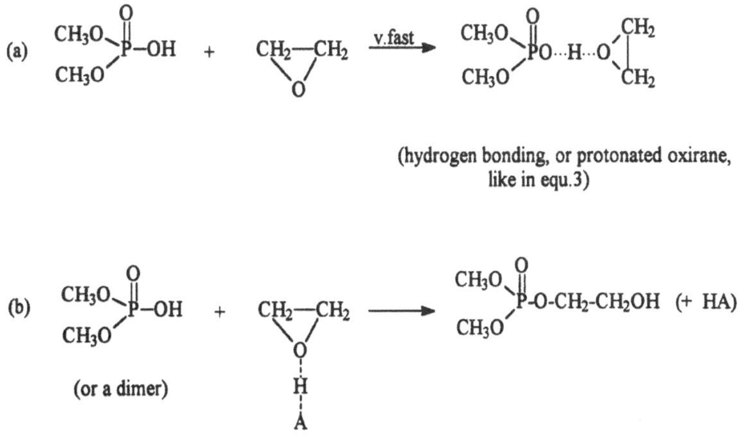

(a)   CH$_3$O–P(=O)–OH / CH$_3$O   +   CH$_2$—CH$_2$(O) $\xrightarrow{\text{v.fast}}$ CH$_3$O–P(=O)···H···O / CH$_3$O (CH$_2$–CH$_2$)

(hydrogen bonding, or protonated oxirane,
like in equ.3)

(b)   CH$_3$O–P(=O)–OH / CH$_3$O   +   CH$_2$—CH$_2$(O) $\longrightarrow$ CH$_3$O–P(=O)–O-CH$_2$-CH$_2$OH / CH$_3$O   (+ HA)

(or a dimer)   H
                                    |
                                    A

(where A = (CH$_3$O)$_2$P(O)O-)   Eq.7

Chains grow then, like in eq.(5), involving an addition of the activated oxirane to the hydroxyl end-group. However, in every initiation step one molecule of acid is consumed. Thus, acid is at the same time an initiator and a catalyst. Finally, when all of the acid is consumed, i.e. when catalyst disappears, no more addition of a monomer to the chain ends is possible, because activation of a monomer can not take place any more and monomer cannot add to the ...-CH$_2$OH ended chain. Thus, somehow, the initiation step (a) is suicidal, because every step of initiation means at the same time consumption of one molecule of a catalyst. For this reason the ratio of the rate of acid addition (in whatever form) to the activated monomer in initiation to the rate of the activated monomer addition to the ...-CH$_2$OH chain end governs finally a number of monomer units in the chains. For H$_3$PO$_4$ or its derivatives it is usually not higher than 3 [4].

These observations have led us to the creation of a system, in which polyaddition either of H$_3$PO$_4$ or of its derivatives to diepoxides leads to the polyesters of phosphoric acid. This polyaddition is related to the much better known polyaddition of the dicarboxylic acids to diepoxides, developed recently by Maréchal [10].

The polyaddition of diepoxides to the acids of phosphorous was described (to the best of our knowledge) for the first time in the open literature in our papers [4,11]. We started with the phosphorous acid (HO)$_2$P(O)H and prepared in this way polymers with M$_n$ up to 3·10^4 [11].

In the present paper we shortly summarize our earlier work on the mechanism and kinetics of polyaddition of H_3PO_4 and its derivatives to mono- and diepoxides, formation of branched polymers or ionic gels and then our more recent work on the formation of reactive cyclic structures in the backbones and functionalization of these polymers by reacting cyclics with alcohols bearing functional groups.

a) formation of the linear/branched polymer

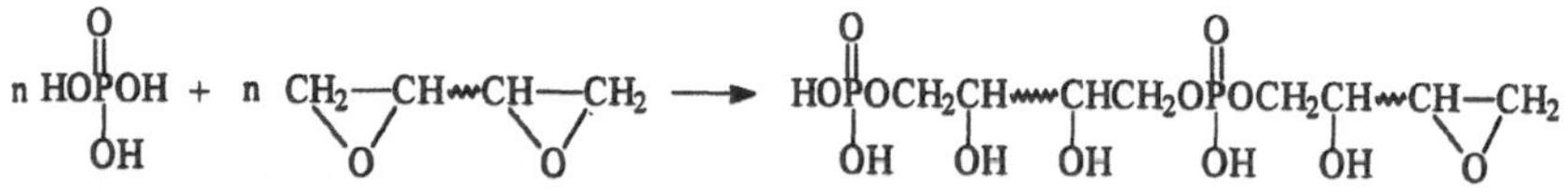

b) cyclization in the backbone

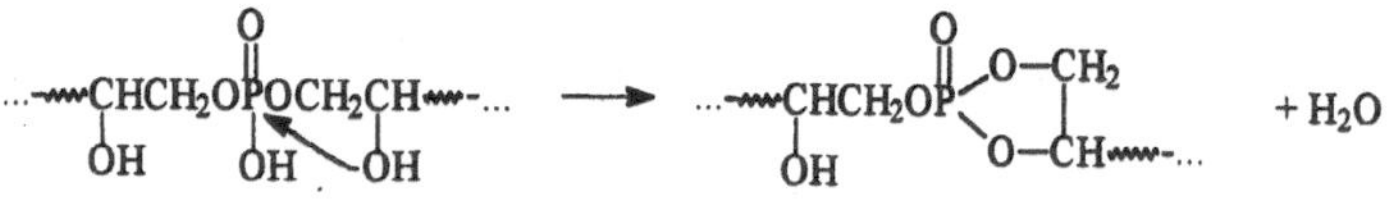

c) functionalization through the rings in the chain with $HO-CH_2CH_2X$

   (where X = functional group)

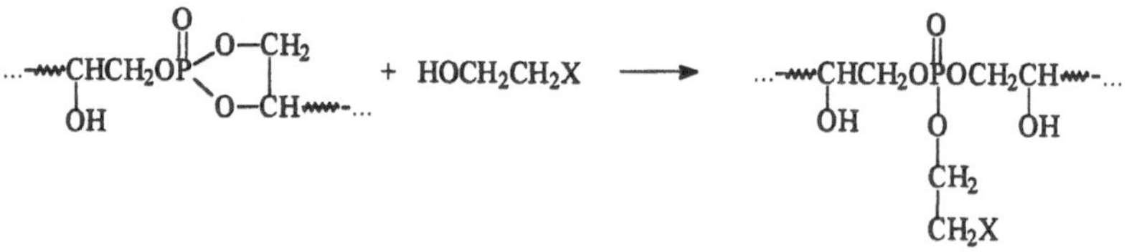

Eq.8

KINETICS AND MECHANISM OF ADDITION OF H_3PO_4 AND ITS DERIVATIVES TO OXIRANES

This paper mostly describes formation of the poly(alkylene phosphates) by a simple addition of the acids of phosphorus to diepoxy compounds. In order to better understand the polyaddition we studied model processes: addition of ethylene oxide (EO) to diethylphosphoric acid (DEPA) and to the phosphoric acid itself.

Analysis of the kinetics of ethylene oxide addition to diethylphosphoric acid:

DEPA   EO   Eq.9

led to conclusion, that reaction proceeds mainly on the hydrogen bonded oxirane. Ionic species play only a minor role, due to the very low ionization degree of DEPA ($\gamma = 0.05\%$ mol) at the studied conditions (1,4-dioxane, 25°C) [5].

This analysis has shown that 85% of additions involves various forms of oxirane activated by hydrogen bonding and only 15% of oxirane activated by protonation.

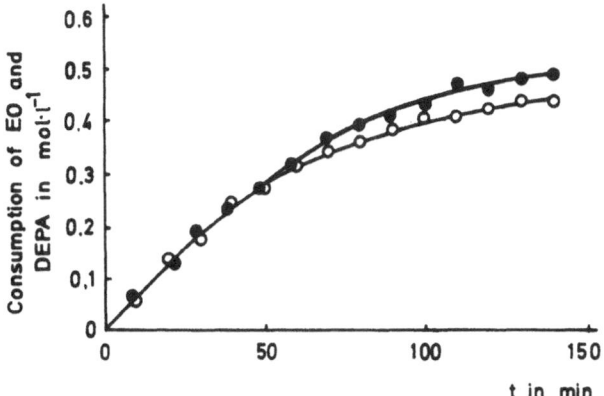

Figure 1. Kinetic curves of consumption of EO (●) and DEPA (○). Conditions: [EO]$_0$ = 1.0 mol·L^{-1}, [DEPA]$_0$ = 0.6 mol·L^{-1}, 1,4-dioxane, 25°C (explanation is given in the text).

$$\triangleright\!\!-\!O\cdots H\text{-}O\text{-}\overset{\overset{\displaystyle O}{\parallel}}{P}(OC_2H_5)_2 \quad \text{vs} \quad \triangleright\!\!\overset{\oplus}{-O}\text{-}H \ , \ A^{\ominus} \qquad\qquad \text{Eq.10}$$

Kinetics of reaction was studied in 1,4-dioxane solution. Below, in Fig.1, a typical curve is given, showing the simultaneous disappearance of DEPA and EO.

The product of reaction, analysed by HPLC, has shown the presence of the series of products of consecutive additions of EO to the hydroxyl groups:

$$\begin{matrix} C_2H_5O \\ \diagdown \\ C_2H_5O \end{matrix}\!\!\overset{\overset{\displaystyle O}{\parallel}}{P}\!-OH \quad + \quad n \ CH_2\!\!-\!\!CH_2 \quad\longrightarrow\quad \begin{matrix} C_2H_5O \\ \diagdown \\ C_2H_5O \end{matrix}\!\!P\!\!-\!\!(OCH_2CH_2)_n OH$$

$$\underset{O}{\diagdown}$$

DEPA   EO

where n = 1,2,3...   Eq.11

When EO was reacted with DEPA without any catalyst added, at 25°C in 1,4-dioxane solvent, then reaction stopped when DEPA was consumed. This happens at a average n=1.5.

Reaction with H$_3$PO$_4$, proceeds in principle similarly, and H$_3$PO$_4$ reacts as an unionized dimer.

Reaction of H$_3$PO$_4$ with EO can be treated as a sequence of parallel-consecutive reactions:

a) reactions with P-OH

$$HO\text{-}\overset{\overset{\displaystyle O}{\parallel}}{\underset{\underset{\displaystyle OH}{|}}{P}}\text{-}OH + \xrightarrow[k_1]{+\ O\triangleleft} HO\text{-}\overset{\overset{\displaystyle O}{\parallel}}{\underset{\underset{\displaystyle OH}{|}}{P}}\text{-}O\frown OH \xrightarrow[k_2]{+\ O\triangleleft} HO\text{-}\overset{\overset{\displaystyle O}{\parallel}}{\underset{\underset{\displaystyle O\frown OH}{|}}{P}}\text{-}O\frown OH \xrightarrow[k_3]{+\ O\triangleleft}$$

$$\longrightarrow \quad HO\frown O\text{-}\overset{\overset{\displaystyle O}{\parallel}}{\underset{\underset{\displaystyle O\frown OH}{|}}{P}}\text{-}O\frown OH \qquad\qquad\qquad \text{Eq.12}$$

As long as there are unreacted P-OH groups present (playing a role of catalyst), reaction with C-OH also takes place.

b) reactions with C-OH

$$HO\text{-}\overset{O}{\underset{OH}{P}}\text{-}O\diagup\diagdown OH \quad \xrightarrow{+O\triangleleft} \quad HO\text{-}\overset{O}{\underset{OH}{P}}\text{-}O\diagup\diagdown O\diagup\diagdown OH$$

Eq.13

This reaction (addition of the activated EO molecule to the -CH$_2$CH$_2$OH groups) is equivalent to the propagation step in the polymerization of EO by the Activated Monomer Mechanism, studied by us earlier [12].

Assuming, that the reactivity of the unreacted P-OH group does not depend on the length of the substitutent at these groups, that had already reacted, i.e. is the same for both species shown in eq.(13), this reaction can still be treated as a sequence of consecutive reactions:

$$HO\text{-}\overset{O}{\underset{OH}{P}}\text{-}OH \xrightarrow{k_1} RO\text{-}\overset{O}{\underset{OH}{P}}\text{-}OR \xrightarrow{k_2} HO\text{-}\overset{O}{\underset{OR}{P}}\text{-}OR \xrightarrow{k_3} RO\text{-}\overset{O}{\underset{OR}{P}}\text{-}OR$$

where R - denotes $-[CH_2CH_2O]_n H$ with n = 1 or 2

Eq.14

Methods of calculation of the ratios of rate constants for the series of consecutive reactions are known [13]. To calculate these ratios, it is sufficient to know the concentrations of all involved species at different stages of reaction.

Thus, the products of reaction between PCA and EO, isolated at different stages of reaction, were analysed by HPLC, ^{1}H- and ^{31}P-NMR. The peaks in HPLC are not well resolved because the diesters with n=1 appear at similar retention volumes as monoester with n=2 and the same is true for the next oligomers of the series. 300 MHz^{1}H-NMR spectra for the same reason, are not fully conclusive. On the other hand, in ^{31}P-NMR spectra, the resolution is sufficiently good and the signals of mono-, di-, and triesters, as well as the signal of acid, are observed separately [14].

Moreover, signals corresponding to mono-, di-, and triesters are separated into two, three, and four lines respectively, which shows, that oligomers with the same degree of substitution but differing in n=1 or 2 can be observed separately. No species with n>2 were observed.

The intensity of signals in ^{31}P-NMR is directly proportional to the concentrations. Thus, from the ratio of integration of signals corresponding to the given population (e.g. mono-, di-, and triesters) to the sum of all signals, the mole fraction of given oligomer could be determined. From these, the ratios of rate constant have been calculated giving the following values:

$$k_2/k_1 \sim 2 \; ; \; k_3/k_1 \sim 4$$

Thus, the reactivity of P-OH group increases with increasing the degree of substitution and the third P-OH group is the most reactive.

Independent of the determination of the ratios of rate constants from the oligomer distribution, we have analysed the reaction kinetics at the early stage of reaction, followed by ^{1}H- and ^{31}P-NMR the rate of disappearance of H$_3$PO$_4$ and EO. It was found, that reaction obeys at this stage the second order kinetics (first order in EO and first order in the P-OH groups). The calculated second-order rate coefficient is equal to $2.4 \cdot 10^{-4}$ mol^{-1}·l·s^{-1} (1,4-di-oxane, 25°C) [5] i.e. is very close to the second-order rate coefficient of reaction between DEPA and EO, determined earlier and equal to $2.7 \cdot 10^{-4}$ mol^{-1}·l·s^{-1} at the same conditions [5]. DEPA is the diethyl ester of H₃PO₄ and for this compound the reactivity of the third remaining P-OH group seems to be the same as the reactivity of the first P-OH group in H₃PO₄. On the other hand, reactivity of the third P-OH group in diester containing HO[-CH₂CH₂O]-$_n$ (n=1,2) groups is ~4 times higher. The observed changes in reactivity indicate, that the reactivity of the P-OH group does not depend on the degree of substitution but on the nature of substituent.

Apparently, the internal hydrogen bond is formed between the oxygen atom in HO[-CH₂CH₂O]-$_n$ substituent and acidic proton in the P-OH group increasing in this way the nucleophilic character of the oxygen atom of the P-OH group. This behaviour is illustrated by the scheme below:

$$
\begin{array}{ccccc}
\text{O-CH}_2\text{-CH}_3 & & \text{OH} & & \overset{\text{CH}_2\text{-CH}_2}{\text{O}\diagdown\text{O—}} \\
| & & | & & \\
\text{O=P-OH} & \sim & \text{O=P-OH} & < & \text{O=P—O—H} \\
| & & | & & \\
\text{O-CH}_2\text{-CH}_3 & & \text{OH} & & \underset{\text{CH}_2\text{-CH}_2}{\text{O}\diagup\text{O—}}
\end{array}
\qquad \text{Eq.15}
$$

Preparation of the linear phosphate (diester) in reaction between H₃PO₄ and EO would be possible only if the third rate constant k_3 were much lower than the second rate constant k_2>. Results presented in this paper show that this is not the case. Rate constant k_3 is about two times higher than k_2. Thus diesters are relatively fast converted into triesters.

Therefore, as it will be shown in the next paragraph, polyaddition of H₃PO₄ and diepoxides involves all three acidic groups and leads to branched products. Nevertheless, when polyaddition is conducted in solvents able to form stronger H-bonds (e.g. glymes) the branching and reticulation are less pronounced, since external hydrogen bonding success-fully competes with the internal one, shown in eq.(15), and levels off the difference between the rate constants discussed above.

POLYADDITION OF H₃PO₄ AND ITS DERIVATIVES TO DIEPOXIDES

Polyaddition of H₃PO₄ or its derivatives to diepoxides provides polymers with different structures and properties, depending on the substrates, their ratio, and conditions of polyaddition.

All of these reactions are performed by directly mixing components in the chosen solvent (e.g. 1,4-dioxane). The progress of reaction was studied by ¹H and ³¹P NMR.

H₃PO₄ is a tribasic acid, and all of the three acidic functions may react with epoxy groups. In this instance reticulated polymers are formed. If, however, conditions are chosen deactivating the acidic groups with conversion, then branched but soluble polymers result. The next explored possibility is related to blocking of the third group and/or applying the dibasic acid or its reactive ester. Thus, three substrates were used:

$$\underset{\underset{\text{OH}}{|}}{\overset{\overset{\text{O}}{\|}}{\text{HOPOH}}} \quad , \quad \underset{\underset{\text{OCH}_3}{|}}{\overset{\overset{\text{O}}{\|}}{\text{HOPOH}}} \quad , \quad (\text{CH}_3)_3\text{SiO}\underset{\underset{\text{OCH}_3}{|}}{\overset{\overset{\text{O}}{\|}}{\text{P}}}\text{OSi(CH}_3)_3$$

$$\text{1} \qquad\qquad \text{2} \qquad\qquad\qquad \text{3} \qquad\qquad\qquad \text{Eq.16}$$

As the diepoxy compounds mostly diepoxybutane or diglycidylethers of oligo(oxyethylene) glycols were applied:

$$\underset{\text{O}}{\overset{/\backslash}{\text{CH}_2-\text{CH}}}-\underset{\text{O}}{\overset{/\backslash}{\text{CH}-\text{CH}_2}} \qquad\qquad \underset{\text{O}}{\overset{/\backslash}{\text{CH}_2-\text{CH}}}-\text{CH}_2\text{O(CH}_2\text{CH}_2\text{O)}_n\text{CH}_2-\underset{\text{O}}{\overset{/\backslash}{\text{CH}-\text{CH}_2}}$$

$$\text{4} \qquad\qquad\qquad\qquad\qquad \text{5 (A, B, or C)}$$

$$(n = 1(\textbf{A}), 2(\textbf{B}), \text{ or } 3(\textbf{C})) \qquad\qquad \text{Eq.17}$$

There is a number of structural units being formed during this cationic polyaddition. In order to comprehend better these structures and the way they were determined, we have given below the ^{31}P NMR spectra of a typical product from $(\text{HO})_2\text{P(O)OCH}_3$ and diglycidyl ether of triethylene glycol, containing the majority of the structural units found in these products, *irrespectively* of the reaction conditions. Later we describe structural differences emerging when either H_3PO_4 or $(\equiv\text{SiO})_2\text{P(O)CH}_3$ are used):

α Polymer repeating units (regioisomers):

$$...-\text{CHCH}_2\text{OPOCH}_2\text{CH}-\text{...} \quad , \quad \text{CHCH}_2\text{OPOCH}- \quad , \quad \text{CHOPOCH}-$$

A1	A2	A3
(α,α–opening)	(α,β–opening)	(β,β–opening)
(tail-to-tail)	(head-to-tail)	(head-to-head)

Eq.18

There are three centres of chirality, one on phosphorus atom and two on the adjacent carbon atoms, giving fine splitting in the spectra, due to the presence of stereoisomers.

Besides these major three repeating units (actually, nine, if chirality is taken into the account), there are also **A1**, **A2**, and **A3** units with epoxy group added to the hydroxyl groups. Without taking into account the discussed above chirality, there are two such products derived from both **A1** and **A3** and three from **A2**. Of course, the probability of formation of some of these products is rather low. As an example the products derived from **A2** are shown below (ramifications **R1** and **R2**):

$$\text{R1} \qquad\qquad\qquad \text{R2} \qquad\qquad\qquad \text{Eq.19}$$

These products are precursors of the reticulated structures, and can be avoided by blocking (in situ) the -OH groups (cf. below).

Finally, there are end-groups. These can either be epoxy or acidic groups, depending on the starting ratio of components. In this work we were interested in having acidic end-groups, and, therefore, only these appear in the spectra:

$$
\begin{array}{ccc}
\underset{\text{E1}}{\underset{\displaystyle \underset{OCH_3}{|}}{HOPOCH_2CH\sim}} & ; & \underset{\text{E2}}{\underset{\displaystyle \underset{OCH_3}{|}}{HOPOCH_2CH\sim}}
\end{array}
\qquad \text{Eq.20}
$$

Besides these units, cyclic structures have also been observed. As it was already discussed in the previous paragraph, the cyclic units are mostly formed by the conformationally enhanced unimolecular esterification and/or transesterification. These reactions proceed on the polymer repeating units and on the end groups. For the end-group **E2** the cyclic group **CE2** forms as shown below:

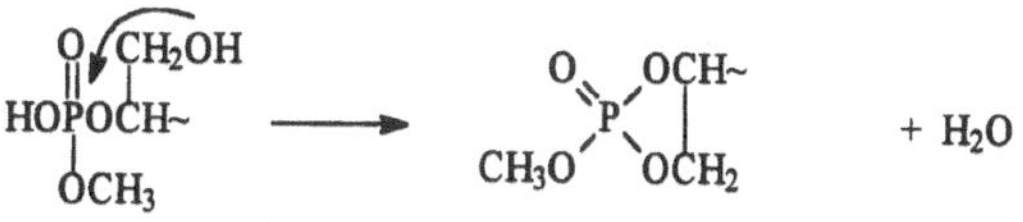

$$
\underset{\text{E2}}{HOPOCH\sim} \longrightarrow \underset{\text{CE2}}{\overset{O}{\underset{CH_3O \ \ OCH_2}{\overset{\diagup OCH\sim}{P}}}} + H_2O
\qquad \text{Eq.21}
$$

The cyclic end group **CE1** as well as the 5-membered cyclics **CA** in the chain are formed similarly, e.g. (from the linear unit **A2**):

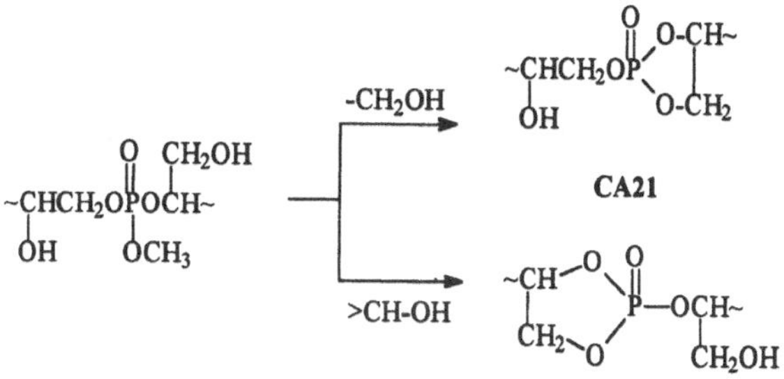

$$
\sim CHCH_2OPOCH\sim \quad \overset{-CH_2OH}{\underset{>CH\text{-}OH}{\Longrightarrow}} \quad \begin{array}{c} \textbf{CA21} \\ \textbf{CA22} \end{array}
\qquad \text{Eq.22}
$$

In Fig.2, in the ^{31}P{^{1}H} NMR spectrum of a model polymer, prepared in such a way, that a number of the end-groups is enhanced, all of the structural units discussed above are indicated.

Several structural differences were observed, when instead of methylphosphoric acid either H$_3$PO$_4$ or ($\equiv$SiO)$_2$P(O)OCH$_3$ are used. Some are discussed in the next paragraphs.

Simple Polyaddition of H$_3$PO$_4$

The course of reaction was studied on the basis of the direct ^{1}H NMR observation of the epoxy groups and the "mobile" protons, i.e. a sum of $\equiv$P-OH protons and protons from hydroxyl groups -CH$_2$OH and >CHOH. The latter sum is constant and only the resulting chemical shift changes, when the proportion of the $\equiv$P-OH groups decreases with the extent of reaction. The dependence of the chemical shift on this proportion was established independently. The spectra of the earlier stages of polyaddition of H$_3$PO$_4$ and diepoxybutane is shown in Fig.3.

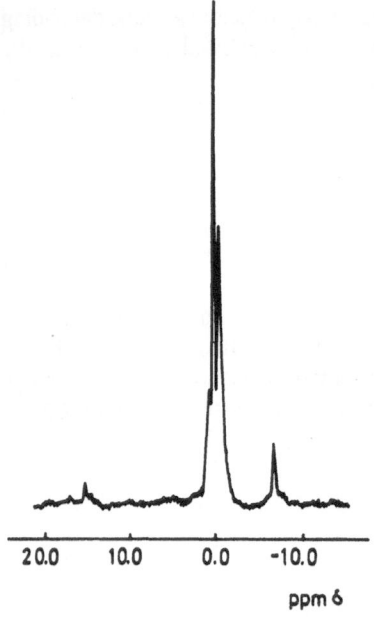

20.0   10.0   0.0   -10.0

ppm δ

Figure 2. $^{31}P\{^{1}H\}$ spectrum of the reaction product of the reaction of $(HO)_2P(O)OCH_3$ and diglycidyl ether of diethylene glycol (explanation is given in the text).

Usually, reaction of H_3PO_4 with diepoxides leads to the formation of reticulated products. These gels can be analysed by a simple titration and by ^{31}P NMR spectra in solution or in solid state by MAS $^{31}P\{^{1}H$ NMR [5]. Although the detailed structure analysis is difficult, because of the variety of isomeric structures, but distinction between the monoesters (end groups), diesters (polymer repeating units) and triester (branching points in the backbones) can clearly be distinguished by both titration and ^{31}P NMR.

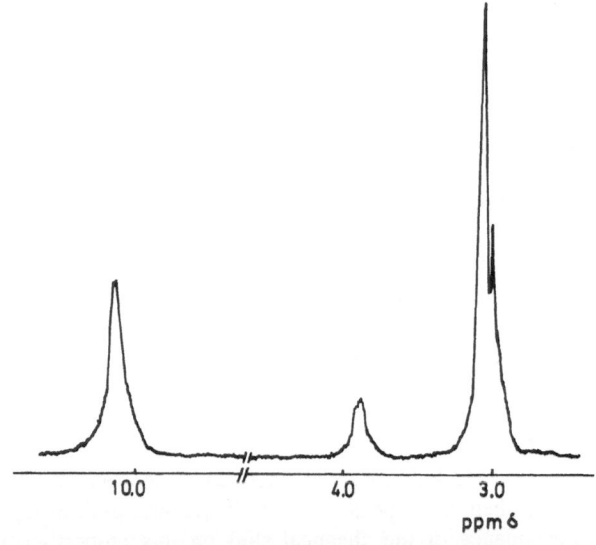

10.0   4.0   3.0

ppm δ

Figure 3. ^{1}H NMR spectra of the reaction mixture of **1** and **2** in 1,4-dioxane-d_8: (a) after 10 min; b) after 20 min; (c) after 30 min; (d) after 45 min (explanation is given in the text).

Moreover, further esterification of the acidic groups with diazomethane gives additional information:

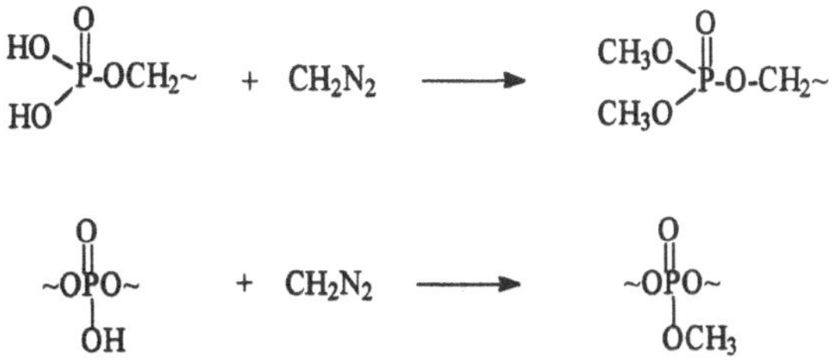

$$\underset{HO}{\overset{HO}{>}}\overset{\overset{O}{\|}}{P}\text{-OCH}_2\sim \ + \ CH_2N_2 \ \longrightarrow \ \underset{CH_3O}{\overset{CH_3O}{>}}\overset{\overset{O}{\|}}{P}\text{-O-CH}_2\sim$$

$$\sim O\overset{\overset{O}{\|}}{\underset{OH}{P}}O\sim \ + \ CH_2N_2 \ \longrightarrow \ \sim O\overset{\overset{O}{\|}}{\underset{OCH_3}{P}}O\sim \qquad\qquad \text{Eq.23}$$

Comparison of these three sources of data showed very good agreement, allowing finally the fine structure to be established [4].

For one of the product, containing altogether approx. 20 phosphorus atoms in the average molecule, the proportions of the corresponding units were: monoesters ~40%, diesters ~30%, triesters ~30%. Structure of such a highly branched polymer is similar to the cut-off from the gel, shown in the next paragraph (eq.27).

Polyaddition in the Presence of Compounds Blocking In-Situ the Hydroxyl Groups

As it was shown in our model studies, approx. 1.5 molecules are needed for the complete esterification of one acidic group. This is because the emerging hydroxyl groups react also with epoxy groups, in the process catalyzed by acidic protons. This reaction is a major source of departing from simply polyesters of phosphoric acid to products having, besides these units, also the ether bonds. One of the ways allowing to avoid this sometimes undersirable reaction is blocking the hydroxyls in situ.

From a number of blocking agents used 2,2,6-trimethyl-4-H[1,3]dioxin-4-one (TMDE) was particularly effective. The original polymer, prepared from (HO)$_2$P(O)OCH$_3$ and diglycidyl ether of triethylene glycol in the presence of TMDE had the structure shown in eq.24.

Table 1. Structural units in gels

Structural unit	HO$\underset{HO}{>}$$\overset{\overset{O}{|}}{P}$-O~ monoester	$\sim$O-$\overset{\overset{O}{\|}}{\underset{OH}{P}}$-O~ diester	$\sim$O-$\overset{\overset{O}{\|}}{\underset{O}{P}}$-O~ triester
% mol	40	30	30

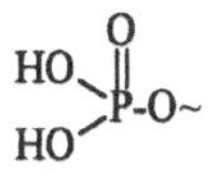

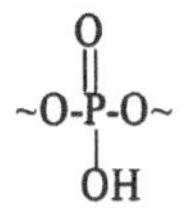

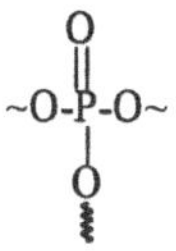

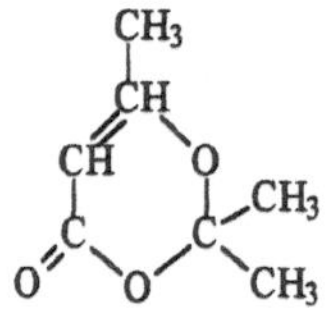

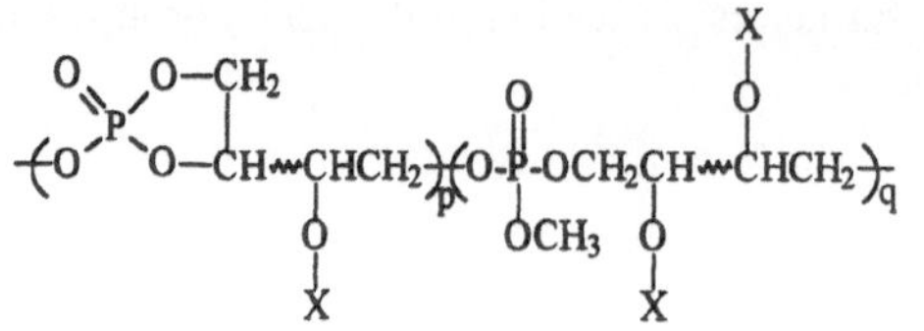

where X = CH$_3$COCH$_2$(O) from TMDE and p/q = 0.55

TMDE

$$\text{—wwww—} \quad = \quad -(CH_2CH_2O)_3$$

Eq.24

This polymer, after the simple deblocking and opening the rings in the chain, gives fairly regular structure.

Polyaddition of Disilyl Ester of Methylphosphoric Acid

Application of the disilyl ester allows the hydroxyl groups to be eliminated. Therefore branching is avoided. Polyaddition proceeds in the following way:

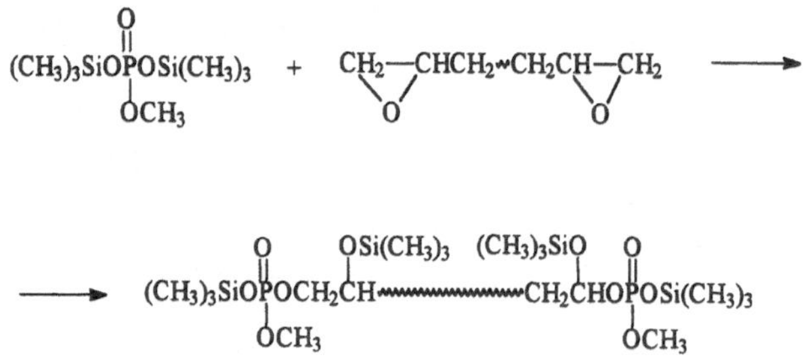

where   wwww   = (CH$_2$CH$_2$O)$_3$

Eq.25

In Fig.4 the ^{31}P{^{1}{H}} NMR spectrum is given for a polymer prepared in dioxane at r.t. with DMAP as catalyst.

In the spectrum (Fig.4) the following chemical shifts, related to structures shown below are observed.

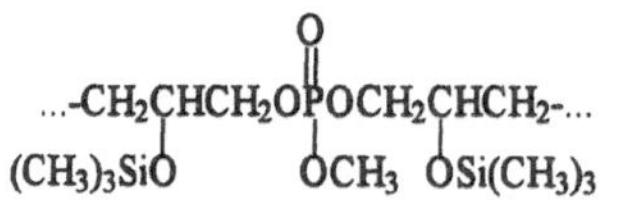

α,α-ring-opening
("tail-to-tail")

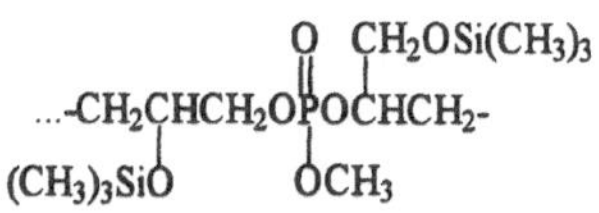

α,β-ring-opening
(tail-to-head)

(in ppm δ)   -0.1   ; ~60%

-0.8   ; ~30%

$$\begin{array}{cc}
\underset{\substack{|\\(CH_3)_3SiOCH_2}}{...\text{-}CH_2CHO}\overset{\substack{O\quad CH_2OSi(CH_3)_3\\\parallel\quad|\\}}{P}\underset{\substack{|\\OCH_3}}{OCHCH_2\text{-}...}
&
\underset{\substack{|\\OCH_3}}{...\text{-}CH_2O}\overset{\substack{O\\\parallel\\}}{P}OSi(CH_3)_3
\end{array}$$

β,β-ring-opening   end-group
(head-to-head)

-8.3 ; ~5%

(in ppm δ)   +0.5 ; ~5%   Eq.26

Besides, small peak at δ = 16.5 indicates the presence of the cyclic group in the chain.

Integration, particularly of the end groups, is not sufficiently precise, but it gives DP_n ~ 20, whereas the DP_n measured by vpo equals ~18.

Recently (in preparation) even better regioselectivity was observed with p-(N,N'-dimethyl)aminopyridine (DMAP) as catalyst.

Structure and Some Properties of Ionic Gels

Reaction of H₃PO₄ with diepoxide, conducted in dioxane at room temperature, usually leads to insoluble products, precipitating out from solution. Remembering, that for every 4-5 epoxy groups reacting with =⁼P-OH approximately one reacts with alcoholic -OH, 20-30% mol. excess of diepoxide over H₃PO₄ was used. Then, in the gel there are mono-, di-, and triesters as well as certain proportion of the ether bonds. Neglecting these latter structures, the cut off structure from the gel is shown below.

It contains altogether 20 P atoms, 10 of the triester structure (branching point), 5 of the diester repeating units and 5 of the =⁼P-(OH)₂ end groups:

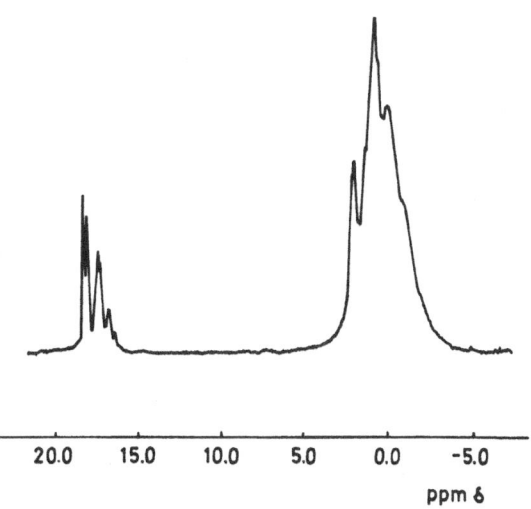

20.0   15.0   10.0   5.0   0.0   -5.0

ppm δ

Figure 4. ³¹P{¹H} NMR spectrum of the reaction product of the reaction of ((CH₃)₃SiO)₂P(O)OCH₃ [1] and diglycicyl ether of diethylene glycol [2]. Reaction conditions: [1] = [2] = 2.1 mol·L⁻¹, 80°C, 2% mol DMAP (explanation is given in the text).

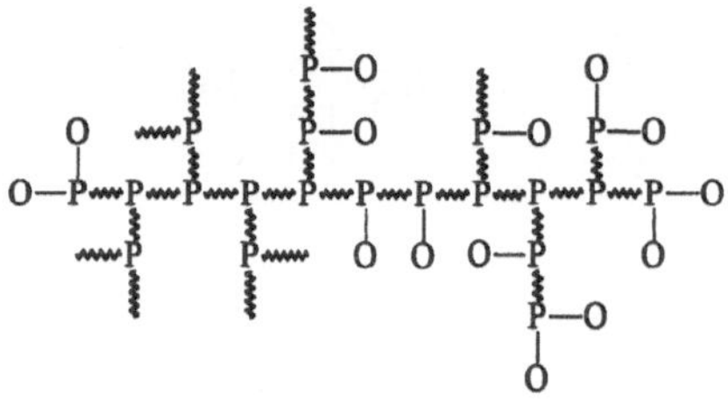

where the symbols given above denote respectively:

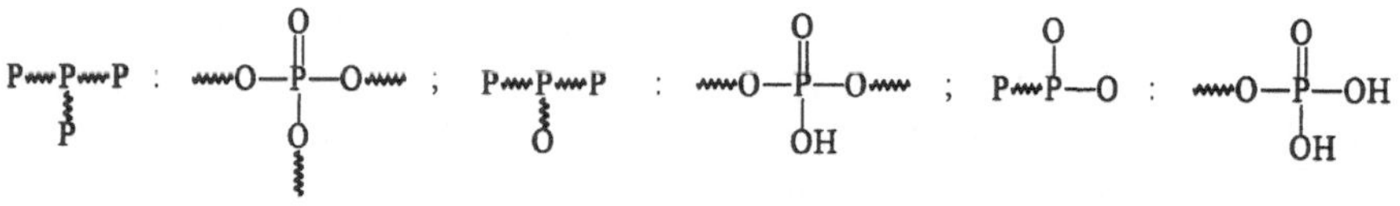

| triester | diester | monoester |

and $P\text{ww}\overset{\text{|}}{P}\text{www}$ or $P\text{ww}\overset{\text{|}}{P}\text{—O}$ describe the points of cut-off   Eq.27

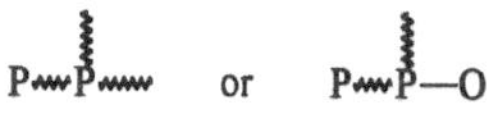

The microstructure of gels was studied and proportions of mono-, di-, and triesters were determined [4]. The proportions depend on the structure of the diepoxide used. For instance, with diepoxybutane up to 80% of epoxy groups reacted with formation of the P-O-C bonds and 20% formed the C-O-C bonds. The proportions of mono-, di-, and triesters were as follows: 26:21:53, i.e. close to the structure of the cut-off described above.

These gels can be made stable or hydrolyzable in water. The time required for solubilization at pH$\approx$2 for a typical gel may be equal to 48-60 hrs at r.t. Swelling ability of gels depends on pH and on their chemical structure, at pH$\approx$2 a gel having the repeating units and branching points derived from triethyleneglycol (eq.28) takes up to 1500% of H_2O [4].

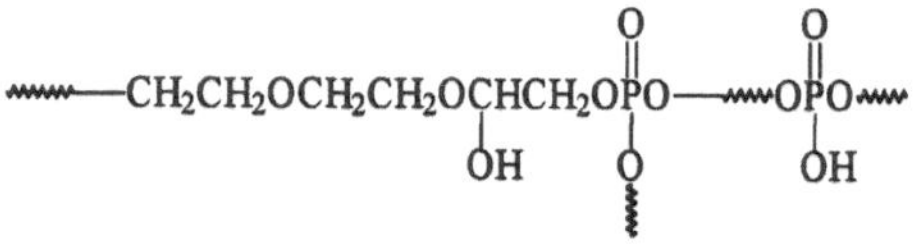

Eq.28

Functionalization of Poly(alkylene phosphates)

Poly(alkylene phosphates) prepared applying the Activated Monomer Mechanism concept, as described in the previous paragraphs, can either be linear/branched or reticulated polymers. At certain conditions these polymers may contain five- and six-membered cyclics as the repeating units in the chains and the same cyclic units at the chain ends. Thus, the general structure is (e.g. from H_3PO_4):

a)

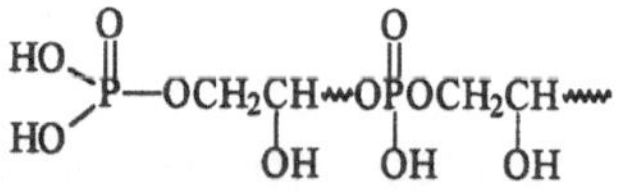

Eq.29a

(branching at CH-OH is not shown). And

b)

$$HO-\overset{\overset{O}{\parallel}}{P}\overset{O-CH_2}{\underset{O-CH}{\diagup\diagdown}}\cdots O\overset{O}{\underset{}{P}}OCH_2CH\cdots O\overset{O}{\underset{}{P}}\overset{O-CH_2}{\underset{O-CH}{\diagup\diagdown}}\cdots$$

$$\quad\quad\quad\quad\quad\quad\quad\quad\quad\quad\quad\overset{}{OH}\;\overset{}{OH}$$

Eq. 29b

We have recently elaborated methods of interconverting the noncyclic polymers (like in eq.29a) into the cyclic ones (eq.29b) and vice-versa. These reactions proceed reversibly and without affecting the chain length, because of the relatively high strain of the five-membered phosphate rings (equal to ~20kJ/mol [16]). Thus, only the endocyclic groups in the cyclic units react and the exocyclic ones do not (their reaction would have led to the chain scission).

The corresponding chemical shifts of these structural units are (^{31}P NMR, in ppm δ):

$$\underset{\sim O}{\overset{O}{\diagdown}}\overset{O-CH_2}{\underset{O-CH\sim}{P}} \quad ; \quad \sim O-\overset{\overset{O}{\parallel}}{P}-OCH_2CH\sim \quad ; \quad \underset{\sim O}{\overset{O}{\diagdown}}\overset{O-CH_2}{\underset{O-CH\sim}{P}}CH_2$$

$$\quad\quad\quad\quad\quad\quad\quad\quad\quad\quad\overset{}{OH}\;\overset{}{OH}$$

| ppm δ: | $16.5 \div 20$ | $3.0 \div -3.0$ | $-6.5 \div 9.5$ |

(only from diepoxybutane)   Eq.30

Simple dissolution of polymers containing cyclics in water restores the linear units. The equally simple heating of thus prepared polymers in boiling benzene/dimethyl sulfoxide mixture gives back cyclics. ^{31}P$\{^1$H$\}$ NMR spectra of cyclic and linear structures (vide supra) differ sufficiently enough to allow controlling of these interconversions. Cyclization of polymers prepared from H$_3$PO$_4$ and diglycidyl ether of tri(ethylene glycol) gives polymers with up to 60% of cyclic repeating units (five-membered rings).

Cyclic units present in the macromolecules are reactive enough and polymers containing these units can simply be functionalized. It has recently been shown, that alcoholysis of the rings proceeds smoothly at room temperature, and various functions can be introduced this way to poly(alkylene phosphates).

Poly(alkylene phosphates) as described in this paper are water soluble, nontoxic and can be used to attach the biologically active substances to the backbones.

Below we have given, as an example, the chemical structure of one of the typical adducts, prepared by a direct addition of Z-serine to the poly(alkylene phosphate) containing cyclic repeating units and prepared from H$_3$PO$_4$ and triethylene glycol in the presence of acetic anhydride, in order to suppress branching:

$$\left(CH_2CHCH_2(OCH_2CH_2)_3OCH_2CHCH_2O\overset{\overset{O}{\parallel}}{P}O\right)_x\left(CH_2CHCH_2(OCH_2CH_2)_3OCH_2CHCH_2O\overset{\overset{O}{\parallel}}{P}O\right)_y$$

$$\quad\; \underset{CH_3}{\overset{|}{OCO}}\quad\quad\quad\quad\quad\quad \overset{|}{OH}\quad \overset{|}{O}\quad\; \overset{|}{OH}\quad\quad\quad\quad\quad\quad \underset{CH_3}{\overset{|}{OCO}}\; \overset{|}{OH}$$

$$\quad\quad\quad\quad\quad\quad\quad\quad\quad\quad\quad\quad CH_2CHCOOH$$

$$\quad\quad\quad\quad\quad\quad\quad\quad\quad\quad\quad\quad NHCCH_2-\langle O\rangle$$

$$\quad\quad\quad\quad\quad\quad\quad\quad\quad\quad\quad\quad\quad\; \overset{}{O}$$

Eq.31

The hydrolytic stability of poly(alkylene phosphates) has recently been studied; it depends on pH, and for simple chains (e.g. poly(methyltrimethylene phosphate) the time required to hydrolyze 10% of bonds in the triester units at 25°C and pH=7 equals several hours [17].

REFERENCES

1. Penczek,S, Sekiguchi,H., and Kubisa,P., in preparation.
2. Tokar,R., Kubisa,P., and Penczek,S., 1994, Cationic polymerization of glycidol: coexistence of the activated monomer and active chain end mechanism, *Macromolecules* 27:320-322.
3. Kazanskii,K.S., Pretula,J., and Kuznetsova,V.I., and Penczek,S., in preparation.
4. Nyk,A., Klosinski,P., and Penczek,S., 1991, Water-swelling, hydrolyzable gels through polyaddition of H₃PO₄ to diepoxides, *Makromol.Chem.* 192:833-846.
5. Biela,T., and Kubisa,P., 1991, Oligomerization of oxiranes in the presence of phosphoric acid. Kinetics of model reactions, *Makromol.Chem.* 192:473-489.
6. Biela,T., Szymanski,R., and Kubisa,P., 1992, Oligomerization of oxiranes in the presence of phosphorus acids, 2. Kinetics of addition of ethylene oxide to phosphoric and phosphorus acid, 1992, *Makromol.Chem.* 193:285-301.
7. Biedron,T., Szymanski,R., Kubisa,P., and Penczek,S., 1990, Kinetics of polymerization by activated monomer mechanism, *Makromol.Chem., Macromol.Symp.* 32:155-168.
8. Bednarek,M., Biedron,T., Kubisa,P., and Penczek,S., 1991, Activated monomer polymerization of oxiranes. Microstructure of polymers vs. kinetics and thermodynamics of propagation, *Makromol.Chem.,Macromol.Symp.* 42/43:475-487.
9. Penczek,S., and Kubisa,P., 1993, Cationic ring-opening polymerization, in Ring-opening polymerization, ed. Brunelle,D.J., 13-86.
10. Madec,P.I, and Maréchal,E., 1985, Kinetics and mechanisms of polyesterifications. II. Reactions of diacids with diepoxides, *Adv.Polym.Sci.* 71:153-228.
11. Klosinski,P., and Penczek,S., 1988, Addition polymerization of H₃PO₃ to diepoxides, *Makromol.Chem.,Rapid Commun.* 9:159-164.
12. Penczek,S., Kubisa,P., and Szymanski,R., 1986, Activated monomer propagation in cationic polymerizations, *Makromol.Chem.,Macromol.Symp.* 3:203-220.
13. Szabo,Z.G., in Comprehensive chemical kinetics, 1969, Bamford,D.H., and Tipper,C.F.H., eds., Elsevier Amsterdam, 69.
14. Biela,T., Szymanski,R., and Kubisa,P., 1992, Oligomerization of oxiranes in the presence of phosphoric acids, *Makromol.Chem.* 193:285-301.
15. Potrzebowski,M., Nyk,A., and Ciesielski,M., in preparation.
16. Sosnowski,S., Libiszowski,J., Slomkowski,S., and Penczek,S., 1984, Thermodynamics of the polymerization of ethylene methyl phosphate, *Makromol.Chem., Rapid Commun.* 5:239-244.
17. Baran,J., and Penczek,S., 1995, Hydrolysis of polyesters of phosphoric acid. I. Kinetics and the pH profile, *Macromolecules*, in press.

21

FUNCTIONAL POLYMERS WITH VARIOUS MACROCYCLIC CHAIN ARCHITECTURES AND WELL-DEFINED DIMENSIONS

Alain Deffieux,[*] Michel Schappacher, and Laurence Rique-Lurbet

Laboratoire de Chimie des Polymères Organiques
ENSCPB-CNRS
Université Bordeaux 1
351, cours de la libération
33405, Talence-cédex
France

INTRODUCTION

Natural macrocycles represent a broad class of medium to large size molecules which actively contribute to numerous biological and chemical processes. Their cyclic architecture, generally associated to a special arrangement of comonomer units allow them to play very specific and sophisticated roles in complex reaction pathways. For example, Valinomycin shown in Figure 1 possesses inner/outer amphiphilic ring chain properties which makes it able to complex organic alkali salts. It may be regarded as one of the natural forerunners of large size crown ethers. Amphotericin, Figure 2a, presents an amphiphilic diblock-type cyclic architecture. The latter allows the formation of ion channels across biological membranes through which salts can be transported. Via this function Amphotericin can manifest antibiotic properties. Other large size macrocycles with a distinct diblock-type amphiphilic structure are also found in living systems. The one shown in Figure 2b, is a constituent of the lipidic membrane of bacteria which grow in very hard conditions (pH=2, 85°C), thus suggesting that cell wall stability can be reinforced by cyclic molecular architecture.

The preparation by living polymerization and the investigation of the properties of cyclic polymers associating a specific function resulting to the presence of functional groups in their backbone to a cyclic or multi cyclic chain architecture has been recently undertaken. After a brief survey of the other strategies used for the preparation of macrocyclic polymers, the specific synthetic procedures developed to prepare mono- and pluri-macrocyclic polymers and to introduce a series of functional or interactive groups,

---

[*] Address correspondence to Dr. Deffieux.

Macromolecular Engineering, Edited by M.K. Mishra et al.
Plenum Press, New York, 1995

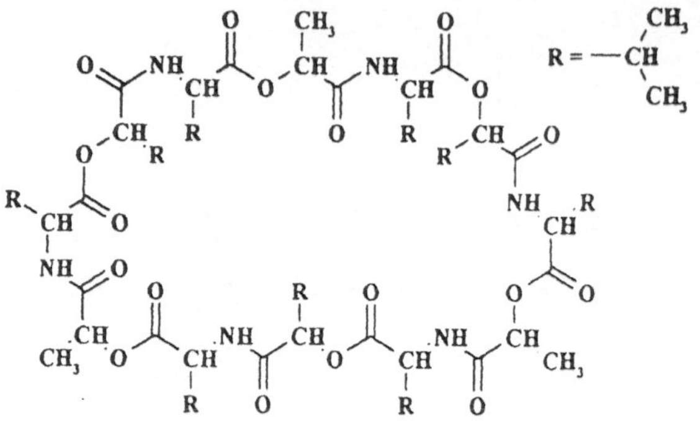

Figure 1. Structure of Valinomycine.

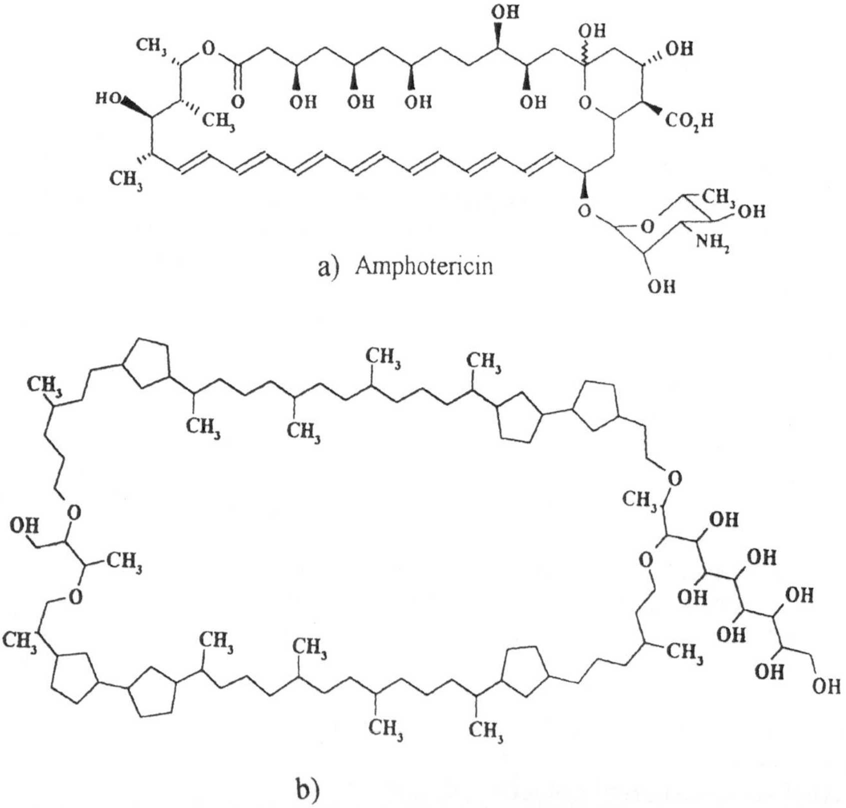

a) Amphotericin

b)

Figure 2. Amphiphilic di-block macrocycles with two distinct structures.

as well as some of the original properties identified on these materials, are presented in this article.

RESULTS AND DISCUSSION

I. Strategies for the Synthesis of Cyclic Polymers

Three main strategies have been used so far for the preparation of polymers with a cyclic chain architecture ;

Formation of Cyclics in Linear-to-Ring Chain Equilibrated Systems. This first method is based on the equilibrated formation of linear and cyclic macromolecules in systems containing reactive functions in their backbone ; polycondensates (polyesters, polycarbonates,..), polymers issued from ring opening polymerization of heterocycles and polyalkenamers belong to this category. Cyclization[1,2] involves the intramolecular attack of one of the reactive functions of the backbone by the active chain-end. Macrocycles obtained in this way have an extremely broad distribution of size, ranging from small to large rings and often contain in admixture a fraction of linear chains. Very efficient fractionation procedures, such as high performance preparative size exclusion chromatography[3], are necessary to recover pure macrocycles of controlled dimensions and narrow polydispersity.

Bimolecular End-To-End Coupling of α,ω-Difunctional Polymers. The second strategy involves the end-to-end coupling of ditelechelic linear polymers in the presence of a bifunctional molecule, in highly diluted conditions. One main advantage of this second approach is the control of the size and the molar masses distribution of the macrocyclic polymers which, in principle, correspond to those of the linear precursor used for the cyclization. Although this cyclization technique does not exclude the use of macromolecules with labile functions in the main chain, studies have been mostly conducted with polymers having non reactive functions in their backbone, typically alkenyl polymers. Polystyrene macrocycles, the most extensively studied system[4-16], but also cyclic poly(butadiene)[14,17], cyclic poly(2-vinylpyridine)[18,19] cyclic poly(isoprene)[20], and more recently cyclic block-copolymers poly(styrene-b-dimethylsiloxane)[21,22], poly(styrene-b-2-vinylpyridine)[22], poly(styrene-b-butadiene)[23], have been prepared. In most cases the experimental cyclization yields are noticeably lower than it could be expected from theoretical calculations based on the Jacobson-Stockmayer theory[24.] Even at the very low concentrations used, cyclic polymers are generally obtained in admixture with a fraction of polycondensates (dimers to polymers) which concurrently forms. An interesting exception concerns the quantitative formation of polyisoprene macrocycles[20]. Some theoretical and experimental reasons which may explain the limited conversions into cyclics have been recently stressed[19,25].

Monomolecular End-To-End Ring Closure of α,ω-Heterodifunctional Polymers. A different route to the synthesis of macrocyclic polymers has been recently investigated[25-28]. The general strategy is presented in Scheme 1. In this approach an heterodifunctional polymer precursor is first prepared by a living polymerization technique, in conventional concentration conditions. The cyclization is then performed in a separated stage, corresponding to highly diluted conditions. The ring closure is achieved, after activation, by the direct attack of one of the precursor active ends onto the other end-function of the same polymer

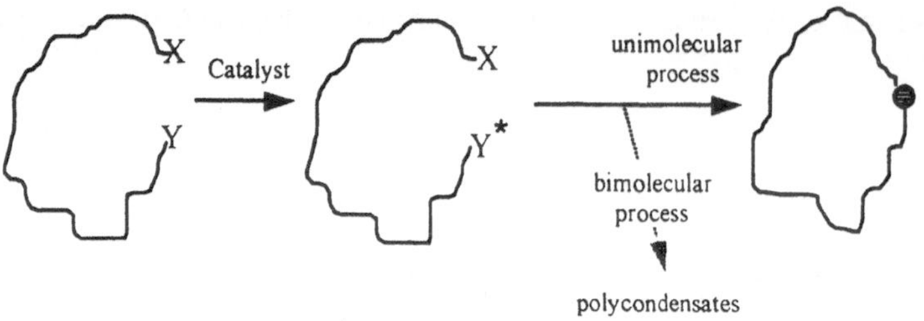

Scheme 1. Formation of macrocyclic polymers by direct end-to-end ring closure. High dilution conditions.

chain. In this mechanism, the ring closure proceeds via an unique unimolecular reaction which is not concentration dependant.

II. Application of Monomolecular End-to-End Ring Closure to the Synthesis of Monocyclic Polymers and Copolymers

a) Poly(Chloroethyl Vinyl Ether)S . The application of monomolecular end-to-end coupling to the synthesis of cyclic poly(chloroethyl vinyl ether), poly(CEVE) has already been described[26,27]. Last refinements in the cyclization experimental procedure and in reaction conditions are indicated in Scheme 2. Briefly, the linear poly(CEVE) precursors with a styrenyl and an acetal termini are added dropwise to a large volume of non-dried methylene dichloride containing a strong Lewis acid as cationogen, preferably $TiCl_4$, at -15°C. In situ cationation of the polymer acetal end is followed by fast reaction onto the α-styrenyl polymer head-group leading to cyclization.

Pure cyclic poly(CEVE)s with narrow molecular weight distribution (<1,1) and molar masses ranging from 1000 to about 10000 are readily prepared by this procedure. The access to cyclic poly(CEVE)s of higher molar masses is mainly limited by the preparation of linear precursors of correct end-functionality.

b) Polystyrene. The unimolecular cyclization procedure has been recently extended to polymer precursors prepared by anionic techniques. This is exemplified by the synthesis of linear α,ω-heterodifunctional poly(styrene)s precursor, Scheme 3, from which it is possible to prepare macrocyclic poly(styrene)s[25-28]. Last refinements in the experimental procedure and the best reaction conditions are also indicated in Scheme 3. The direct slow addition of the α-acetal, ω-styrenyl terminated polystyrene precursor into a large volume of chlorinated solvent containing an appropriate activating agent, either a strong Lewis acid ($SnCl_4$) or trimethylsilyltriflate (route II, Scheme 3) leads to end-to-end ring closure. In these conditions, the formation of polycondensates can be avoided and the cyclization of polystyrenes with molar masses ranging from 1000 to about 10000-15000 is quantitative. Typical GPC chromatograms of linear polystyrene precursors and of the corresponding crude cyclic products are shown in Figure 3a. For polystyrene with molar masses higher than 15000 it should be mentioned however, that the cyclization yields tend to decrease due to uncomplete coupling, see Figure 3b. It is not clear yet whether this situation results from an uncomplete functionalization of the precursor or to the cyclization step itself.

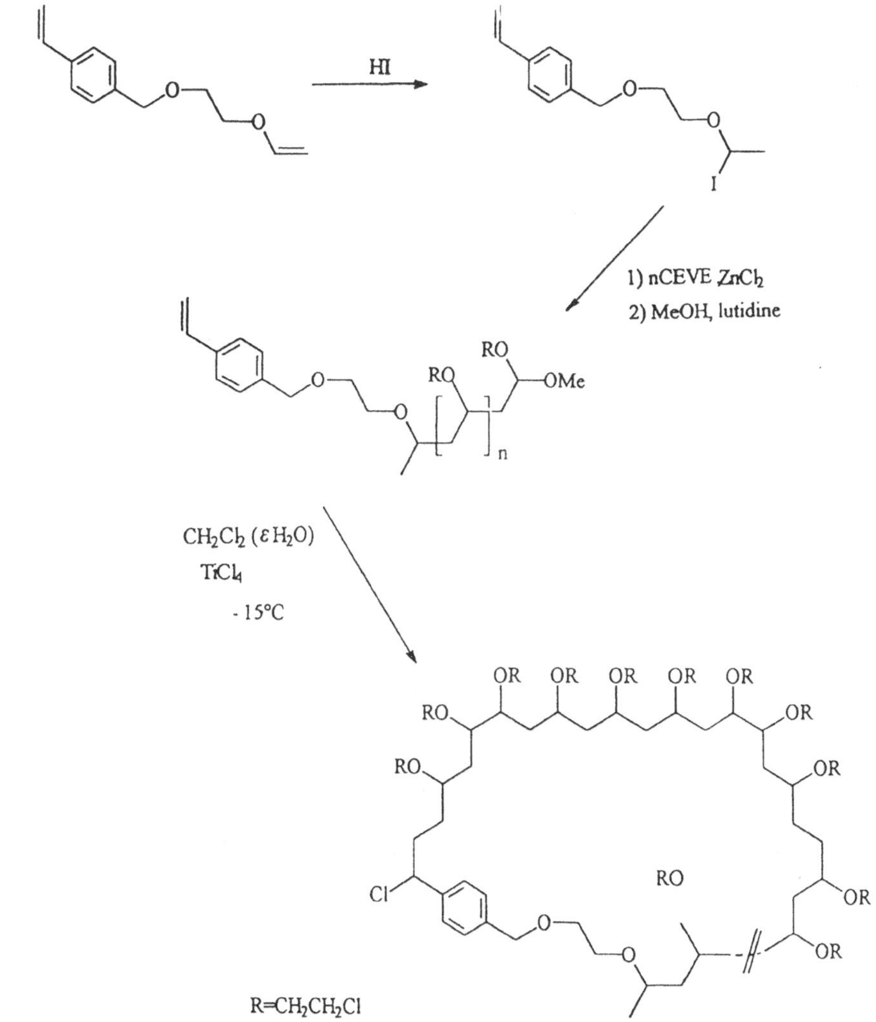

Scheme 2. Synthesis of poly(CEVE) macrocycles.

c) Poly(Styrene-b-Chloroethyl Vinyl Ether) Macrocycles. The synthesis of poly(styrene-b-vinyl ether) diblock copolymers with different chain topologies has been recently investigated[27] in our group. The two architectures indicated in Schemes 4 and 5 have been successively prepared.

In the case of macrocycles constituted of both styrene and vinyl ether blocks, type **I**, a polystyrene precursor with an α-acetal and an ω-styrenyl end is first prepared and used as a macroinitiator. Then the CEVE polymerization is initiated from the polystyrene diethyl acetal end, after it has been derivatized into an α-iodo ether end by the action of trimethyl silyl iodide (TMSI). ZnCl$_2$ is used as polymerization cocatalyst. At the end of the vinyl ether polymerization, addition of ammoniacal methanol regenerates an acetal terminus. The linear heterodifunctional di-block copolymer is then submitted to end-to-end cyclization, yielding macrocyclic poly(styrene-b-CEVE) copolymer. Typical GPC chromatograms of the polystyrene first block, the linear poly(styrene-b-CEVE) and the corresponding macrocyclic di-

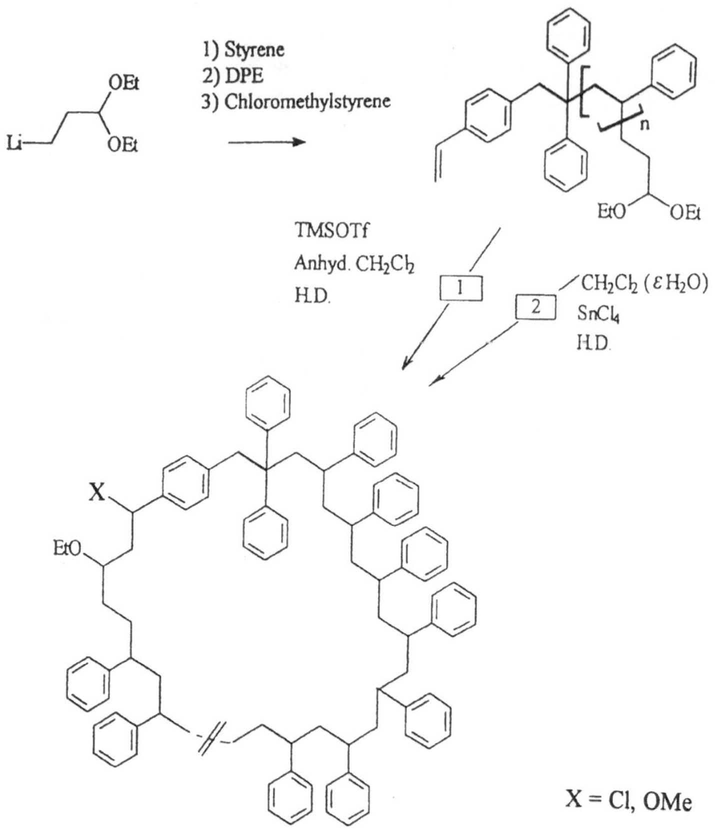

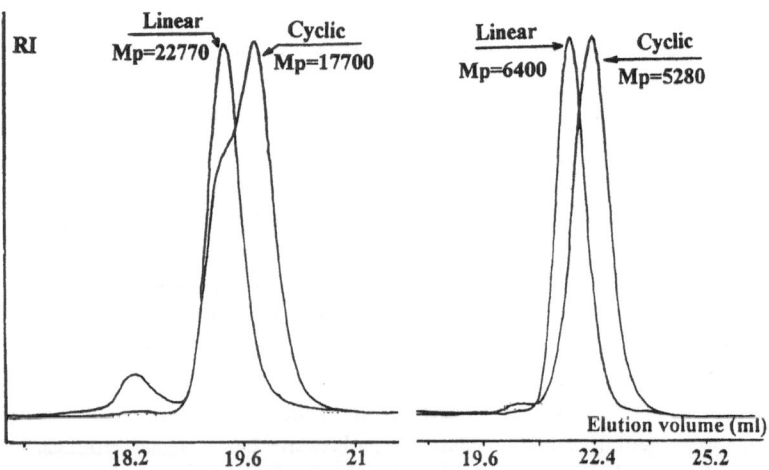

Scheme 3. Synthesis of polystyrene macrocycles.

Figure 3. GPC chromatograms of cyclic polystyrenes obtained by unimolecular end-to-end cyclization of α–styrenyl, ω-acetal polystyrenes in the presence of SnCl₄ as catalyst ; Mp= apparent peak molecular weight.

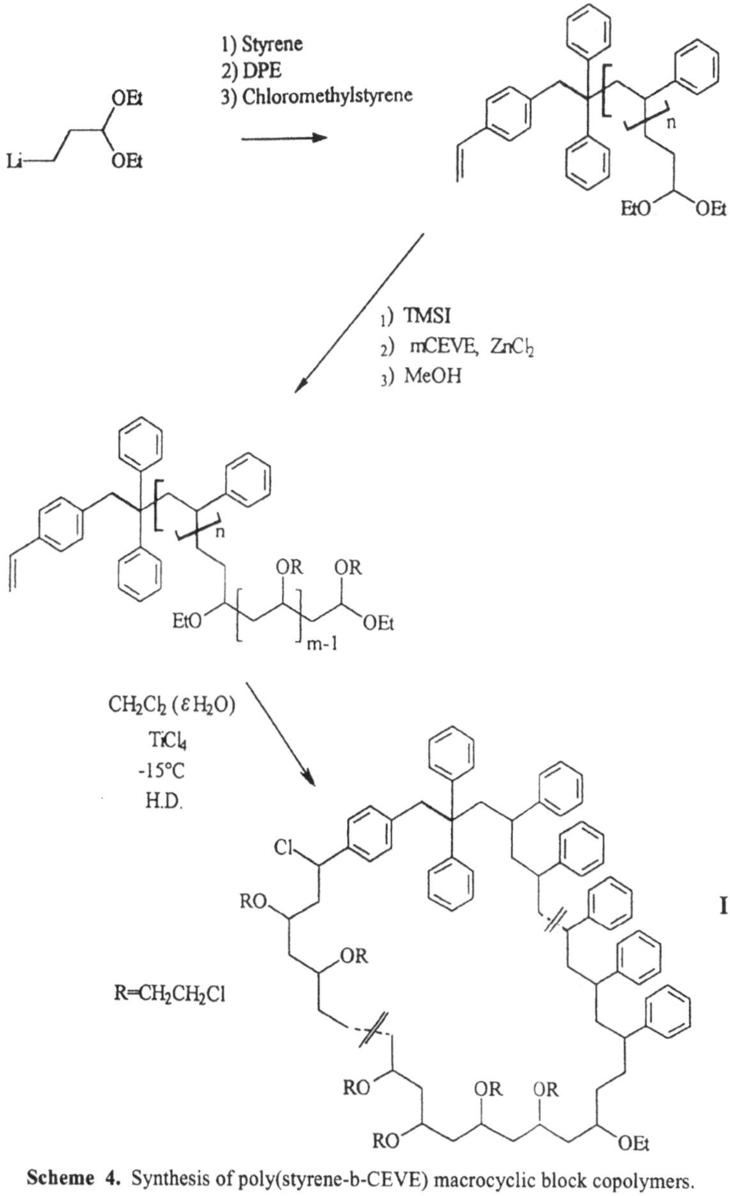
1) Styrene
2) DPE
3) Chloromethylstyrene
OEt
Li
OEt
n
EtO
OEt
1) TMSI
2) mCEVE, ZnCl2
3) MeOH
n
OR
OR
EtO
OEt
m-1
CH2Cl2 (ε H2O)
TiCl4
-15°C
H.D.
Cl
RO
OR
R=CH2CH2Cl
RO
OR
OR
RO
OEt

I

Scheme 4. Synthesis of poly(styrene-b-CEVE) macrocyclic block copolymers.

blocks are presented in Figure 4. The dimensional characteristics of the different polymers and copolymers are collected in Table 1.

For the preparation of block copolymers constituted of a vinyl ether macrocycle and a linear styrene tail, type **II** Scheme 5, the diethyl acetal-end group of an α-functional polystyrene precursor is first transformed into a cyclic acetal terminus by transacetalisation. The addition of TMSI onto the cyclic acetal polymer terminus results both in the initiation of the vinyl ether polymerization, in the presence of $ZnCl_2$, and in the formation of a trimethyl silyl ether group. After hydrolysis, at the end of the polymerization, the latter group yields a pendant hydroxyle located in between the styrene and the CEVE blocks. Typical GPC

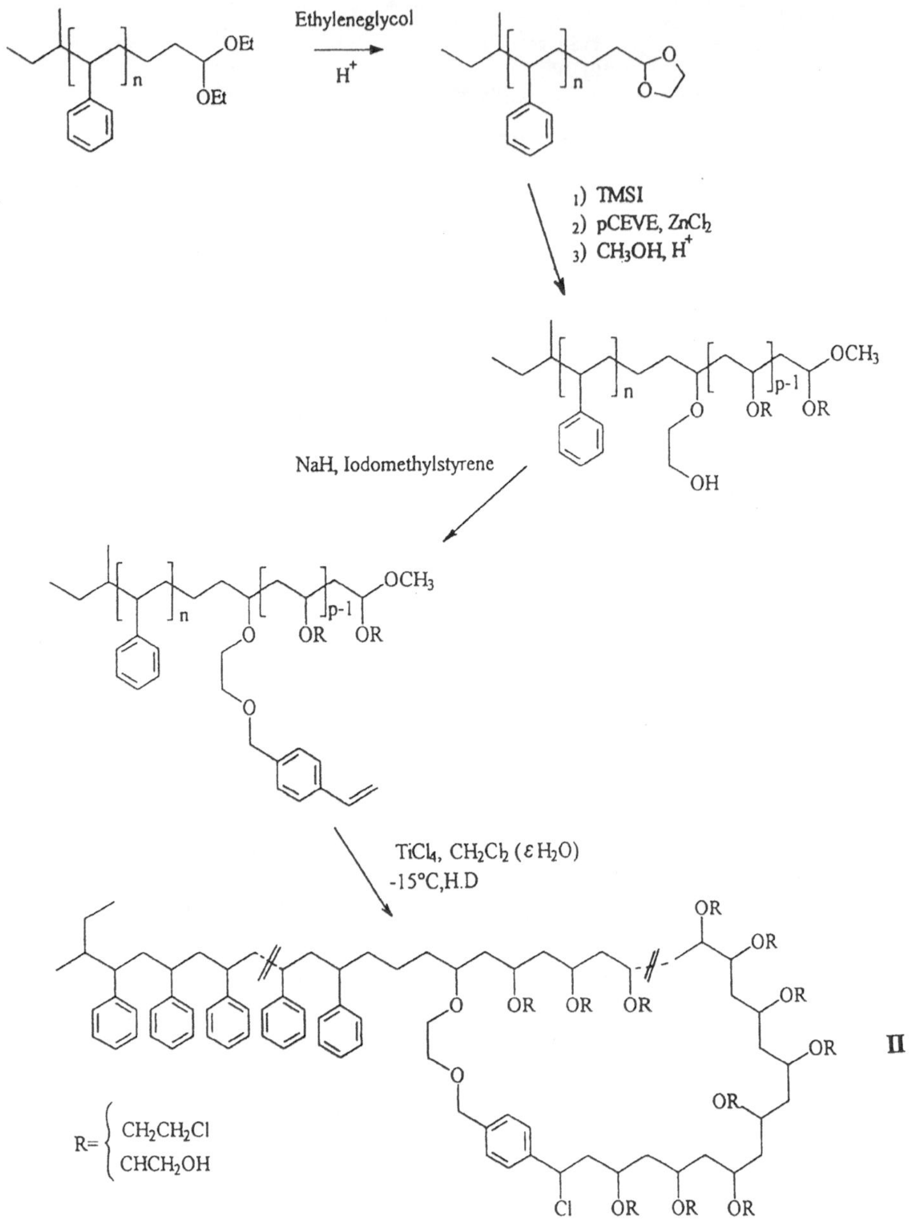

Scheme 5. Synthesis of linear poly(styrene)-b-cyclic poly(CEVE) block copolymers.

chromatograms of the polystyrene first block and of the linear poly(styrene-b-CEVE) are shown in Figure 5.

The hydroxyl group located in the center of the chain is then used to anchor a styrenyl group by reaction with p-iodomethylstyrene. The ^{1}H NMR spectrum of the di-heterofunctional di-block copolymer bearing an ω-acetal and a μ-styrenyl group (where μ denote a function located in the middle of the chain) is shown in Figure 6. The proportion between the styrene and CEVE units in the one hand as well as the relative intensities of the signals

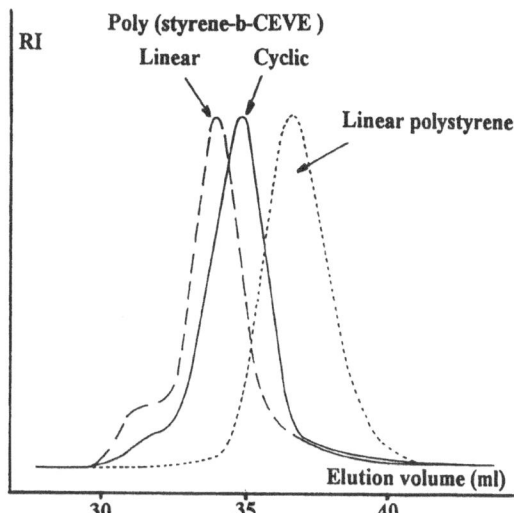

Figure 4. GPC chromatograms of a cyclic poly(styrene-b-CEVE) (M_p = 4350) obtained by unimolecular end-to-end cyclization of α–styrenyl, ω-acetal poly(styrene-b-CEVE) (M_p = 2200) in the presence of TiCl$_4$ as catalyst ; polystyrene first block, $\overline{M}_n$ = 1800, $\overline{M}_W/\overline{M}_n$ = 1,07 ; α–styrenyl, ω-acetal poly(styrene-b-CEVE),$\overline{M}_n$ = 4420, $\overline{M}_W/\overline{M}_n$ = 1,08.

characteristic to the styrenyl and acetal functions in the other hand are in good agreement with the expected copolymer structure.

The di-heterofunctional copolymer is then used as linear precursor in the synthesis of macrocyclic compounds of structure **II**, see Scheme 5. Typical GPC chromatograms of the linear poly(styrene-b-CEVE) and the corresponding macrocyclic di-block are presented in Figure 5. On the basis of NMR spectra and GPC chromatograms we may assume that diblock macrocycles of type **II** are almost quantitatively formed.

The reduction of the hydrodynamic volume <G_{exp}> of the chains resulting from cyclization of the linear block copolymers can be determined from the ratio of the respective apparent peak molecular weights (Mp). <G_{exp}> for structures of types **I** and **II** are indicated in Tables 1 and 2 respectively. Results are in agreement with previous data concerning polystyrene and poly(CEVE) macrocycles, each one close to 0,8, and with the two expected macrocyclic block copolymer structures **I** and **II** ; a <G_{exp}> value of 0.81 is observed for the poly(styrene-b-CEVE) macrocycles whereas for the poly(chloroethyl vinyl ether) macrocycles with a polystyrene tail, <G_{exp}> varies with the relative

Table 1. Characteristics of linear α,ω-heterodifunctional poly(styrene-b-CEVE)s and the corresponding cyclic poly(styrene-b-CEVE)s

Chain architecture	$\overline{DP}_n{}^a$ (NMR)	$\overline{M}_n$ (NMR)	Mp[b]	$\overline{M}_W/\overline{M}_n$	<Gexp>[c]
Linear polystyrene[d]	18	2300	2200	1.07	
Linear poly(styrene-b-CEVE)	18 and 20	4416	4350	1.08	
Cyclic poly(styrene-b-CEVE)	18 and 20	4420	3520	1.16	0.81
Linear polystyrene[d]	23	2820	2515	1.02	
Linear poly(styrene-b-CEVE)	23 and 33	7390	6400	1.15	
Cyclic poly(styrene-b-CEVE)	23 and 33	7384	5200	1.24	0.81

[a]Determined on the basis of end-groups.
[b]Apparent peak molecular weight determined by gel permeation chromatography using linear polystyrene standards.
[c]<Gexp> = M_{pcycl}/M_{plin}.
[d]α,ω-heterodifunctional polystyrene used as chain initiator.

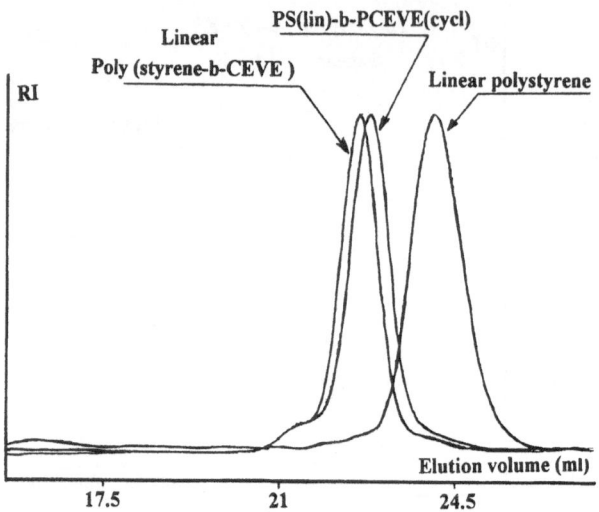

Figure 5. GPC chromatograms of a linear poly(styrene)-b-cyclic poly(CEVE) ($M_p = 4300$) obtained by unimolecular end-to-end cyclization of μ–styrenyl, ω-acetal poly(styrene-b-CEVE) ($M_p=$) in the presence of $TiCl_4$ as catalyst ; polystyrene first block, $\overline{M}_n = 2680$, $\overline{M}_W/\overline{M}_n = 1{,}03$; α-styrenyl, ω-acetal poly(styrene-b-CEVE), $\overline{M}_n = 4580$, $\overline{M}_W/\overline{M}_n = 1{,}04$.

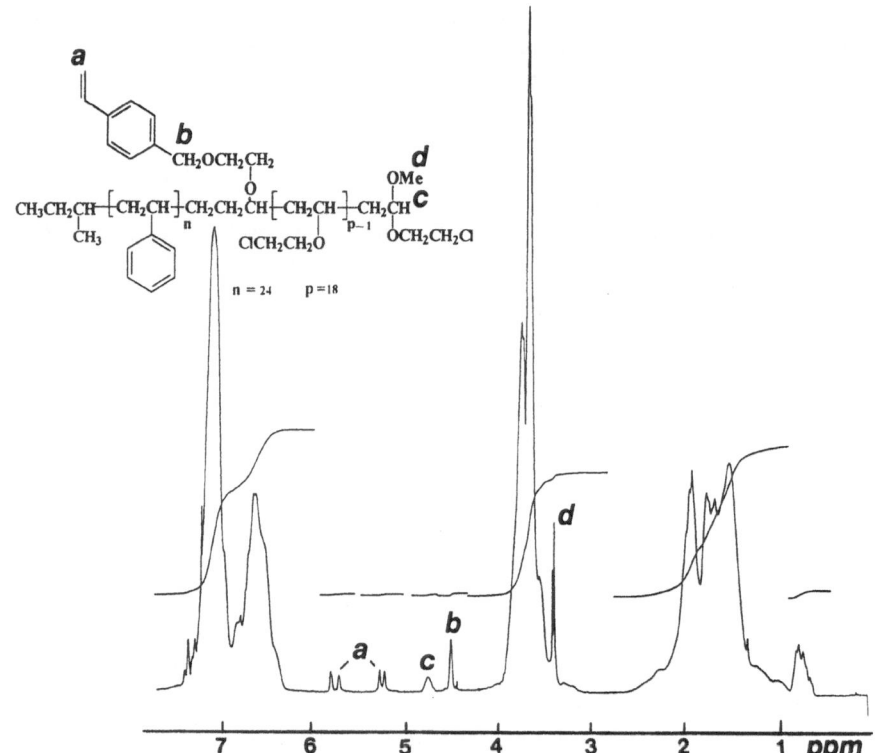

Figure 6. 200 MHz ^{1}H NMR spectrum of an μ–styrenyl, ω-acetal poly(styrene-b-CEVE) used for the synthesis of linear poly(styrene)-b-cyclic poly(CEVE), type **II** and attribution of the resonances of the reactive chain end functions.

Table 2. Characteristics of linear heterodifunctional poly(styrene-b-CEVE)s and the corresponding poly(lin.styrene-b-cyc.CEVE)s

Chain architecture	$\overline{DP}_n{}^a$ (NMR)	$\overline{M}_n$ (NMR)	Mp[b]	$\overline{M}_w/\overline{M}_n$	<Gexp>[c]
Linear polystyrene[d]	24	2684	2460	1.03	
Linear poly(styrene-b-CEVE)	24 and 18	4587	4700	1.04	
poly(lin.styrene-b-cycl.CEVE)	24 and 18	4591	4300	1.05	0.91
Linear polystyrene[d]	38	4128	4624	1.03	
Linear poly(styrene-b-CEVE)	38 and 16	5818	6820	1.04	
poly(lin.styrene-b-cycl.CEVE)	38 and 16	5822	6547	1.05	0.96
Linear poly(styrene-b-CEVE)	38 and 27	7003	7839	1.02	
poly(lin.styrene-b-cycl.CEVE)	38 and 27	7007	7141	1.03	0.91

[a]Determined on the basis of end-groups.
[b]Apparent peak molecular weight determined by gel permeation chromatography using linear polystyrene standards.
[c]<Gexp> = M_{pcycl}/M_{plin}.
[d]α-acetal polystyrene used as first building block.

dimensions of the linear and cyclic blocks; for example it is close to 0.9 when the two blocks have the same $\overline{DP}_n$.

d) Functional Macrocycles by Chemical Modification. The pendant chlorines of chloroethyl vinyl ether units can be exchanged using nucleophilic substitution. Their replacement under mild conditions by various functions or organic groups can be achieved in very good yield, almost quantitatively in many cases, without any noticeable influence of the substitution conditions on the chain architecture and polymer dimensions[29,30]. Some examples of functional side groups which have been anchored onto poly vinyl ether chains are presented in Scheme 6. In the same way, a large series of organic groups able to generate specific polymer functions (liquid crystal, complexing, biological, etc..properties), can also be introduced in the chain[30,31].

According to the nature and the proportion of the lateral functions, the polarity, the solubility and the hydrophobic/hydrophylic balance of the poly(vinyl ether) segments can be drastically modified. Macrocyclic poly(chloroethyl vinyl ether) can be converted into macrocyclic random (chloroethyl vinyl ether-co-hydroxyethyl vinyl ether) copolymers through partial substitution of the chloride groups with acetate or benzoate followed by alkaline hydrolysis. Total substitution of chlorine can also be readily achieved by increasing the proportion of nucleophile, yielding an hydrophylic macrocyclic poly(hydroxyethyl vinyl ether).

This approach was used to synthesize linear and macrocyclic amphiphilic poly (hydroxyethyl vinyl ether-b-styrene) corresponding to types **I** and **II**. Results obtained in the case of di-blocks of type II are presented in some detail to illustrate the efficiency of this functionalization technique.

Replacement of chloro-2 ethyl pendant groups by acetoxy-2 ethyl ones is first achieved by reacting the poly(styrene-b-CEVE) block copolymers, linear and cyclic, with sodium acetate in the presence of tetrabutyl ammmonium hydrogenosulfate as catalyst. Final transformation into hydroxy ethyl group is then realized by treatment of the esterified polymer with sodium hydroxide. The NMR spectra of the poly (acetoxyethyl vinyl ether-b-styrene) and of the poly (hydroxyethyl vinyl ether-b-styrene) are presented in Figures 7a and 7b. The CH_2Cl resonance of the initial poly(CEVE), located at 3.5 ppm (Figure 6), is replaced by new signals corresponding to protons of the methylene and

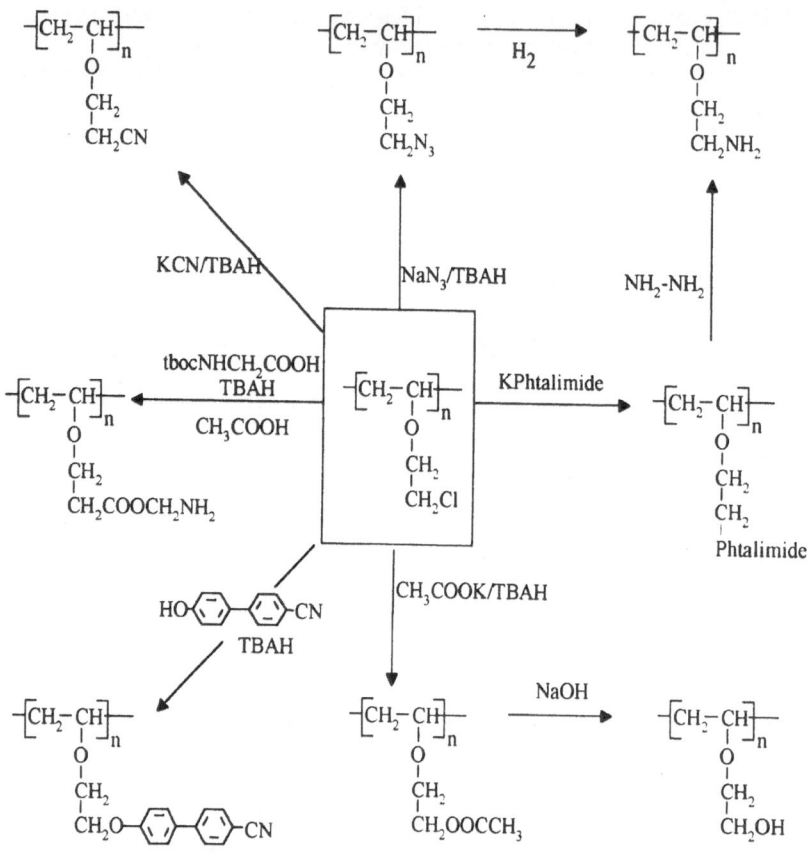

Scheme 6. Derivatizations of chloroethyl vinyl ether units into various functional groups.

methyl of the new pendant C$\underline{H}_2$-OCOC$\underline{H}_3$ groups respectively at 4.2 and 2.1 ppm. The acetoxy groups in turn are quantitatively converted into CH$_2$-OH characterized by a distinct O$\underline{H}$ resonance, located at 4,6 ppm in DMSO. Besides, the GPC chromatograms of the different polymer derivatives, do not show any noticeable change in the chromatogram profiles which remain very narrow ($\overline{M}_W/\overline{M}_n$ < 1,1), in agreement with a clean chemical modification process.

We have recently started an investigation of the amphiphilic properties of the linear and macrocyclic di-block copolymers. Preliminary results show that these properties are strongly influenced by the chain architecture. For example, macrocyclic copolymers of type **II** give Langmuir-Blodgett films[32] much more stable than their linear homologues, suggesting different organization capacities of the macrocyclic block copolymers.

III. Application of Unimolecular End-To-End Ring Closure Technique to the Synthesis of Polymers with a Plurimacrocyclic Chain Architecture

Because the synthesis of well-defined mono-macrocyclic polymers was not satisfactorily controlled until recently, the preparation of polymers with a pluricyclic chain architecture remained for long a limited research domain. Unimolecular end-to-end polymer cyclization procedures have opened interesting possibilities in this area.

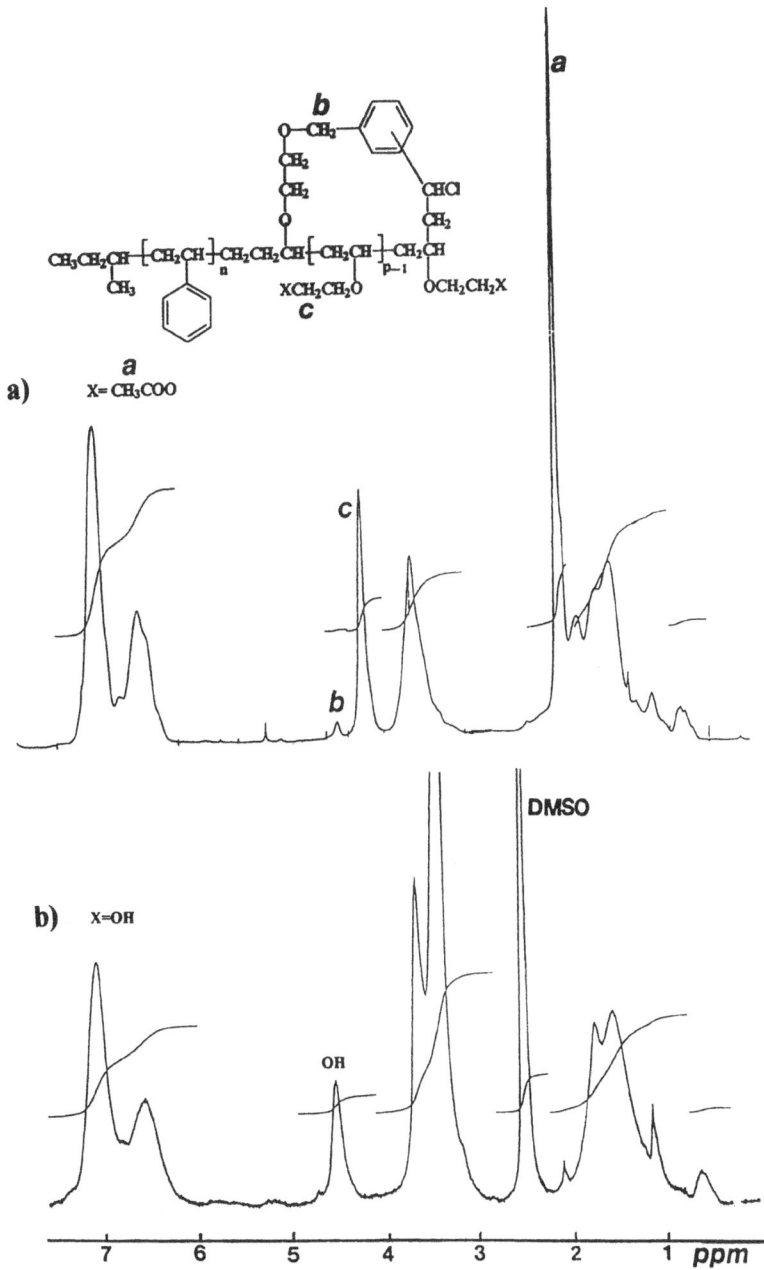

Figure 7. 200 MHz ^{1}H NMR spectrum of ; a) a linear poly(styrene)-b-cyclic poly(acetoxyEVE) obtained by chemical modification of the corresponding chlorinated copolymer; b) a linear poly(styrene)-b-cyclic poly(hydroxyEVE) resulting from the hydrolysis of a).

a) Synthesis of Bicyclic "Eight-Shaped" poly(CEVE). First direct attempts to prepare polymers with a bicyclic "eight-shaped" chain-architecture have been reported by Antonietti[33], in 1988 and very recently by El Madani[34]. The strategy applied by the two groups is derived from the synthesis of mono-macrocycles through bimolecular end-to-end coupling. The use of a tetrafunctional compound, SiCl₄, instead of a difunctional molecule,

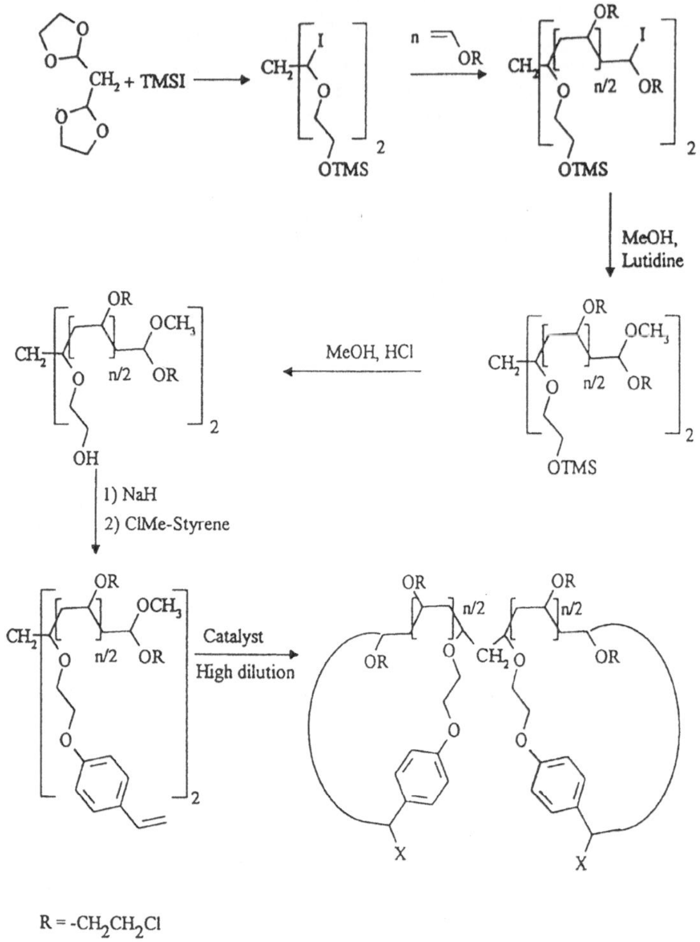

R = -CH₂CH₂Cl

Scheme 7. Synthesis of bicyclic eight-shaped poly(CEVE).

allowed them to react two bifunctional polymer chains per one coupling molecule. Indeed, although the low molar mass fraction was identified as the bicyclic polymer, several other populations of polymers including polycondensates were obtained in important proportions.

Using a different strategy inspired from the unimolecular ring closure process developed for mono-macrocycles, well-defined bicyclic eight-shaped poly(chloroethyl vinyl ether)s have been recently synthesized[35]. The general approach consists in the synthesis of a linear polymer precursor bearing two pairs of dual functions, acetal and styrenyl, which can be coupled under appropriate conditions. The detailed reaction pathway used for the synthesis of bicyclic eight-shaped poly(CEVE)s is given in Scheme 7. A typical GPC chromatogram of a linear precursor and of the corresponding crude bicyclic polymer are presented in Figure 8. As may be seen, a characteristic shift of the GPC signal towards the low molecular weight polymer range, corresponding to a ratio $Mp_{lin}/Mp_{bi-cycl}$ close to 0,8 is observed in agreement with a reduction of the hydrodynamic volume of the cyclized macromolecule. The low polydispersity of the bicyclic polymer, the absence of polycondensation products, as well as structural characterization, such as polymer end-group analysis,

PCEVE

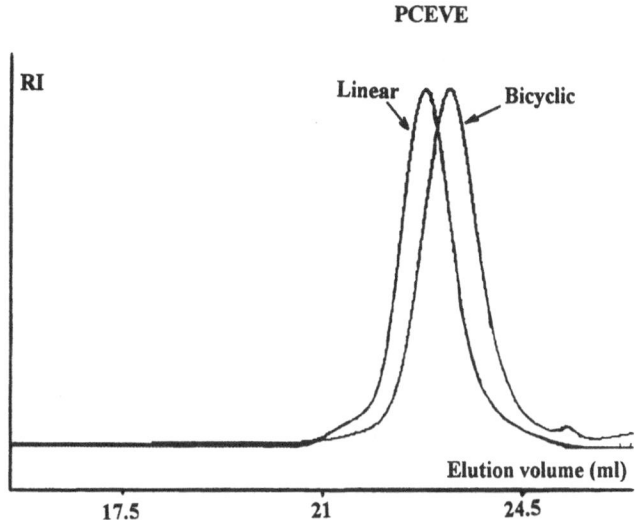

Figure 8. SEC curves of linear α,α'- distyrenyl, ω,ω' diacetal poly(CEVE) precursor ($\overline{M}_n$ = 3990, $\overline{M}_w/\overline{M}_n$ = 1,06) and the corresponding bicyclic eight-shaped poly(CEVE) ($\overline{M}_w/\overline{M}_n$ = 1,05) obtained by unimolecular coupling in the presence of $TiCl_4$ as catalyst ; crude reaction product without any fractionation.

support the almost quantitative formation of bicyclic eight-shaped poly(CEVE) of controlled dimensions.

Further studies of the solution and bulk behaviour of this new family of macromolecules of original architecture and interesting functionalization capacities are of great interest both for theoretical and applicative aspects.

b) Macrotricyclic Poly(Vinyl Ether)S. Finally, the possibility to prepare macromolecules of controlled size with other types of pluricyclic architectures has also been explored[36]. An interesting example of such bulky

macromolecular objects is the synthesis of macrotricyclic poly(chlorethyl vinyl ether)s, MTC, a cage-like macromolecule which exhibits very interesting complexing properties.

The general strategy followed for the synthesis of poly(chloroethyl vinyl ether) macrotricycles is summarized in Scheme 8. It involves the use of a di-heterotrifunctional polymerization initiator from which a three-armed star-shaped polymer is grown. The cyclization of each arm is achieved under highly diluted conditions by further activation of the active polymer ends in order to allow their addition onto the styrenyl head groups, as previously reported. The main structural and physico-chemical polymer characteristics, which have been determined at the different steps of the synthesis, are in close agreement with the indicated reaction pathway, thus supporting the predominant formation of polymer with a tricyclic architecture. A strong decrease of the apparent molar masses of the polymer (M_{tricyc}/M_{lin} = 0.75) is observed after cyclization while the polydispersity of the final polymer remains narrow (<1,3) and the amount of polycondensates formed is negligible, Figure 9.

Very intriguing complexation properties[29] between MTC and various organic compounds of different size and chemical structure, including dyes and porphyrins are observed, whereas any interaction could be detected when MTC is replaced by linear, monocyclic and star poly(CEVE) in similar conditions.

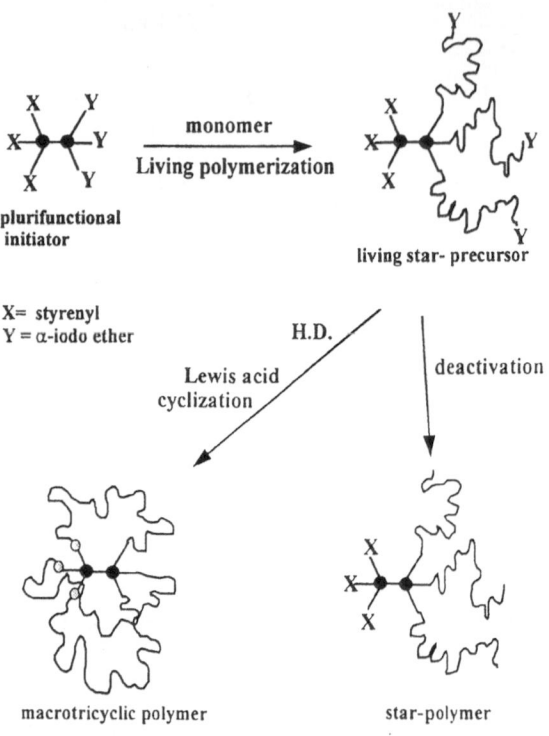

Scheme 8. Strategy for the synthesis of tricyclic poly(CEVE).

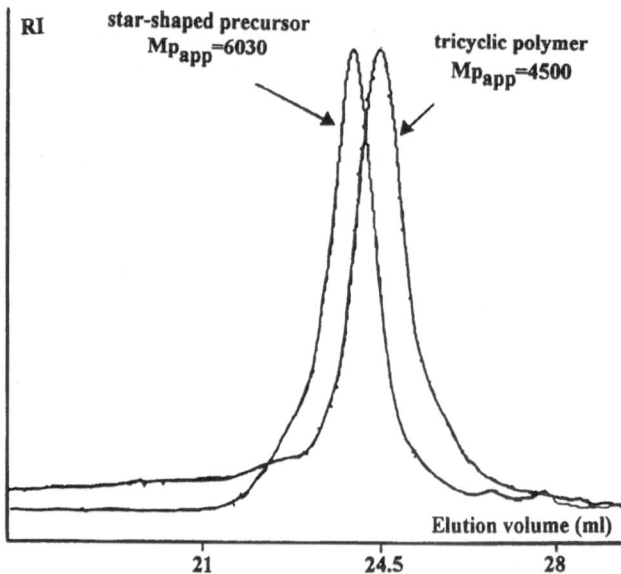

Figure 9. SEC curves of a star α,α',α''- tri-styrenyl, ω,ω',ω'' tri-acetal poly(CEVE) precursor ($\overline{DP}_n$ = 15,7x3, $\overline{M}_w/\overline{M}_n$ = 1,25) and the corresponding macrotricyclic poly(CEVE) obtained by unimolecular coupling ; crude reaction product without any fractionation.

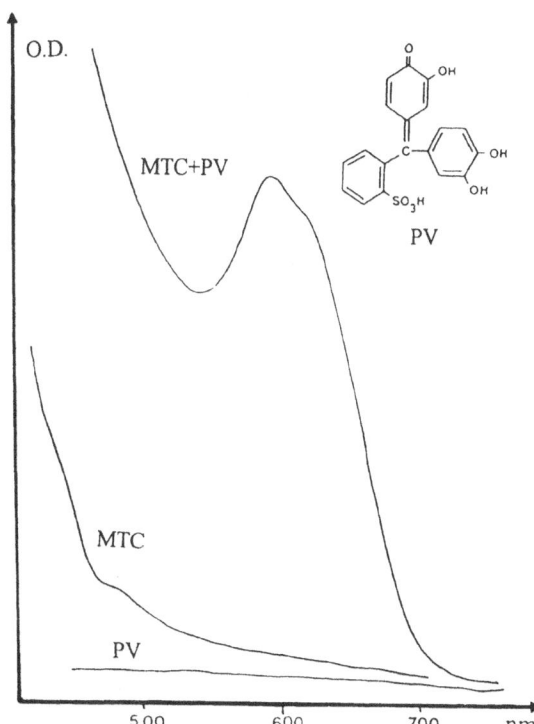

Figure 10. UV-Visible spectra in CH_2Cl_2 of Pyrochatecol Violet (PV), Macrotricyclic Poly(CEVE) (MTC) of $\overline{DP}_n = 15,7 \times 3$ and their mixture 1:1.

Pyrochatecol Violet which is insoluble in methylene dichloride, is readily transferred into the organic phase in the presence of MTC. The UV absorption spectra of the dye, in CH_2Cl_2, in the absence and in the presence of MTC are presented in Figure 10. A similar effect is observed also with Hemin, which becomes soluble in CH_2Cl_2 on addition of MTC, Figure 11. Further evidences for selective complexation have been obtained by NMR[29]. Since, molecular recognition generally implies specific interactions with an host molecule of controlled dimensions and spatial shape, the particular architecture of the tricyclic polymer which may be regarded as a cage-like macromolecule might explain this very specific behaviour.

CONCLUSION

Synthetic mono- and pluri-macrocyclic polymers can be considered as a special class of macromolecules which possess as common feature the capacity to develop very specific properties, in direct or indirect relation with their cyclic chain architecture. Some of the solution and bulk properties of macrocyclic homopolymers have been already identified ; lower chain mobility, higher glass transition of oligomers, lower viscosity, etc...

The results presented in this paper show that other original and very specific properties can be developed by increasing the complexity of the chain architecture. Far ahead in this direction, natural macrocycles are interesting examples which are able to participate to very complex cascade reactions in biological processes.

The recent progress in the living polymerization techniques and in the cyclization procedures now allow the controlled synthesis of polymers with complex macrocyclic

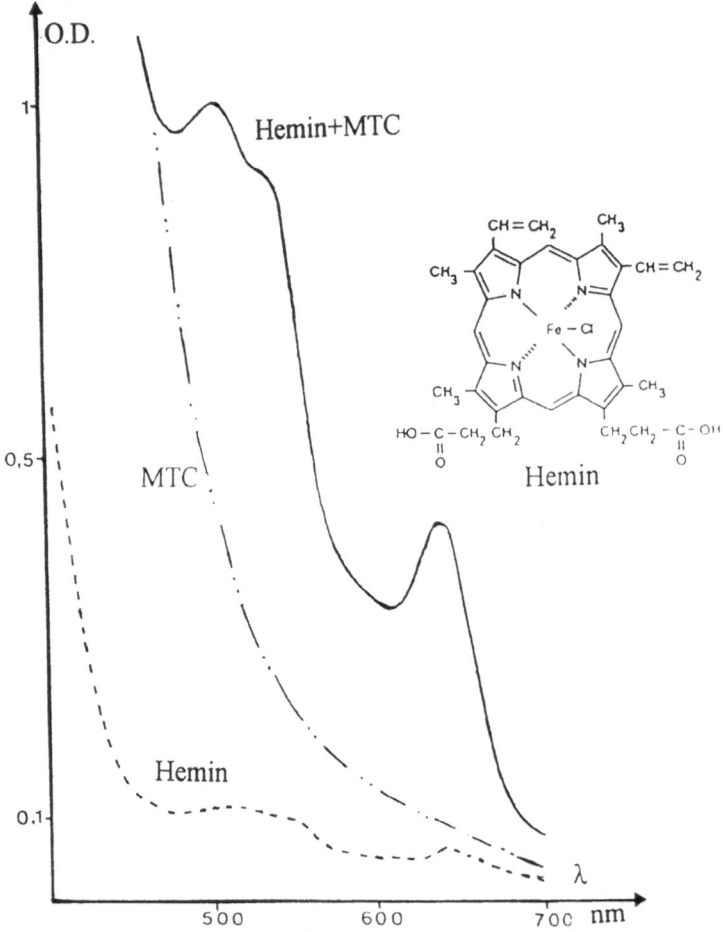

Figure 11. UV-Visible spectra in CH_2Cl_2 of Hemin, Macrotricylic Poly(CEVE) (MTC) of $\overline{DP}_n$ = 15,7x3 and their mixture 1:1.

architectures and adjustable chain dimensions. Future aims of our research group in this area will be focused on the elaboration and the study of simplified amphiphilic synthetic macrocyclic models of these natural macrocycles in order to investigate their potential as substitutes in biological processes.

REFERENCES

1. J. A. Semlyen ed., *Cyclic polymers*, Elsevier Applied Science, London, 1986.
2. S. Penczek and S. Slomkowski, *Comprehensive Polymer Science*, Vol 3, 725, Pergamon Press, Oxford, 1989.
3. K. Dodgson, D. Sympson and J. A. Semlyen, *Polymer*, 1285, **19** (1978).
4. G. B. Mc Kenna, G. Hadziioannou, P. Lutz, G. Hild, C. Strazielle, C. Straupe, P. Rempp and A. J. Kovacs, *Macromolecules*, **20**, 498 (1987).
5. G. B. McKenna., B. J. Hostetter, N. Hadjichristidis, L. J. Fetters and D. J. Plazk, *Macromolecules*, **22**, 1834 (1989).
6. G. Hild, H. Kohler and P. Rempp, *Eur. Polym. J.*, **16**, 525 (1980).

7. G. Geiser and H. Höcker, *Polym. Bull.*, **2**, 591 (1980); ibid. *Macromolecules*, **13**, 653 (1980).
8. B. Vollmert and J. X. Huang, *Makromol. Chem., Rapid Comm.*, **2**, 467 (1981).
9. J. Roovers and P. M. Toporowski, *Macromolecules*, **16**, 843 (1983).
10. G. Hild, C. Strazielle and P. Rempp, *Eur. Polym. J.*, **19**, 721 (1983).
11. J. Roovers, *Macromolecules*, **18**, 1359 (1985).
12. P. Lutz, G. B. McKenna, P. Rempp, C. Strazielle, *Makromol. Chem., Rapid Comm.*, **7**, 599 (1986).
13. P. Lutz, C. Strazielle, P. Rempp, *Recent Advances in Anionic Polymerization*, 403, T.E Hogen-Esch and J. Smid ed., Elsevier Sci. Pub.,1987.
14. R. P. Quirk and J. J. Ma, *Polym. Prep.*, **29**, 2, 10 (1988).
15. X. Liu, D. Chen, Z. He, H. Zhang and H. Hu, *Polym. Comm*, 32, 4, 123 (1991).
16. J. Sundararajan and T. E. Hogen-Esch, *Polym. Prep.* **32**, 3, 604 (1991).
17. J. Roovers and P. M. Toporowski, *J. Polym. Sci., part B, Polymer Physics*, **26**, 1251 (1988).
18. T. E. Hogen-Esch, J. Sundararajan and W. Toreki, *Polym. Prep.*, **29**, 2, 17, (1988).
19. T. E. Hogen-Esch, J. Sundararajan and W. Toreki., *Makromol. Chem., Macromol. Symp.*, **47**, 23 (1991).
20. A. El Madani, J. C. Favier, P. Hémery and P. Sigwalt, *Polym. Internat.*, **27**, 353 (1992).
21. R. Yin and T. E. Hogen-Esch, *Polym. Prep.* **33**, 1, 239 (1992); *Macromolecules*, **26**, 6952 (1993).
22. Y. D. Gan, J. Zöller, R. Yin and T. E. Hogen-Esch, *Macromol Symp.*, **77**, 93 (1994).
23. J. Ma, *Polym. Prep.*, **34**, 2, 626, (1993).
24. H. Jacobson and W. H. Stockmayer, *J. Chem. Phys.* **18**, 1600 (1950).
25. L. Rique-Lurbet, M. Schappacher and A. Deffieux, *Macromolecules*, 27, 6318 (1994).
26. M. Schappacher and A. Deffieux, *Makromol. Chem., Rapid Comm.*, **12**, 447 (1991).
27. A. Deffieux, M. Schappacher and L. Rique-Lurbet , *Polymer*, **35**, 21, 4562 (1994).
28. A. Deffieux, M. Schappacher and L. Rique-Lurbet , *Polym. Prep.*, **35**, 2, 494 (1994)
29. M. Schappacher and A. Deffieux, submitted to *Polym. Adv. Tech.*.
30. E. Papon, A. Deffieux, F. Hardouin and M. F. Achard, *Liquid Crystals*, **11**, 6, 803 (1992)
31. M. Sanchez, A. Deffieux, M. Fontanille, C. Baquey and L. Bordenave, *Clinical Materials*, **15**, 253 (1994)
32. C. Mingotaud, A. Deffieux, M. Schappacher and S. Beinat to be published.
33. M. Antonietti and K. J. Fölsch, *Makromol. Chem., Rapid Comm.*, **9**, 423 (1988).
34. P. Hémery, J. M. Boutillier, A. El Madani, J. C. Favier and P. Sigwalt, *Polym.Prep.*, **35**, 2, 478 (1994).
35. M.Schappacher and A. Deffieux, submitted to *Macromolecules*.
36. M.Schappacher and A. Deffieux, *Macromolecules*, **25**, 6744 (1992).

22

FUNCTIONALIZED POLYMERS
Synthesis and Modification

Helmut Ritter

Bergische Universität-GH Wuppertal
FB 9, Organische Chemie und Makromolekulare Chemie
Gaußstr. 20
D-42097 Wuppertal
Germany

INTRODUCTION

Polymers bearing reactive groups are of great scientific and economical interest. Even since the beginning of macromolecular chemistry, functional groups of biopolymers, e.g. of proteins, cellulose or rubber, have been modified chemically to control e.g. processibility, mechanical properties or solubility. Additionally, in the last several decades, many synthetic polymers bearing functional groups have been partially responsible for the rapid development of markets for tailor made organic materials. Recent demands to increase the degree of polymer recycling have also enhanced some developments of polymers containing functional groups for enzymatical or chemical degradation.

Thus, it should be of great interest to extend the structural variety of functionalized polymers systematically. Functional groups can be attached, as usual, within a linear polymer chain but they can also be combined in different manner with a branched or comb-like polymer (Scheme 1; A-D). Typically, comb-like polymers contain a linear main chain of different chemical types and at least 4 C-atoms in the side chains. Numerous papers appeared dealing with this type of polymers. Although comb-polymers were investigated in general because of their interesting thermal and solution properties, up to now only a relative moderate knowledge exists about the characteristics of functionalized combs (1).

The article deals with some new types of functionalized polymers and points out some relationships between molecular architecture and properties of the polymers. The reactivity of a functionalized polymer with comb-like structure can be controlled in a significant manner by the crystalline order of the side chains. In this connection, as an example, the air drying process of a modified polybutadiene system with comb-like structure was measured and correlated with the side chain order. Also the photosensibility of combs bearing cinnamic acids or mesoions in the side chains was investigated to find out some correlation between structures and reactivities.

Synthetic methods to prepare functionalized polymers may also include the use of enzymes as catalysts. Thus, some examples of chemoenzymatic synthesis and modifications

Macromolecular Engineering, Edited by M.K. Mishra et al.
Plenum Press, New York, 1995

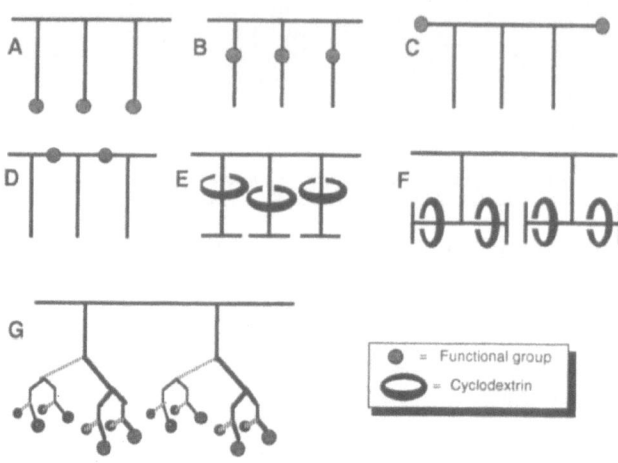

Scheme 1. Structures of functionalized comb-polymers.

of new functionalized monomers and polymers are briefly mentioned. In this connection, the construction of comb-like polymers containing non covalently anchored cyclodextrins threaded around linear and branched side chains has been realized (Scheme 1; E,F). This new class of polyrotaxanes is characterized by a supramolecular (2) architecture. The use of an enzyme to degrade the attached cyclodextrin rings was investigated. Additionally, polymerizable chiral dendrimers containing ester functions have been prepared recently to evaluate the possibility of enzymic modification (Scheme 1; G).

PHOTOSENSIBLE COMB-LIKE POLYMERS

Recently, comb-like polymers containing azobenzene and cinnamic acid moieties in the side chains were synthesized (3). As expected, it was clearly demonstrated that, for example, the E/Z-isomerization of the azogroup of the homopolymer is much faster in solution than in the condensed phase. Therefore, an amorphous copolymer with MMA was prepared and a significant increase of mobility of the azobenzene function in the solid phase could be observed.

The temperature depending quantum yield of the (2+2)-cycloaddition of combs bearing cinnamic acid components in the side chains decreases exactly above the melting point of the side chains (Scheme 2). In this connection, we also synthesized aromatic polyamides containing cinnamic acids in the side chains to produce asymmetric photocrosslinkable membranes. It was shown that the mechanism of the photoreaction is controlled by the position of the cinnamic double bonds. (4)

A new class of photosensitive polymers containing mesoionic groups in the side chains was prepared by radical polymerization of a corresponding styrene derivative as shown in Scheme 3. As an important result, the mesoionic starting material is not soluble in toluene while the irradiated product (bis-ß-lactam) becomes soluble in this solvent. Additionally, it was observed that the photoreactivity decreases with the length of the alkyl side chains (5). Recently, the synthesis of photosensitive polymers containing mesoionic groups in the main chain was successfully achieved (Scheme 3a).

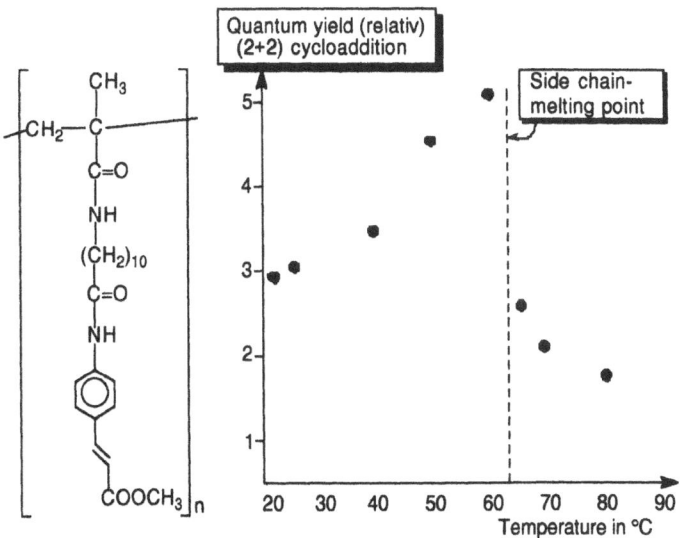

Scheme 2. Influence of side chain order on kinetics of photodimerization of cinnamic acid component

DIELS-ALDER AND ENE-REACTIONS

Modification of AIBN with furan derivatives was performed to produce new radical initiator which were used for the synthesis of telechelic comb polymers containing furan endgroups (s. Scheme 1,C). By a similar way, C,C-splitting initiators were also used to prepare oligomeric systems with furan endgroups. As an example, stearyl acrylate was

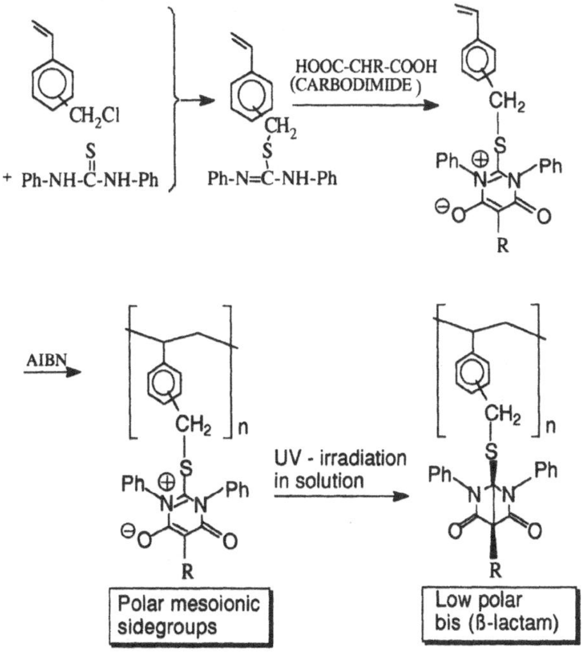

Scheme 3. Mesoionic groups in polymers

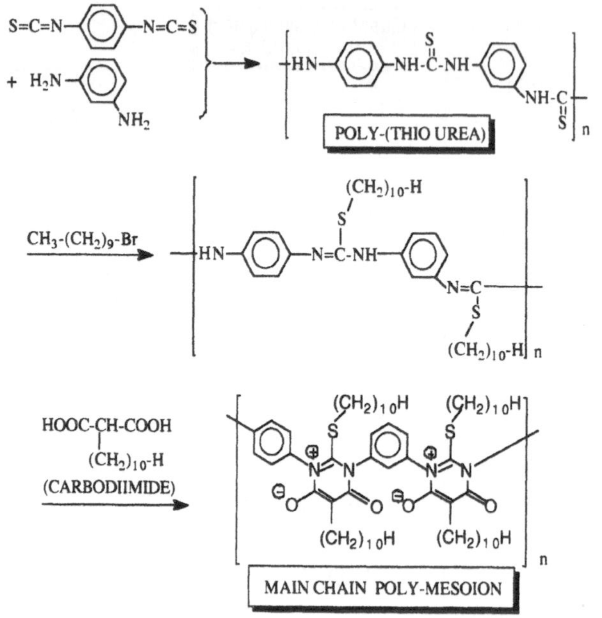

Scheme 3a. Mesoionic groups in polymers.

polymerized with this modified C,C-labile initiator yielding bifunctional oligomers with comb structure. According to Scheme 4, these telechelic systems were used for crosslinking of unsaturated polyesters prepared from maleic anhydride and diols via Diels-Alder additions (6). The formation of the network was followed by measuring the increase of pendulum hardness.

In this connection, comb polymers were prepared containing furan groups at the end of the side chains (s. Scheme 1, A). Kinetics of Diels-Alder reactions with an alkine derivative were found to depend on the length of the spacer groups and on hydrogen bond interactions between the neighboring side groups (Scheme 5).

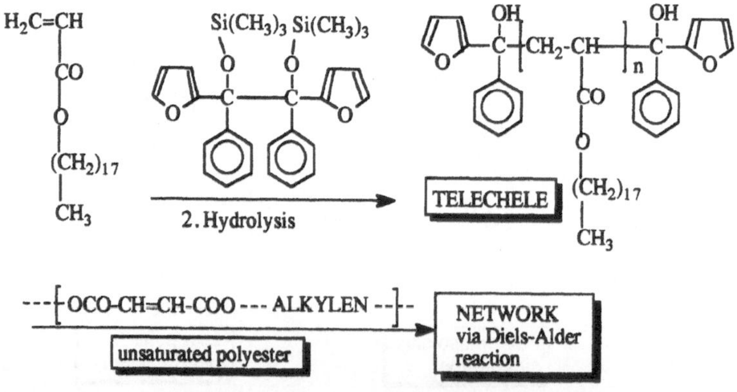

Scheme 4. Synthesis of a telechelic comb-oligomer

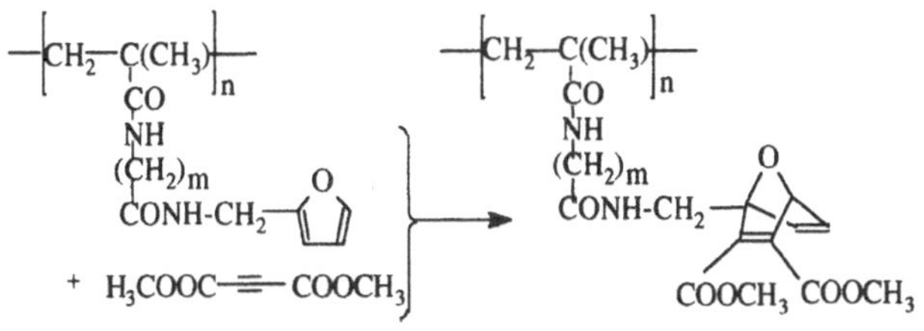

Scheme 5. Diels-Alder addition at furan containing comb-polymer

Unsaturated comb polymers (s. Scheme 1, D) were obtained via ene reaction of stearylacrylate with polybutadiene (Scheme 6). As an important result, the oxidative crosslinking reaction was found to increase significantly above the melting point of the side chains (7).

ENZYMES

Enzymes act as biological catalysts, and enhance the rate of chemical reactions normally taking place within cells. The reactants in enzyme-catalyzed reactions are termed as substrates and each enzyme is more or less specific in acting on those functional groups of the substrates that can interact effectively with its active center. Enzymes are extensively applied as selective catalysts in low molecular weight organic chemistry (8). The realization that enzymes can act not only in aqueous media, but also in dry organic solvents has opened a major field of enzymic catalysis. This means that for example acrylmonomers and even high molecular weight compounds which are only soluble in organic solvents can also become potential substrates for enzymes. In this connection, the degradation of suitable polymers, e.g. peptides, polyesters or cellulose with enzymes is well known. In contrast, only a few papers have appeared describing the anchoring of low molecular weight com-

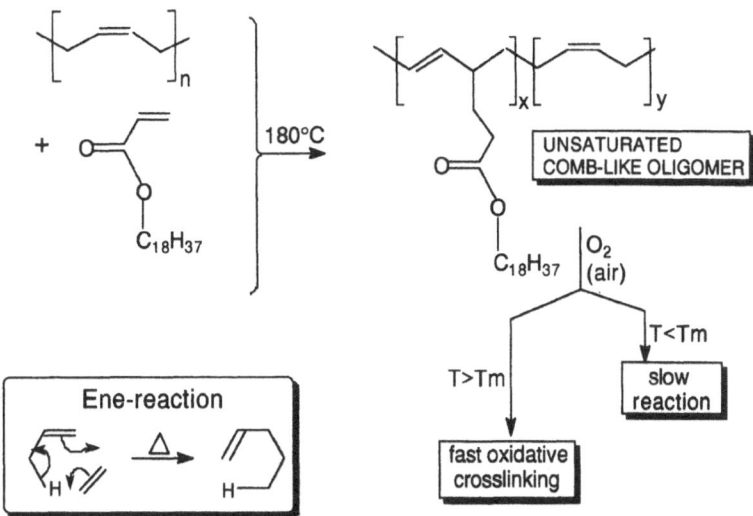

Scheme 6. Ene reaction of stearylacrylate with polybutadiene

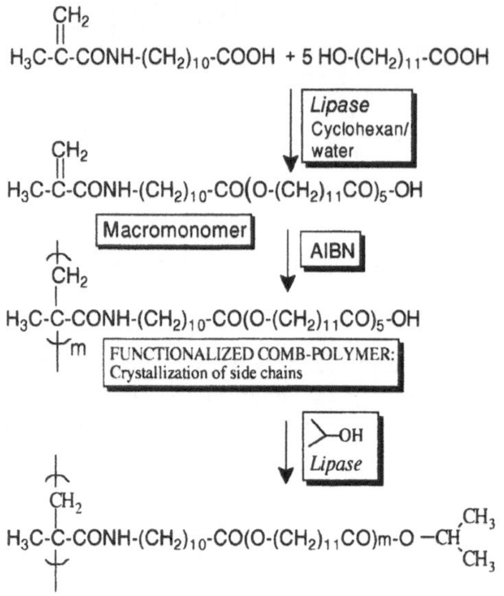

$$CH_2$$
$$\|$$
$$H_3C\text{-}C\text{-}CONH\text{-}(CH_2)_{10}\text{-}COOH + 5\ HO\text{-}(CH_2)_{11}\text{-}COOH$$

↓ | *Lipase* Cyclohexan/ water |

$$CH_2$$
$$\|$$
$$H_3C\text{-}C\text{-}CONH\text{-}(CH_2)_{10}\text{-}CO(O\text{-}(CH_2)_{11}CO)_5\text{-}OH$$

| Macromonomer |

↓ | AIBN |

$$CH_2$$
$$|$$
$$H_3C\text{-}C\text{-}CONH\text{-}(CH_2)_{10}\text{-}CO(O\text{-}(CH_2)_{11}CO)_5\text{-}OH$$
$$]_m$$ | FUNCTIONALIZED COMB-POLYMER: Crystallization of side chains |

↓ | ⟩-OH *Lipase* |

$$CH_2$$
$$|$$
$$H_3C\text{-}C\text{-}CONH\text{-}(CH_2)_{10}\text{-}CO(O\text{-}(CH_2)_{11}CO)m\text{-}O\ -CH\begin{smallmatrix}CH_3\\CH_3\end{smallmatrix}$$

Scheme 7. Chemo-enzymatic synthesis and modification of a macromonomer

pounds to synthetic polymers (9). For example, we recently demonstrated a lipase catalyzed acetylation of a high molecular weight comb-like polymer bearing racemic OH-groups at the end of the side chains (10) up to 40 % yield. A low enantioselectivity of this esterification was proved by measuring the optical rotation. The construction of methacryl-macromonomers with the use of lipases has also been realized. The crystallizable polymer could be esterified enzymatically with isopropanol in cyclohexane solution (9). A slight degradation of the oligoester side chains by transesterification with isopropanol was observed (Scheme 7).

In a solvent free system containing the components 11- methacryloyl-aminoundecanoic acid (1) and methyl - a - D - glucopyranoside (2a) the lipase from *Candida antarctica* was added and this mixture was stirred at 75°C for 24 h yielding the corresponding ester 3a (Scheme 7a). Analogously, the monomeric ester 3b was prepared enzymatically from 1 and 3 - O - methyl - a- D - glucose (2b). The isolation of the esters was performed by preparative high pressure liquid chromatography in a yield of about 20 % and 5% resp.. In contrast to that, the enzymic esterification of the free carboxylic group of 1 with e.g. cyclohexanol or cholesterol was realized in a quantitative yield after 24 h. It is interesting to note that the enzymatically catalyzed condensation of 1 with the anomeric ß-glycoside (methyl -ß- D - glucopyranoside) could not be realized under the same conditions. This observation indicates a high stereo-specifity of this enzymic reaction. (Scheme 7a). (11).

Recently, the chemo-enzymatic epoxidation of methacryl monomers containing chalkone components was realized in the presence of glucose, *glucose oxidase* and oxygen in a preparative scale. The methacrylic function was not attacked under the applied conditions (Scheme 8). The similar enzyme-glucose system was also used to degrade water soluble polymers (12).

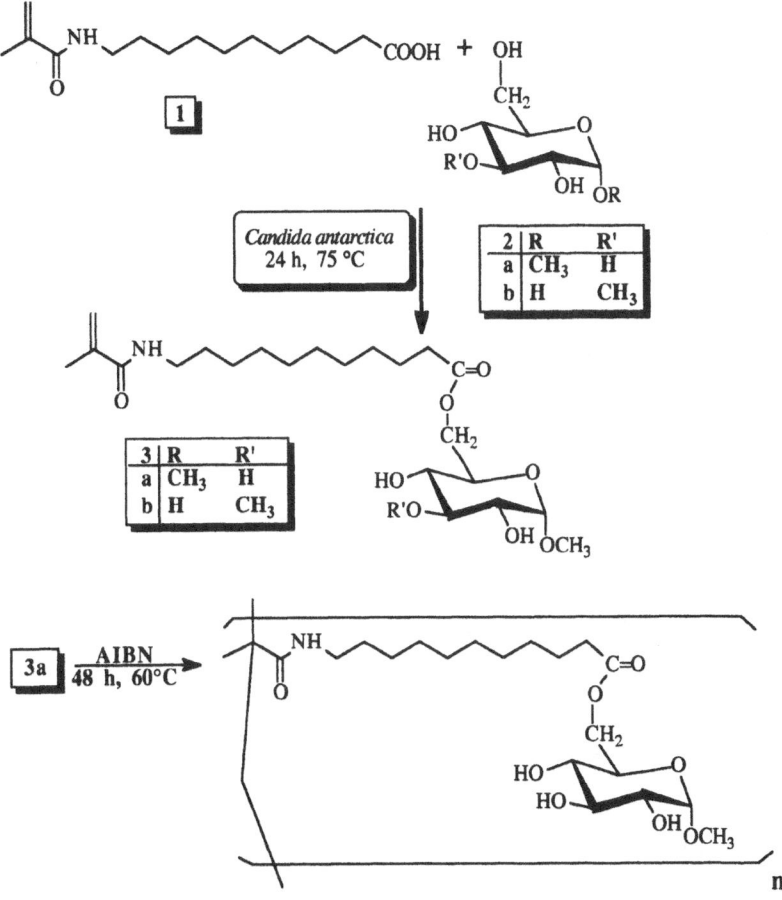

Scheme 7a. Esterification of a methacylmonomer with glucose derivatives

In this connection, the production of a phenoxy substituted polyester using the multi-enzymic system of the bacteria *Pseudomonas oleovorans* was successfully performed. The maximum growth rate of the bacterial culture was determinated by measuring the optical density of the dispersion. Surprisingly it was found that, from a series of different feed components only 11-phenoxy undecanoic acid or a fluorinated derivative was accepted from the bacteria. Furthermore, the phenoxy derivatives with a reduced chain length (n<10) are not accepted from *p.o.* . Only a prolongation of the alkylene chain seem to be slightly suitable for a bacterial polyester formation. The characterization was performed with the extracted material obtained from freeze dryed bacteria. It was proved by IR-spectroscopy that the extracted material shows the expected estergroup at 1730 cm^{-1}. In contrast, the starting material shows the typical acid group adsorptions at 1700 cm^{-1}. For a complete assignment of the NMR-signals the polyester was methanolized and the obtained hydroxyacids were identified by mass spectrometry. It was clearly shown, that the the bacteria have partially fragmentated the alkylene chain of the starting material and introduced a chiral ß-hydroxy group which is important for the formation of the polyester. As a main result, the

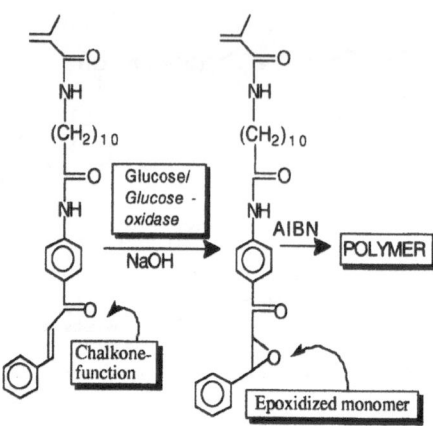

Scheme 8. Chemo-enzymatic synthesis of a epoxidized methacrylmonomer

existence of the monomers 3-hydroxy-5-phenoxypentanoate and 3-hydroxy-9-phenoxy-nonanoate was proved by MS. From the NMR-spectrum a 1:6 composition of the copolyester was calculated. In principle, the obtained copolyesters (Scheme 9) are comb polymers containing a functional phenoxygroup (13).

ROTAXANES AND DENDRIMERS

Recently, the preparation of comb-like polymers containing rotaxanes in the side chains has been achieved (s. Scheme 1, E). For example, a low molecular weight guest molecule N-(4-aminobutanoyl)-4-triphenyl methylaniline was shown to form a stable complex with 2,6-dimethyl-ß-cyclodextrin or triacetyl-ß-cyclodextrin. This type of complex has

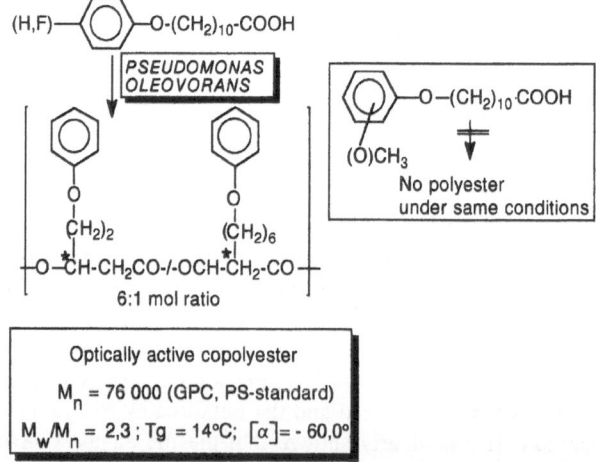

Scheme 9. Bacterial production of a polyester bearing phenoxygroups in the side chains

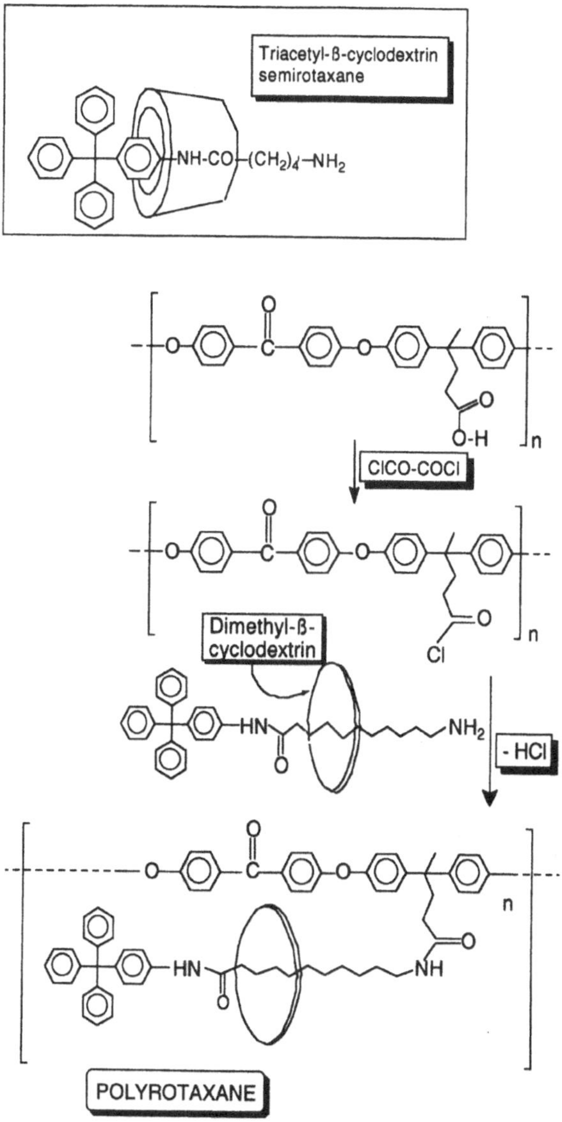

Scheme 10. Structure of a triacetyl-ß-cyclodextrin semirotaxane and synthesis of a side chain polyrotaxane with a polyether-ketone backbone

the structure of a semirotaxane and can be characterized e.g. by FAB-mass spectrometry, thin layer chromatography and NMR-spectroscopy. From NMR-spectroscopy, a selective spatial formation of a triacetyl-ß-cyclodextrin semirotaxane is highly indicated. This means that the wide part of the ring molecule is close to the phenyl barrier group while the narrow part is nearer to the amino function (Scheme 10) The aminogroup of the semirotaxane is still nucleophilic and can react with an activated comb polymer in a nealy quantitative yield (Scheme 10). The structure of the resulting polyrotaxane was proved by a characteristic broadening and shift of the NMR-signals of the 2,6-aniline protons (14).

Recently, the synthesis of a "Tandem-polyrotaxane" according to Scheme 1.-F was performed via similar polymer analogous reaction (Scheme 11). It is interesting to note that

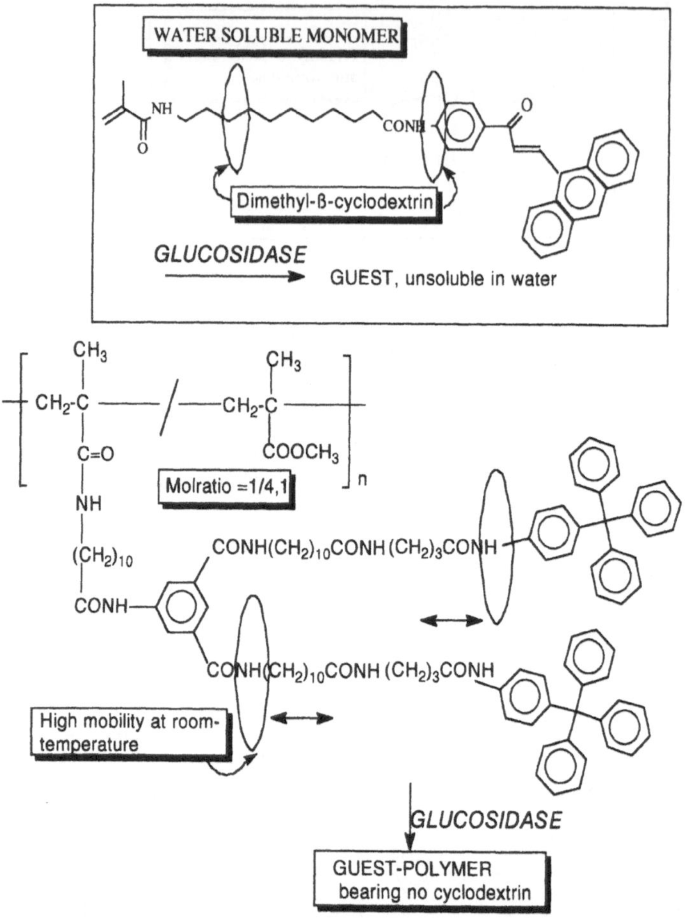

Scheme 11. Enzymatic degradation of cyclodextrin host-components from a chalkon-monomer and a tandem-polyrotaxane.

the enzymatically catalyzed degradation of the cyclodextrin rings from the polyrotaxane was demonstrated yielding the pure guest polymer (15). Furthermore, a new water soluble chalkone-monomer was prepared by threading two cyclodextrin rings per mol of the originally hydrophobic methacryl-derivative. According to preliminary experiments, it was clearly demonstrated that *amyloglucosidase* was able to remove the cylcodextrin rings from the guests via hydrolyzation of the glucoside bonds. As expected, the resulting monomer was unsoluble in water.

The synthesis of chiral (meth)acrylic monomers with a dendrimeric structure bearing 4 or even 8 ester groups, has been performed to produce polymers with highly branched side chains. It was demonstrated, that these types of monomers can be polymerized by radical mechanism. According to mass spectrometry (MALDI) the existence of series of oligomers up to heptamers was detected. This means that, for example, this heptamer contains 56 ester groups. Regarding to molecular modeling calculation, a helical conformation of this highly branched system seems likely (15, Scheme 12).

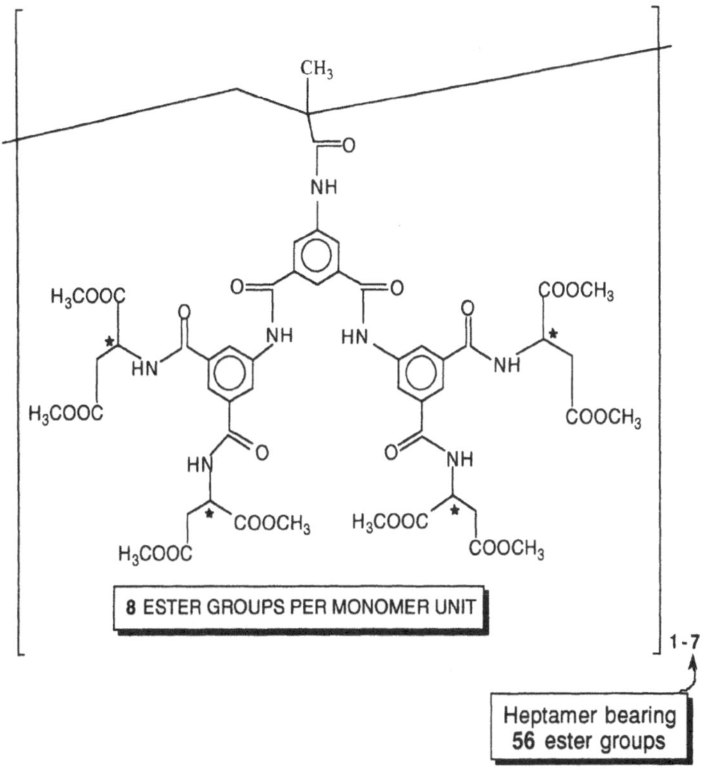

Scheme 12. Oligomerizable dendrimer.

REFERENCES

1. a) K. K. Kapellen, R. Stadler, 1994, Synthesis and characterization of amphiphilic comb-polymers via ring-opening metathesis polymerization of exo-exo-5,6 bis(alkoxymethyl)-7-oxabicyclo(2.2.1)hept-2-enes, Polym. Bull. 32: 3-10; b) V.V. Volkov, A.G. Fadeev, N.A. Plate, N. Amaya, Y. Murata, A. Takahara, T.Kajiyama, 1994, Effect of thermal motion on pervaporation behaviour of comb shaped polymers with fluorocarbon side groups, Polym. Bull. 32: 193-200; c) W. H. Daly, D. Poche, I. I. Negulescu, 1994, Poly(g-alkyl-a,L-glutamate)s derived from long chain paraffinic alcohols, Progr.Polym. Sci. 19: 79-135; d) S. Grutke, J.H. Hurley, W. Risse, 1994, Poly (phenylene oxide) macromonomers for graft copolymer synthesis via ring opening olefin metathesis polymerization, Macromol. Chem. Phys., 195: 2875-2885; e) H. Ritter, A. Stock, 1994, Synthesis of thermoreversible polymers by aminolysis of poly- (methyl-2-(acrylamino)-2-methoxyacetate): Correlation of the lower critical solution temperatures (LCST) with the side group structures and salt concentration in aqueous systems, Macromol. Rapid Commun., 15: 271-277
2. F. Vögtle, 1992, Supramolekulare Chemie, 2d. ed., B. G. Teubner Stuttgart,
3. a) F. Ciardelli, O. Pieroni. A. Fissi, C. Casrlini., A. Altomare, 1989, Photoresponsive Optically Active Polymers - A Review, British Polymer Journal, 21: 97-106; b) M. Niemann, H. Ritter, 1993, Comb-like methacrylamide polymers containing condensates of amino acids and azobenzene moieties in the side chains, Makromol. Chem., 194: 1169-1180; c) J. Stumpe, O. Zaplo, D. Kreysig, M. Niemann, H. Ritter, 1992, Photochemical behaviour of cinnamoyl-containing crystalline polymers with comb-like structure, Makromol. Chem., 193: 1567-1578
4. a) H. Ommer, H. Ritter, 1993, Photocrosslinkable polymers for membranes: Polyamides from N-cin-namoyl-5-aminoisophthalic acid and aromatic diamines Makromol. Chem, 194: 767-776; b) H. Ommer, H. Ritter, 1995, Photocrosslinkable comb-like polymers for membranes: Polyamides from 5-aminoiso-phthalic acid and aromatic diamines, submitted

5. a) H. Ritter, R. Sperber, C. M. Weisshuhn, 1995, Polymerizable mesoionic 4,6-dioxo-1,3-diazines: 1. Synthesis and photochemical behaviour of polystyrenes bearing mesoionic 4,6-dioxo-1,3-diazines as pendant groups Macromol. Chem., Phys. 195: in press; b) Th. Deutschmann, H. Ritter, R. Sperber, Polymeric mesoionic 4,6-dioxo-1,3-diazines: 2, in preparation

6. a) D. Edelmann, H. Ritter, 1993, Synthesis of telechelics with furanyl endgroups: 2. Radical polymerization with C-C splitting initiators and network formation with unsaturated polyesters via Diels-Alder addition, Makromol. Chem., 194: 2375-2384; b) H. Ritter, R. Sperber, C. M. Weisshuhn, 1993, Reactive comb-like polymers. Kinetic studies of the Diels-Alder reaction of furan-containing comb-like polymers with dimethyl butynedioate by means of H-NMR spectroscopy, Makromol. Chem., 194: 1721-1931

7. J. Luchtenberg, H. Ritter, 1994, Synthesis of crystallizable comb-like oligomers via ene reaction and the influence of side chain order on the network formation by autoxidation Macromol. Chem. Phys., 195: 1623-1632

8. C. H. Wong, G. M. Whitesides, 1994, Tetrahedron Organic Chemistry Series, Vol. 12, Elsevier Science, Ltd, Ed. J.E. Baldwin, FRS & P D Magnus, FRS, Pergamon;

9. a) H. Ritter, 1993, The use of enzymes in polymer science, Trends in Polymer Science, 6: 171-173; b) H. Uyama, S. Kobayashi, 1994, Lipase catalyzed polymerization of divinyladipate with glycols to polyesters, Chem. Lett.: 1687

10. K. Pavel, H. Ritter, 1992, Enzymes in polymer chemistry. 6. Lipase catalyzed acylation of comb like methacrylic polymers containing OH-groups in the side chains, Makromol. Chem., 193: 323-328

11. U. Geyer, D. Klemm, K. Pavel, H. Ritter, 1995, Enzymes in polymer chemistry. 8. Chemoenzymatic synthesis of polymerizable 11-methacryloyl-amino-undecanoic ester of 1- and 3-O-methyl-a-D-glucose in 6-O- position, paper submitted

12. a) Ch. Goretzki, H. Ritter, 1995, Chemoenzymatic synthesis of epoxidized methacrylamides involving glucoseoxidase/glucose, Macromol. Reports, accepted; b) M. D. Cho, Y. Okamoto, 1994, Enzymatical chain scission of water soluble polymers by the glucose-glucose oxidase reaction Macromol., Rapid Commun. 15: 629-631

13. a) H. Ritter, A. Spee, 1994, Bacterial production of polyesters bearing phenoxy groups in the side chains: Poly(3-hydroxy-5-phenoxypentanoate-co-3-hydroxy-9-phenoxynonanoate) from Pseudomonas oleovorans, Macromol. Chem. Phys.,195: 1665-1672; b) O. Kim, R. A. Gross, 1994, Opto-active polymers obtained by microbial transformations Polym. Preprints Div. Polym. Chem. ACS 35, no. 2: 627-628

14. M. Born, T. Koch, H. Ritter, 1994, Side chain polyrotaxanes: 2. Functionalized polysulfone with non-covalently anchored cyclodextrins in the side chains, Polym. Acta, 45: 68-72

15. a) M. Born, H. Ritter, 1995, Side chain poly-rotaxanes with tandem structure based on cyclodextrins and a polymethacryl main chain, Angewandte Chemie, accepted, b) Ch, Goretzki, H. Ritter, 1995, to be published

16. G. Draheim, H. Ritter, 1995, Polymerizable Dendrimers. Synthesis of a symmetrically branched methacryl-derivative bearing eight ester groups, Macromol. Chem. Phys., accepted

23

POLYMERS WITH TRIAZENE UNITS IN THE MAIN CHAIN

Application for Laser-Lithography

Oskar Nuyken,[1] Jürgen Stebani,[1] Alexander Wokaun,[2] and
Thomas Lippert[2]

[1] Lehrstuhl für Makromolekulare Stoffe
TU München
Lichtenbergstr. 4
D-85747 Garching
Germany
[2] ETH Zürich and Paul-Scherrer-Institut
CH-5232 Villigen
Switzerland

ABSTRACT

Several monomeric and polymeric triazenes have been synthesized and characterized by common methods such as ^{1}H NMR-, IR-, Raman-, UV-spectroscopy, DSC, and GPC. Their thermolysis, photosensitivity, and the effect of H_3O^+ on their stability have been studied in detail, since those characteristics are very important for an application of these compounds in microlithography.

In laser ablation experiments, polymers containing triazene units in the main chain (**TP1 - TP9**) exhibited several advantages over systems in which common polymers (e.g. PMMA) are doped with low molecular triazenes (**T1 - T14**), especially the fact that their ablation craters have clean contours and sharp edges is of great interest for this type of microstructering.

1. INTRODUCTION

"Triazenes" (R_1-N=N-N(R_2)(R_3)) are rather "old" compounds from the organic chemists viewpoint. It was as early as 1862 that Griess described a suitable way for the synthesis of 1,3-diphenyltriazene [1]. Since one could not find any application for triazenes at that time, these compounds have been ignored for many decades. Moreover, they have been considered an undesired sideproduct in the azo-dye synthesis.

Unsubstituted triazenes are unstable under normal conditions. Substituted triazenes, however, can be rather thermally stable. In newer time, some attention was paid to substituted

Macromolecular Engineering, Edited by M.K. Mishra et al.
Plenum Press, New York, 1995

303

triazenes, especially to 1-aryl-3,3-dialkyl-triazenes, which were synthesized for the first time in 1875 [2] because some of them show activity as insecticides and anti-tumor properties [3]. This is also the main reason for more than 1200 references on triazenes in the last 20 years.

A rather modern aspect, and the main reason for our interest in triazenes, is their potential in the information storage technology, mainly due to their high photosensitivity [4,5]. Therefore we have studied the ground state and excited state properties of selected 1-aryl-3,3-dialkyl-triazenes but also of polymers containing triazene groups as part of the main chain in some detail.

2. SYNTHESIS AND GROUND STATE PROPERTIES OF TRIAZENES

The most suitable route for the synthesis of triazenes is given in the following scheme:

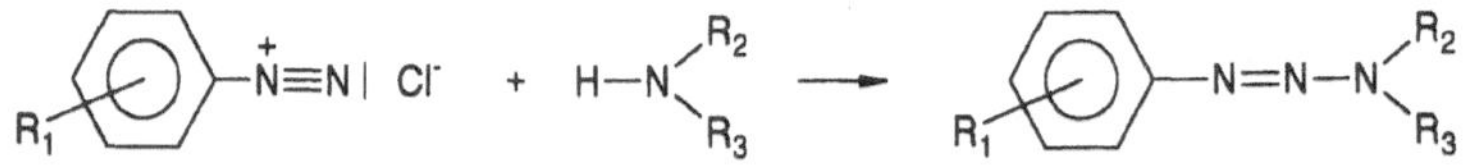

Triazenes are photosensitive, thermosensitive, and sensitive against protons. Further-more, their stability is strongly influenced by their substitution pattern. Thermolysis and photolysis yield radicals as intermediates, meanwhile the retrosynthetic route to diazonium ions and amines is forced by protons.

From X-ray it is known that the most stable isomer has a trans- conformation at the N=N-bond [6].

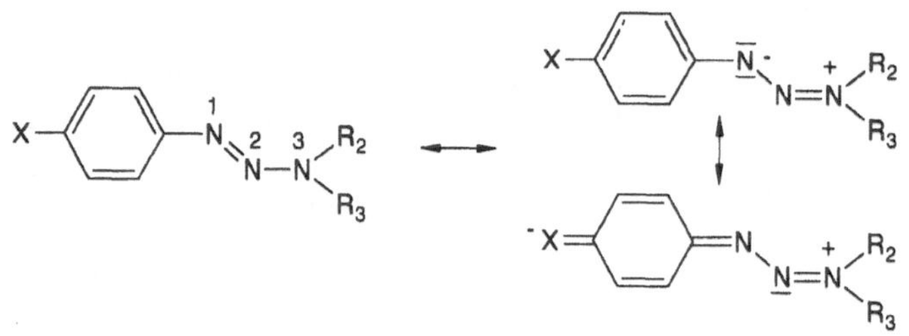

The 1,3-dipolar structures become increasingly important when the aromat is substi-tuted with electron acceptor groups such as NO_2 or CN. In this case the rotation barrier around the N^2-N^3 bond increases [7].

To support the existence of this rotation barrier we carried out ^{1}H NMR experiments for selected triazenes in d_4-methanol at different temperatures. As shown in Table 1, the highest coalescence temperatures were determined for compounds with electron withdraw-ing substituents on the aromatic ring(T1,T2). A typical NMR pattern in these experiments is given in Fig.1:

The Hammett-plot (Fig.2) of the rate constants of the rotation around the N^2-N^3 bond at 273 K, obtained from the NMR experiments, results in ρ= -1.95 ± 0.21 which is in good agreement with literature values [8,9]. The negative value is a clear indication for a rotation barrier due to the π -overlapping of azo and amine orbitals. Furthermore it supports the postulated 1,3-dipolar structure.

Table 1. Coalescence temperature for model triazenes of
the structure R_1-Ph-N=N-N(R_2)(R_3) ($R_2 = R_3 = C_2H_5$)

Compound	R_1	T (K)
T1	4-NO$_2$	308
T2	4-CN	302
T3	4-COOH	283
T5	H	255
T7	4-OCH$_3$	238
T9	3,5-di(COOH)	280
T11	3-COOH	284
T13	3-COOH*	274
T14	3-COOH**	285

* $R_{2,3}$ = n-propyl; ** $R_{2,3}$ = isopropyl

3. THERMOANALYSIS OF TRIAZENES

From thermolysis data one can get information about the thermostability, which is important not only from the academic view point but also from the view point of safety and application. Therefore selected triazenes have been studied by DSC (differential scanning calorimetry) to determine the decomposition temperatures and the decomposition rate constants [compare ref. 10].

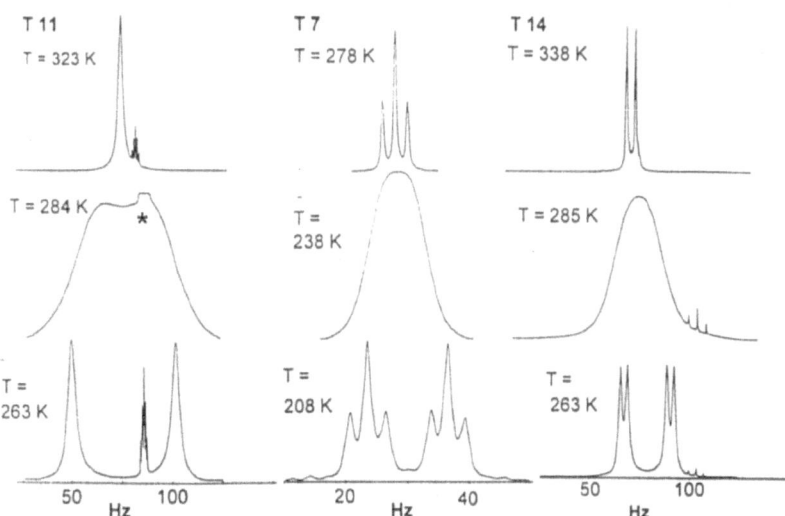

Figure 1. ^{1}H NMR of methyl protons (**T11, T7**) and methin protons (**T14**) of triazenes at different temperatures.

abbrev.	R_1	R_2	R_3
T11	3-COOH	C$_2$H$_5$	C$_2$H$_5$
T7	4-OCH$_3$	C$_2$H$_5$	C$_2$H$_5$
T14	3-COOH	C$_2$H$_5$	CH(CH$_3$)$_2$

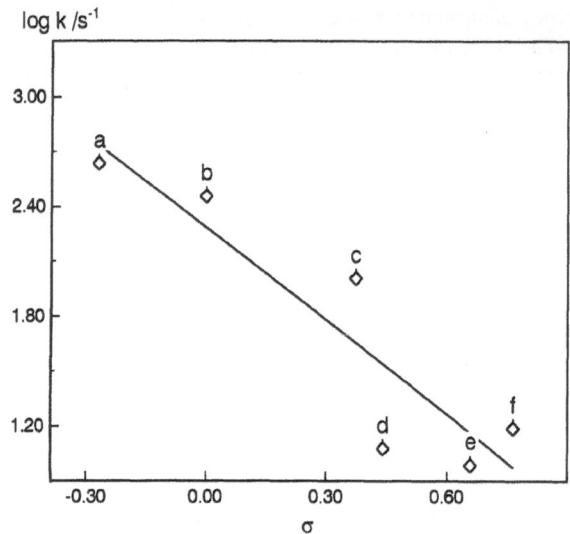

Figure 2. Hammett-plot for the rate constants for the hindered rotation around N^2-N^3 bond in different triazenes at 273 K; a) **T7**, b) **T5**, c) **T12**, d) **T3**, e) **T2**, f) **T1**.

A typical result (arrhenius plot of the rate constants at different temperatures, determined by DSC) for triazenes is shown in Fig.3 for **T9**.

The thermolysis experiments quite clearly showed, that triazenes having electron withdrawing substituents in para-position of the aromatic ring stabilize the compound. Some of the models do not decompose at T<200°.

The thermolysis of triazenes has been studied several times [11-13]. Aromatic triazenes (1,3-diaryltriazenes) decompose thermally via radicalic intermediates under rapture of the N^2-N^3-bond.

The products depend strongly on the solvent used, e.g. decomposition in benzene yields diaryls, decomposition in alcohols yields ethers [14]. The intermediate radicals were

Table 2. Decomposition temperatures T_{max} and E_a for different triazenes (measured by DSC in bulk), structures R_1-Ph-N=N-N(R_2)(R_3)

Compound	R_1	R_2, R_3	T_{max} (°C)	E_a (kJmol⁻¹)
T3	4-COOH	di-C_2H_5	148	242
T9	3,5-di(COOH)	di-C_2H_5	151	282
T11	3-COOH	di-C_2H_5	113	282
T13	3-COOH	di-n-propyl	128	221
T14	3-COOH	di-isopropyl	136	—*
T15	4-OCH$_3$	CH$_3$, $(CH_2)_{17}CH_3$	270	—*
T16	4-CN	CH$_3$, $(CH_2)_{17}CH_3$	308	—*

* not determined

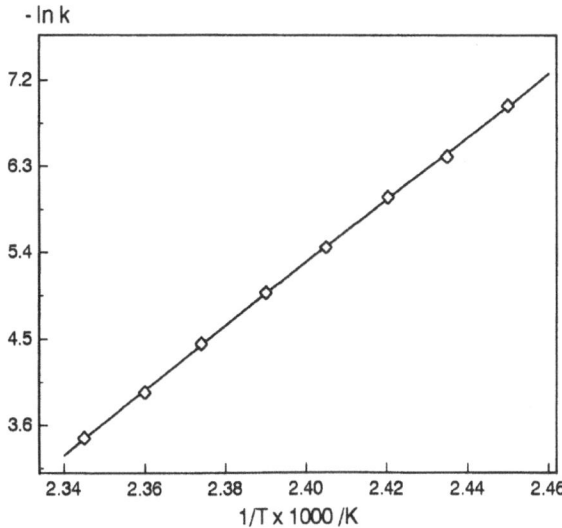

Figure 3. Arrhenius-plot of the decomposition of **T9** in bulk, measured by DSC.

found to be able to initiate a free radical polymerization, e.g. of butadiene, a copolymerization of acrylonitrile with vinylpyridine, and others [11-13, 15].

An interesting observation was that addition of acids reduce the decomposition temperatures of triazenes [16]. Therefore we found it necessary to study the effect of protons on the decomposition of triazenes in some detail.

4. PROTOLYSIS OF TRIAZENES

It is known from earlier work that triazenes decompose in the presence of acids. Di- and trialkyltriazenes [17], 3-alkyl-1-aryltriazenes [18,19], and 3-alkyl-1,3-diaryltriazenes [20,21] are studied in detail. Surprisingly little is known, however, for the protolysis of 1-aryl-3,3-dialkyltriazene [22-24].

Typical experiments were carried out in buffer -methanol mixtures in the presence of a 5 fold concentration of NaCl compared to the triazene ($8 \cdot 10^{-5}$ mol l^{-1}). The decrease of

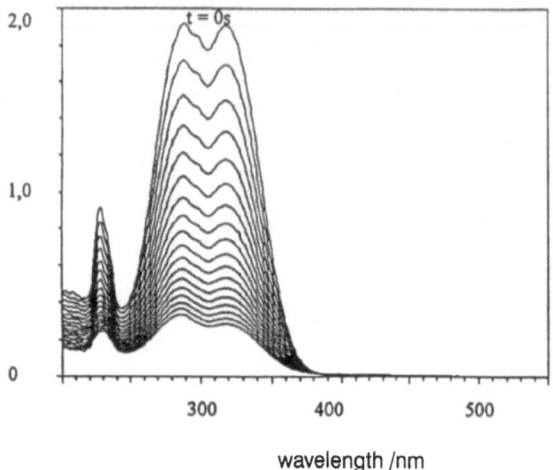

Figure 4. Protolysis of R_1-Ph-N=N-N(C_2H_5)$_2$ (R_1 = Cl), followed by UV-spectroscopy, pH = 4.7, RT, intervals: 60 min.

triazene concentration with time was measured at room temperature in different buffer solutions. A characteristic example for a protolysis experiment followed by UV-measurements is presented in Fig 4.

The effective rate constants as a function of pH values results in straight lines for all triazenes investigated in this work (Fig. 5). Meanwhile **T8** (R_1 = 4-N(CH$_3$)$_2$) decomposes already at pH = 10 with a half life time of only 100s, **T1** (R_1 = 4-NO$_2$) is very resistant against the retrosynthetic decomposition.

From the decomposition scheme for the protolysis of triazenes results (E1):

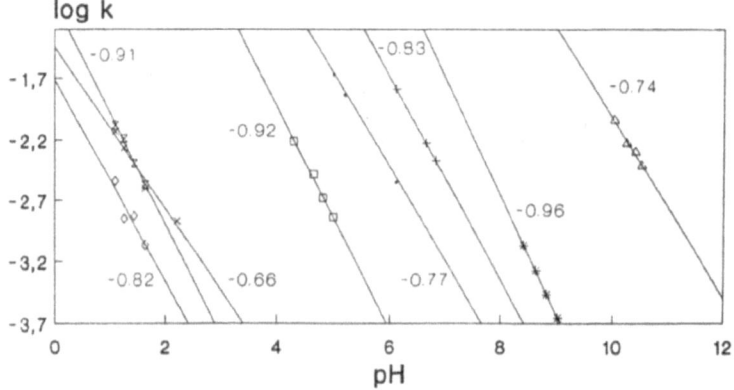

Figure 5. Dependence of the effective first order rate constants of triazene decomposition on solution pH. For each curve the slope of the corresponding linear regression is included in the figure. Structures according to Table 1, R_1=H (■), CH$_3$ (+), OCH$_3$ (∗), Cl (□), CN (x), NO$_2$ (⊖), N(CH$_3$)$_2$ (Δ), m-NO$_2$ (⊠).

$$k_a^1 = \frac{[TH^+]}{[T][H_3O^+]} \tag{E1}$$

Ka^1 equilibrium constant of protonation
$[T]$ concentration of the triazene in equilibrium
$[TH^+]$ concentration of the protonated triazene in equilibrium

If unimolecular decomposition of $[TH^+]$ is rate determined then (E2) is valid:

$$-\frac{d[T]}{dt} = k_1 [TH^+] = k_1 k_a^1 [T][H_3O^+] \tag{E2}$$

In this case the effective rate constant depends on H_3O^+ concentration and the decomposition reaction has the characteristics of a specific acid catalysis.

Systematic variation of the buffer concentration and of the buffer type (pH = 2 for glycine or citrate buffer) shows that "specific acid catalysis" is a suitable description for the protolysis of triazene, since the reaction depends not on the chemical nature of the buffer but on the actual H_3O^+ concentration only. This view is also supported by deuterolysis: D_2O was used for protolysis instead of H_2O. The isotope effect (differences in the rate constants) was determined to be $k_H/k_D = 0.432$. Since D_3O^+ is a stronger acid than H_3O^+ the reaction is faster in D_2O than in H_2O.

As already mentioned above, the protolytic stability is dramatically influenced by the substituents at the aromatic ring of the triazenes. A comparison of the rate constants for the decomposition of triazenes at pH=7 is given in Table 3. The Hammett-plot for these results (Fig. 6) is a straight line, resulting in $\rho = -4.70$ which is quite similar to a value given in the literature for 1-aryl-3,3-dimethyltriazene ($\rho = -4.03$) [24].

The linear dependency of the triazene decomposition on the H_3O^+ concentration opens the chance to control the rate by variation of the pH-value. An additional possibility to produce protons is given by irradiation of photo acids e.g. sulfonium salts. Those investigations are in progress.

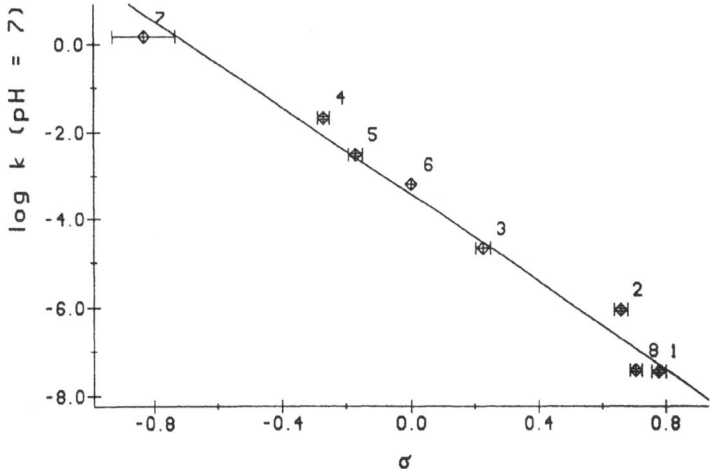

Figure 6. Hammett-plot for the protolysis of triazenes **T1 - T9** at room temperature, pH=7; 1=**T1**, 2=**T2**, 3=**T4**, 4=**T7**, 5=**T6**, 6=**T5**, 7=**T8**, 8=**T12**.

Table 3. Extrapolated rate constants $k_{exp.}$ at pH = 7 for
different triazenes R_1-Ph-N=N-N$(C_2H_5)_2$

Triazene	Substituents	- log $k_{exp.}$	$t_{1/2}$
T1	4-NO$_2$	- 7.470	39 years
T12	3-NO$_2$	- 7.430	36 years
T2	4-CN	- 6.068	2 years
T4	4-Cl	- 4.676	23 days
T10	3-OCH$_3$	- 3.629	2 days
T5	H	- 3.197	18 h
T6	4-CH$_3$	- 2.522	4 h
T7	4-OCH$_3$	- 1.697	35 min
T8	4-N(CH$_3$)$_2$	0.170	28 s

5. PHOTOLYSIS OF TRIAZENES

Among the 1200 papers published on triazene chemistry in recent years there are only few dealing with the photochemistry of these compounds and those mostly are related to 1,3-diaryltriazenes and their application as doping material for laser ablation of polymethylmethacrylate [25-27].

Irradiation of triazenes with UV-light in general yields radicals and nitrogen:

$$\langle O \rangle - N=N-N \overset{R_2}{\underset{R_3}{\diagup}} \overset{h\nu}{\longrightarrow} \langle O \rangle^\bullet + N_2 + {}^\bullet N \overset{R_2}{\underset{R_3}{\diagup}} \longrightarrow \text{products}$$

The radical intermediates have been detected by ESR-spectroscopy [28]. A typical photolysis experiment is shown in Fig. 7.

The observed isosbestic points (1-3) are a clear indication for a well-defined decomposition mechanism, which is suitable described by a first order kinetics [29,30]. The uniformity of the decomposition is strongly supported by evaluation in Mauser plots according to (E3) (Fig.9) [31-34] in which straight lines were observed for most triazenes.

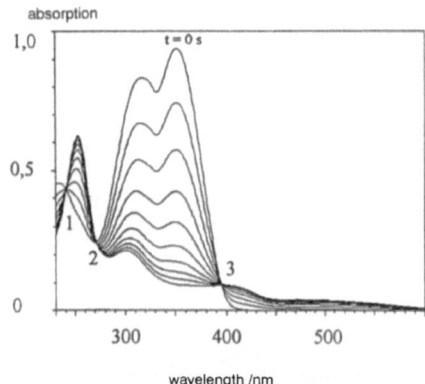

Figure 7. Photolysis of **T7** in THF, RT, followed by UV-spectroscopy, 0.2s intervalls; lamp: Xe-high pressure.

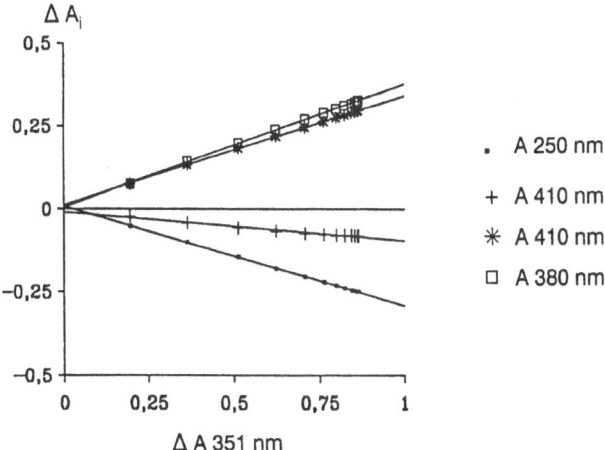

Figure 8. Mauser plot for the photolysis of 4-N(CH$_3$)$_2$-Ph-N=N-N-(C$_2$H$_5$)$_2$ (**T8**) in THF at room temperature for four different wavelengths; lamp: Xe-high pressure.

$$\Delta A_{i_t} = f(\Delta A_t) \tag{E3}$$

ΔA_{i_t} = difference between absorption at t = 0 (A$_0$) and absorption at t=t for a certain wavelength

ΔA_t = difference between A$_0$ and A$_t$ for a reference wavelength.

Since the uniformity of the photolysis is confirmed one can determine the actual rate constants. A Hammett-plot of these rate constants results in a straight line, from which it was possible to determine ρ = -2,13. The advantage of the availability of Hammett-plots is that with known σ-values one can predict values of δ for substituents which were not measured.

In order to get information about the excited state of triazenes some experiments were carried out in the presence of oxygen which is known to function as triplet quencher [35]. However, no change was observed in these experiments compared with those carried out in an argon atmosphere. Therefore the triplet excited state was considered to be unlikely. In contrast, experiments in the presence of pyrene, which is known to be an effective singulett quencher, have shown strong effects, e.g. decrease of the quantum yield by a factor of 5 to 10.

UV-spectra taken from selected triazene support the view, that the 1,3-dipole character is favored when the aromatic ring is substituted with electron withdrawing functions such as NO$_2$ or CN (**T1, T2**), indicated by absorption at high wavelength with high absorption coefficient (Table 4).

6. IR- AND RAMAN SPECTROSCOPY

A combination of IR and Raman spectra was used for the assignment of the N=N frequency. A survey on the position of the N=N signals of different triazenes is given in Table 5 .

For para-substituted triazenes the frequency ν (N^1=N^2) decreases with increasing electron withdrawing character of the substituent. Meanwhile, the frequency of the N^2-N^3 single bond increases, however, to a smaller extent. This behavior is expected for a 1,3 dipolar resonance structure.

Table 4. UV-absorptions of selected triazenes 4-R_1-Ph-N=N-N(C_2H_5)$_2$

Triazene	R_1	λ_1 (nm)	ε_1 (lmol^{-1}cm^{-1})	λ_2 (nm)	ε_2 (lmol^{-1}cm^{-1})	λ_3 (nm)	ε_3 (lmol^{-1}cm^{-1})
T1	NO_2	242	8400	—	—	371	23700
T2	CN	231	9500	—	—	328	22400
T4	Cl	227	10200	289	17900	320	15500
T5	H	227	8400	285	13900	308	13100
T6	CH_3	228	9300	287	14500	316	13400
T7	OCH_3	222	10300	291	15900	328	14000

Ab initio and semiempirical calculations (AM1, PM3, SCAMP [36]) also support the 1,3 dipolar structure of the triazenes: the highest degree of N^1=N^2-double bond character is calculated for R_1 = OCH_3 and the lowest for R_1 = NO_2, and corresponding with those results the single bond character for N^2-N^3 increases from R_1 = NO_2 to R_1 = OCH_3.

7. DOPANT INDUCED LASER ABLATION OF PMMA AT 308 NM

Since the early reports in 1982 [39,40] ablative photodecomposition of polymers became a field of intense investigation. Recently promising results have been published for the ablation of PMMA by a 308 nm excimer laser, using diphenyltriazene as dopant [41].

The following figures show undoped PMMA after irradiation with 308 nm excimer laser (fluence F > 6,5 Jcm^{-2}). In this case PMMA is damaged but not structured (Fig. 10). Better results were received by doping of PMMA with triazenes (Fig. 12): the film surface is now clearly structured. For these experiments films were cast from THF solution of PMMA doped with different amounts of 4-NC-Ph-N=N-N(C_2H_5)$_2$ (**T2**) of 200µm thickness. These films were irradiated with a 308 nm excimer laser with a repetition rate of 2 Hz and a pulse length of 30 ns (fluence between 0,5 and 11,2 J cm^{-2}). Results of the ablation experiments performed on two doped PMMA samples of different molar masses are presented in Fig. 11. The triazene (**T2**) concentration was varied between 1 and 2 wt %.

Doubling the triazene concentration roughly halves the limiting etch depth per pulse. Fig.11 shows also a dependence of the etch depth on the molar mass, indicating that the etch depth is always lower for higher molar masses under comparable conditions.

The ablation quality is also a function of the triazene concentration - sharper, higher quality structures are received when higher triazene concentrations are applied (Fig. 12).

Table 5. Effect of the substituents R_1 in 4-R_1-Ph-N=N-N(C_2H_5)$_2$
on the vibrational frequencies of the triazene group in cm^{-1}

Compound	R_1	ν (N^1=N^2) Raman	ν (N^1=N^2) IR	ν (N^2-N^3) Raman	ν (N^2-N^3) IR
T8	$N(CH_3)_2$	1417/1407	1416/1406	1255	1237
T5	H	1414	1414	1252	1239
T7	OCH_3	1408	1407	1253	1245
T6	CH_3	1403	1403	1254	1237
T3	COOH	1397	1401	1252	1239
T2	CN	1386	1384	1260	1243
T1	NO_2	1391	1383	1259	1246

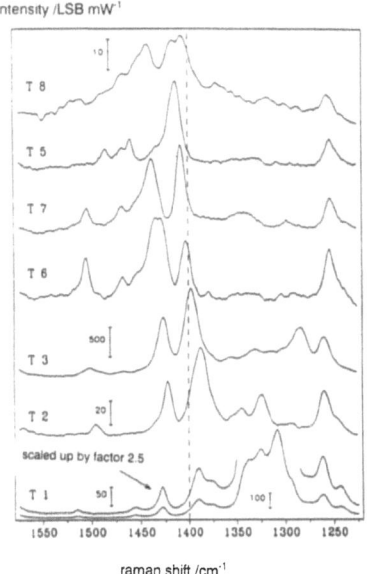

intensity /LSB mW⁻¹

raman shift /cm⁻¹

Figure 9. Raman spectra of para-substituted triazenes.

The much better ablation results on the doped material might be best explained on the basis of a "driving gas" effect of the nitrogen formed during irradiation. However, a large quantity of ejected material is seen around the edges of the crater. Furthermore, both the walls and the bottom of the crater exhibit an irregular structure.

These studies demonstrate that non-absorbing polymers can be efficiently doped for 308 nm eximer laser ablation by adding small amounts of substituted 1-phenyl-3,3-dialkyl triazenes. Crater formation is induced by a photochemical ablation mechanism on the basis of photolytic fragmentation of the dopant. Nitrogen acts as driving gas for the ejection of the polymer material. Ablated depth per pulse as high as 80μm have been observed.

The dependence of the etch rate on laser fluence exhibits two regimes, i.e. a linear dependence on the logarithm of fluence $\{\ln (F/F_o)\}$ followed by a plateau region. In the linear

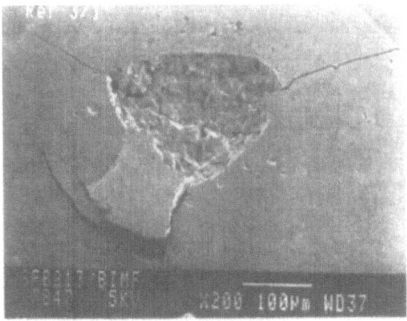
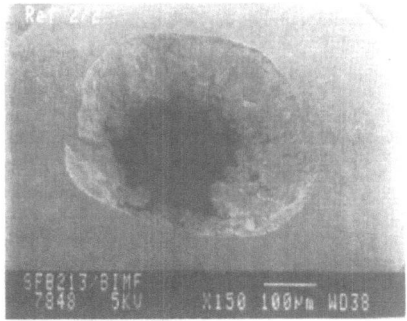

Figure 10. Electron micrograph of PMMA film, irradiated with a 308 nm excimer laser, 1 and 2 pulses (F = 6.5 and 11.7 Jcm⁻²).

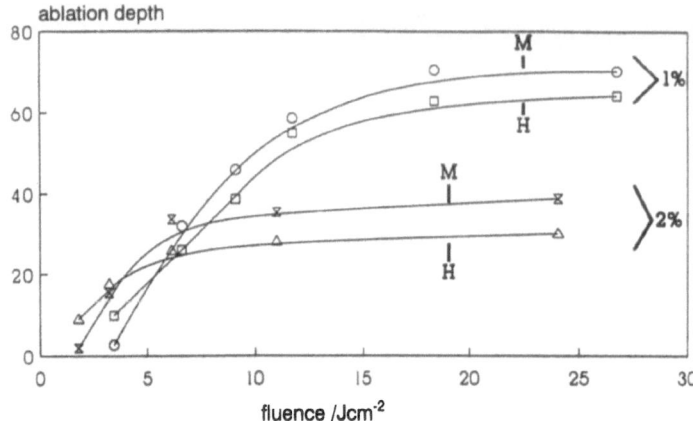

Figure 11. Influence of the molar mass on the ablation characteristics of PMMA doped with 1 and 2 w% of 4-NC-Ph-N=N-N(C$_2$H$_5$)$_2$ (**T2**); M = PMMA (M$_w$ = 97000); H = PMMA (M$_w$ = 500 000).

regime, the ablated depth per pulse is proportional to the quantum yield of the photochemical decomposition and follows the 1/c dependence. In the plateau a thermal mechanism, i.e. the absorption of photons followed by the thermolization of the energy, provides the dominant contribution to the ablation. Therefore the photolysis quantum yield QY appears to be no longer of importance.

8. EXCIMER LASER ABLATION OF TRIAZENE POLYMERS

The alternate route to chromophore doped polymers as described above is the tailored synthesis of photopolymers containing a suitable chromophore group, e.g. triazene polymers of the following structure:

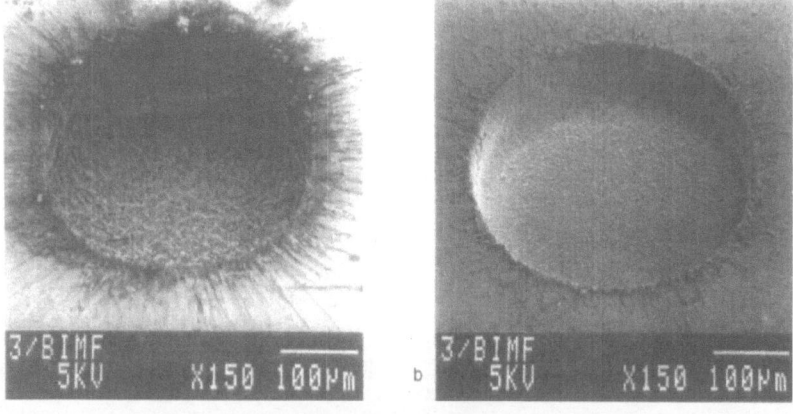

Figure 12. Electron micrograph of laser ablated PMMA doped with **T6**; a) 2mol%, b) 5mol% (F = 5.8 Jcm^{-2}).

Table 6. Structural formulas of investigated triazene polymers

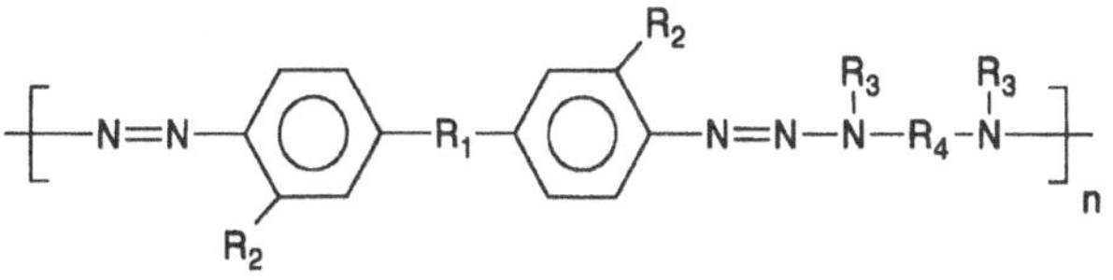

Polymer	R_1	R_2	R_3	R_4
TP1	O	H	CH_3	C_6H_{12}
TP2	O	H	CH_3	C_2H_4
TP3	—	OCH_3	CH_3	C_6H_{12}
TP4	p-SO_2	H	CH_3	C_6H_{12}
TP5	m-SO_2	H	CH_3	C_6H_{12}
TP6	CO	H	CH_3	C_6H_{12}
TP7	O	H	CH_3	C_3H_6
TP8	O	H	C_2H_5	$(CH_2-CH=)_2$
TP9	CO	H	C_2H_5	$(CH_2-CH=)_2$

The polymers were synthesized according to the following general procedure: A bis(4-aminophenyl) ether, ketone, or sulfone (R_1 = O, CO, SO_2) was diazotized with the nitrite in hydrochloric acid. The resulting diazonium salt was reacted in situ directly with a substituted α,ω-diaminoalkane to yield the desired product in a polycondensation reaction. The polymers **TP1** through **TP9** were characterized by GPC, NMR spectroscopy, DSC, and TGA.

Molar masses of **TP1** to **TP9** are in the range of 50,000 to 120,000 g/mol (M_w), measured by GPC. From TGA-studies one can conclude that the N_2 and the R_3-N-R_4-N-R_3 unit are released in one step between 210 and 300°C. Some data are summarized in Table 7.

From these polymers films were prepared by casting from THF-solution. After drying films of 100 - 200 μm thickness were received. Technical details about the laser ablation procedure are described elsewhere [41,43].

Polymer **TP8** was chosen as a representative example to illustrate the structures resulting from excimer laser irradiation. A scanning electron micrograph of an ablation crater

Table 7. Physicochemical reference data of triazene polymers

Polymer	M_w (gmol^{-1})	λ_{max} (nm)	ε_1/lmol^{-1}cm^{-1}at 308nm (in solution)	photolysis quantum yield (%)
TP1	71,000	330	27,700	0.26
TP2	46,000	353	24,900	0.39
TP3	53,000	367	11,300	0.17
TP4	107,000	336	23,200	0.12
TP5	207,000	293	22,500	0.16
TP6	62,000	352	17,200	0.13
TP7	19,000	331	23,900	0.74
TP8	9,000	334	22,400	0.64
TP9	69,000	357	18,400	0.16

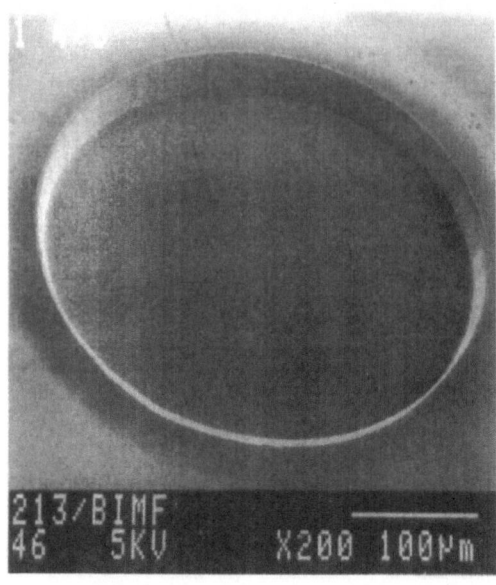

Figure 13. Electron micrograph of **TP8**, irradiated with a 308 nm laser, 100 pulses, fluence = 1.2 J/cm^{-2}.

produced is shown in Fig.13. The sample was irradiated with 100 pulses of 1.2 Jcm^{-2} fluence at 308 nm. Very sharp contours and steep edges were observed for the irradiated area.

Similar sharp contours have been found for all other polymers **TP1** to **TP9**. This favorable behavior which has rarely been observed with any other materials, appears to be an unique property of triazene polymers. For triazene polymers the ablation is fully governed by the photolytic decomposition of the triazene groups.

A striking feature that distinguishes the triazene polymers from other photosensitive materials used in laser ablation is the absence of ejected material deposited around the crater. A possible reason for the absence of solid ablation products, the socalled "debris", is the well-defined fragmentation of the polymers. Small gaseous products, in particular N_2, are acting as a driving gas which promotes at least the initial stages of the ablation.

For all triazene polymers the total ablated depth at given fluence (laser pulse energy per irradiated area) increases linearly with the number of pulses. In the low-pulse energy regime the fluence dependence is well described by the equation:

$$d(F) = \frac{1}{\alpha_{eff}} \ln\left(\frac{F}{F_0}\right)$$

(E4)

$d(F)$ = ablated depth per pulse
F = laser fluence
F_0 = threshold fluence
α_{eff} = effective absorption coefficients for ablation
(typical data: α_{eff} = 2 x 10^{4}cm^{-1}, $F_0 \cong$ 100 mJ/cm, F_0 x $\alpha_{eff} \cong$ 2 kJ/cm)

A common feature of the triazene polymers is the absence of any simple relation between the effective absorption coefficients obtained from the ablation data and the molar absorption coefficients at 308 nm in solution. This lack of correlation is not unusual in

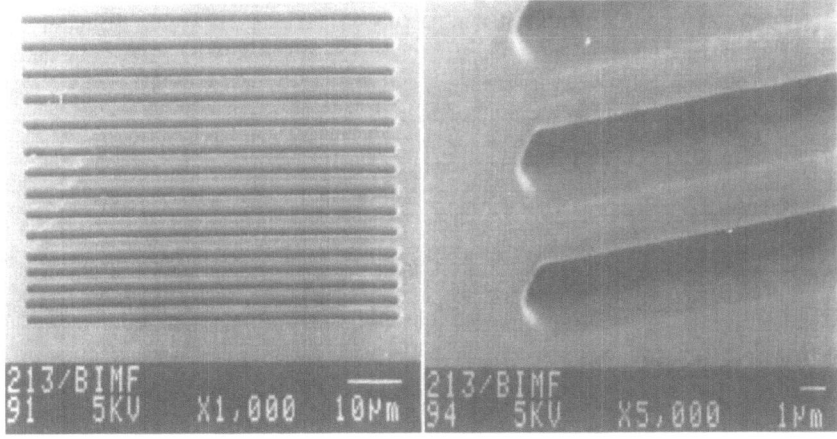

Figure 14. Electron micrograph of laser ablated **TP1** with slit mask applied.

polymer ablation [44] and has been ascribed to nonlinear effects (saturation of absorption, multiphoton effects) and to absorption by laser induced plasma photoproducts [45].

9. RESOLUTION TESTS

The achievable structural resolution by laser ablation was tested for material **TP1**. A rectangular slit mask was imaged onto the sample (coated as described above) to illuminate a stripe of $\approx 1 \mu m$ width. The sample was irradiated with 10 pulses of 0.2 Jcm^2 fluence. The results are shown in Fig. 14.

Again, an unique feature is the absence of any solid polymer deposits around developed structures. Furthermore the high resolution tests with the triazene polymers material yield resolutions near to 0.4 µm. These both facts render the triazene polymers very suitable for laser ablation structuring in microlithography.

CONCLUSION

Low molecular triazenes, added to common polymers, and triazene polymers are both suitable for laser ablation. The differences between PMMA doped with low molecular triazenes (A) and polymers containing triazene units in the main chain (B) are as follows:

(A)
- ablation depth ~ 70 µm/pulse
- ablation depth ~ 1/[triazene]
- low level of doping necessary for large effects
- photochemical produced N_2 is driving gas for the ablation
- change from photochemical into photothermical ablation as a function of laser fluence

(B)
- ablation depth ~2-3 µm/pulse

- ablation depth ~ $1/\alpha_{eff}$
- unusual proportionality between photolysis and quantum yield
- sharp contours
- resolution 500A
- no deposits around the ablation crater

REFERENCES

1. P. Griess, Justus Liebigs Ann. Chem. *121*, 258 (1862).
2. A. V. Baeyer, C. Jaeger, Ber. dtsch. Chem. Ges. *8*, 148 (1875).
3. T. Giraldi, T. A. Connors, G. Carter (Ed. s), "Triazenes - Chemical, Biological and Clinical Aspects" Plenum Press, New York, 1990.
4. J. Stebani, PhD Thesis, Bayreuth, 1993.
5. T. Lippert, PhD Thesis, Bayreuth, 1993.
6. F. R. Fronczek, C. Hansch, S. F. Watkins, Acta Cryst. *C 44*, 1651 (1988).
7. M. H. Akhtar, R. S. McDaniel, M. Feser, A. C. Oelschlager, Tetrahedron *24*, 3899 (1968).
8. N. P. Marullo, C. B. Mayfield, E. H. Wagener, J. Am. Chem. Soc. *90*, 510 (1968).
9. L. Lunazzi, G. Cerioni, E. Foresti, D. Macciantelli, J. Chem. Soc.Perkin Trans. II, 686 (1978).
10. O. Nuyken, J.Gerum, R. Steinhausen, Makromol. Chem. *180*, 513 (1979).
11. C. Koningsberger, G. Salomon, J. Polym. Sci. *1*, 200 (1946).
12. US Patent 2.313.233 (1941), B. F. Goodrich Co., Inv. C. F. Fryling, C.A. *37*, 5284[4] (1943).
13. US Patent 2.643.990 (1953), Chemstrand Co., Inv. G. E. Ham, C.A. *47*, 9025i (1953).
14. R. L. Hardie, R. H. Thomson, J. Chem. Soc., 1268 (1958).
15. US Patent 2.376.015 (1945), B. F. Goodrich Co., Inv. W. L. Semon, C. A. *39*, 5545[8] (1945).
16. P. A. Vinogradov, Z. Obsc. Chim. *26*, 2882 (1956).
17. R. H. Smith, B. D. Wladkwski, A. F. Mehl, M. J. Cleveland, E. A. Rudrow, G. M. Chmurny,C. J. Michejda, J. Org. Chem. *54*, 1036 (1989).
18. J. Iley, R. Moreira, E. Rosa, J. Chem. Soc. Perkin Trans. II, 81 (1991).
19. N. J. Isaacs, E, Rannala, J. Chem. Soc. Perkin Trans. II, 899 (1974).
20. O. Pytela, P. Svoboda, M. Vecera, Coll. Czech. Chem. Commun. *46*, 2091 (1981).
21. O. Pytela, T. Nevecna, J. Kavalek, Coll. Czech. Chem. Commun. *55*, 2701 (1990).
22. V. Zverina, J. Divis, M. Remes, M. Matrka, Chem. Prum. *22*, 454 (1972).
23. J. R. Barrio, N. Satyamurthy, H. Ku, M. E. Phelps, J. Chem. Soc., Chem. Commun. 443 (1983).
24. Y. Hashida, H. Endo, K. Matsui, Nihon Kagakku Kaishi *8*, 1433 (1975)
25. J. Baro, D. Dudek, K. Luther, J. Troe, Ber. Bunsenges. Phys. Chem. *87*, 1155, 1161 (1983).
26. M. Julliard, M. Scelles, A. Guillemonat, G. Vernin, J. Metzger, Tetradedron Lett., 375 (1977).
27. M. Julliard, G. Vernin, J. Metzger, Helv. Chim. Acta *63*, 456, 467 (1980).
28. A. Stasko, V. Adamcik, J. Dauth, O. Nuyken, T. Lippert, A. Wokaun, Makromol. Chem. *194*, 3385 (1993).
29. T. Lippert, J. Stebani, O. Nuyken, A. Stasko, A. Wokaun, J. Photochem. Photobiol. A: Chem. *78*, 139 (1994).
30. O. Nuyken, J. Stebani, T. Lippert, A. Wokaun, A. Stasko, Macromol. Chem. Phys. *196* (1995), in print.
31. H. Mauser, Z. Naturforsch. *23b*, 1025 (1968).
32. H. Mauser, V. Starrock, H. J. Niemann, Z. Naturforsch. *27b*, 1354 (1972).
33. H. Mauser, G. Gauglitz, Chem . Ber. *106*, 1985 (1973).
34. J. Polster, Z. Phys. Chem. *NF 97*, 113 (1975).
35. N. J. Turro, Modern Molecular Photochemistry, The Benjamin / Cummings Publ., Menlo, 1978.
36. M. J. S. Dewar, E. V. Zoebisch, E. F. Healy, J. J. P. Stewart, J. Am. Chem. Soc. *107*, 3902 (1985).
37. J. J. P. Stewart, J. Comput. Chem. *10*, 209, 211 (1989).
38. T. Clark, SCAMP 4.30, Universität Erlangen, 1990.
39. Y. Kawamura, K. Toyoda, S. Namba, Appl. Phys. Lett. *40*, 374 (1980)
40. R. Srivivasan, V. Mayne-Banton, Appl. Phys. Lett. *41*, 576 (1982)
41. M. Bolle, K. Luther, J. Troe, J. Ihlemann, H. Gerhardt, Appl. Surf. Sci. *46*, 279 (1990)
42. J. Stebani, O. Nuyken, T. Lippert, A. Wokaun, Makromol. Chem. Rapid Commun. *14*, 365 (1993)
43. T. Lippert, J. Stebani, J. Ihlemann, O. Nuyken, A. Wokaun, J. Phys. Chem. *97*, 12296 (1993)
44. S. Lazare, J. Gramer, Laser Chem. *10*, 25 (1989)
45. S. Lazare, J. Gramer, Chem. Phys. Lett. *168*, 593 (1990)

24

SYNTHESIS AND PROPERTIES OF POLY(ARYL ETHER KETONE) POSSESSING CROSSLINKING GROUPS

Application for Electronic Device

Yoshihiro Taguchi,[1] Hiroshi Uyama,[1] and Shiro Kobayashi,[1*] and Katsuhisa Osada[2]

[1] Department of Molecular Chemistry and Engineering
Faculty of Engineering
Tohoku University
Aoba
Sendai 980-77
Japan
[2] Product Department 1
Tohoku Alps Co., LTD
Wakuya-Cho
Toda-Gun
Miyagi-Ken 987-01
Japan

INTRODUCTION

Fully aromatic poly(ether ketone)s (PEKs) are widely used as high-performance engineering plastics in various industrial fields because of their excellent thermal and chemical stabilities, and mechanical properties at high temperature (May, 1986). A non-substituted PEK is generally insoluble in organic solvents at room temperature because it shows highly crystalline property. This often restricts the use of the PEK for some applications.

The introduction of a substituent onto the PEK backbone is able to suppress crystallinity, and hence, improve its solubility. Recently, PEK derivatives soluble in organic solvents have been synthesized by using an alkyl- or aryl-substituted hydroquinone as monomer (Risse and Sogah, 1990; Ueda, *et al.*, 1994; Wang, *et al.*, 1993). These polymers may be utilized as coating materials or matrix resins with inorganic compounds. However, the solubility was not high enough to prepare their concentrated solution.

---

[*] Address correspondence to Dr. Kobayashi.

Macromolecular Engineering, Edited by M.K. Mishra et al.
Plenum Press, New York, 1995

319

Soluble aromatic polymers generally do not show good chemical resistance. The introduction of a crosslinkable group at chain ends of these polymers has been investigated in order to improve their chemical stability (Hergenrother, 1986). The polymer becomes insoluble by the curing treatment and enhances chemical resistance. As a crosslinkable group, acetylene group is often used because of its easy introduction into polymer terminals and of using no catalyst for curing. (Bennett and Farris, 1994; Hergenrother et al., 1994; Núñez et al., 1992; Percec and Auman, 1984). Furthermore, no volatiles are evolved during curing of such polymers.

The first part of this article describes the synthesis of PEK containing more than two substituents per one polymer unit. The resulting polymer was highly soluble in organic solvents. The second part deals with synthesis and properties of crosslinkable PEK bearing acetylene groups at chain ends. The acetylene group was reacted by curing to produce insoluble products exhibiting high chemical stability.

ALKYL-SUBSTITUTED POLY(ARYL ETHER KETONE)

PEK with more than two alkyl substituents per one unit was synthesized by the aromatic nucleophilic reaction of (1,1'-(p-phenylenedioxy)bis[2-methyl-4-(4-fluoroben-zoyl)benzene] (1) with aromatic diol (2) using anhydrous potassium carbonate as base (Fig. 1) in a mixture of N,N-dimethylacetamide (DMAc) and toluene. The polymerization was performed at 130 °C for 1 h, followed by heating at 170 °C for 2 h. The polymer was purified by reprecipitation procedure (DMAc as good solvent, aqueous methanol as poor solvent). The aromatic diols used in this study were hydroquinone (2a), methyhydroquinone (2b), t-butylhydroquinone (2c), 2,5-di-t-butylhydroquinone (2d), and resorcinol (2e).

Polymerization results are shown in Table 1. In all cases, alkyl-substituted PEK 3 was obtained in a high yield. The polymer structure was confirmed by ^{1}H and ^{13}C NMR spectroscopies. Molecular weight of the polymer was evaluated by gel permeation chroma-

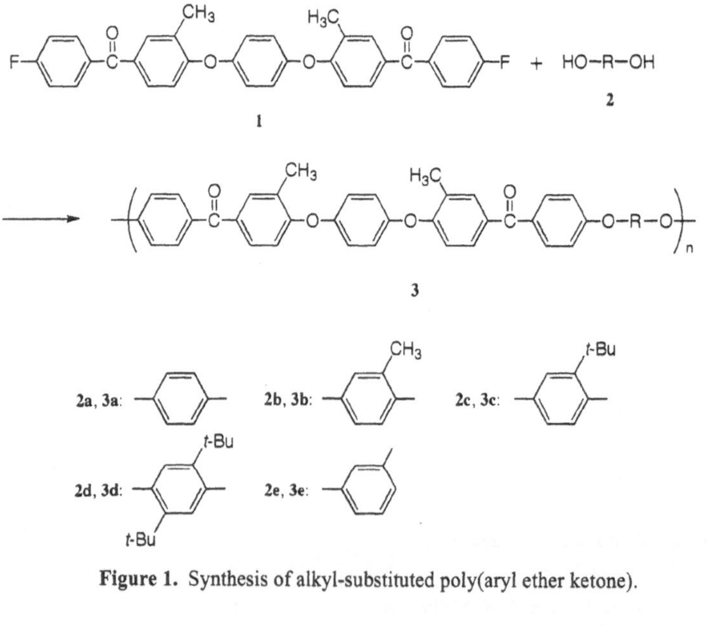

Figure 1. Synthesis of alkyl-substituted poly(aryl ether ketone).

Table 1. Preparation and solubility of alkyl-substituted poly(aryl ether ketone) **3**

| | | | | | Polymer | | | | | | |
| | | | | | | | Solubility[b] | | | | |
Aromatic Diol	Structure	Yield (%)	M_n[a]	M_w/M_n[a]	Chloroform	DMAc	NMP	Toluene	Triglyme	Acetone	Methanol
2a	3a	82	43000	2.8	++	++	++	-	±	-	-
2b	3b	90	33000	2.2	++	++	++	±	++	-	-
2c	3c	85	40000	1.9	++	++	++	++	++	-	-
2d	3d	92	35000	2.2	++	++	++	±	+	-	-
2e	3e	82	43000	2.8	++	++	++	+	++	-	-

a) Determined by GPC.

b) ++: soluble; +: partly soluble; ±: swelling; -: insoluble.

tographic (GPC) analysis using polystyrenes as standard. The number-average molecular weight was around 4×10^4 and hardly affected by the nature of the substituent in **2**.

The solubility of **3** toward organic solvents has been examined (Table 1). Non-substituted PEK is scarcely soluble in common organic solvents (May, 1986). On the other hand, **3** was soluble in chloroform, DMAc, and N-methyl-2-pyrrolidone (NMP). PEK **3c** obtained by using t-butylhydroquinone as aromatic diol was still more soluble in toluene and 2-methoxyethyl ether (diglyme). The solubility limit of **3** in chloroform at room temperature was as follows: **3a**, 0.7 % (weight); **3b**, 6.3 %, **3c**, 9.8 %; **3d**, 3.2 %; **3e**, 1.4 %. From these data, PEK from the substituted aromatic diol shows higher solubility toward chloroform than from non-substituted aromatic diol. It is to be noted that the solubility limit of **3c** is quite high, as compared with that of fully aromatic polymers. Therefore, **3c** can be used as solution of relatively high concentration, leading to new applications of PEK.

Glass transition temperature (Tg) was measured by using differential scanning calorimeter (DSC) at 10 °C/min heating rate under nitrogen. All the polymers prepared showed Tg of more than 130 °C (Table 2). Tg of PEK **3e** from meta-benzenediol was lower

Table 2. Thermal properties of PEK **3**

Polymer	T_g[a] (°C)	T_{di}[b] (°C)	T_{d10}[c] (°C)
3a	148	445	475
3b	146	437	451
3c	154	447	464
3d	178	448	460
3e	135	438	466

a) Glass transition temperature.

b) Onset temperature of the initiation of thermal decomposition.

c) Temperature of 10% loss of intial weight.

than that from para-isomer (**3a**). This may be because the meta-linkage induces inhibition of the packing of the polymer chain. **3b** from methyl-substituted hydroquinone showed lower Tg than **3a** from non-substituted one, on the other hand, PEK **3c** from *t*-butylhydrpquinone possessed higher Tg. Furthermore, the use of disubstituted monomer (**2d**) resulted in the formation of PEK **3d** with higher Tg, compared with that obtained by using mono-substituted monomer **2c**. These data indicate that the introduction of more bulky substituent increases Tg of PEK.

Thermal stability was evaluated by thermogravimetric analysis (TGA) under nitrogen. The temperature of initial decomposition was around 440 °C, indicating that all the samples show high thermal stability under air. The structure of the aromatic diol monomer affected the decomposition temperature very little.

CROSS-LINKABLE POLY(ARYL ETHER KETONE) BEARING ACETYLENE GROUPS AT CHAIN ENDS

Synthesis

In the first part, we have shown the reaction of **1** with the alkyl-substituted aromatic diol to produce PEK **3** showing high solubility property toward organic solvents. This property, in some cases, implies low chemical stability. In order to improve its chemical resistance, an acetylene group has been introduced into the polymer terminal. Crosslinking reaction of the acetylene group by curing affords an insoluble product with high chemical stability.

At first, the reaction of **1** with 3-ethynylphenol was performed as a model reaction for the introduction of ethynyl group into the polymer terminal (Figure 2). The reaction proceeded under the similar condition of the synthesis of **3** to give **4** quantitatively. The structure of **4** was confirmed by NMR and IR spectroscopies.

Acetylene-terminated PEK (**6**) was synthesized according to Figure 3. Excess of **1** was reacted with hydroquinone (**2a**) to give PEK (**5**) having fluoro groups at both chain ends, followed by the reaction with 3-ethynylphenol to give **6**. Polymerization results are shown in Table 3.

Polymerization in the various feed ratio was carried out to obtain PEK of different molecular weight. The molecular weight was determined by GPC and NMR analyses. Figure 4 shows ^{1}H NMR spectrum of **6a** in CDCl$_3$. Besides multiplet peaks due to aromatic protons, two singlet peaks are observed. Peak A at δ 2.3 is due to methyl protons attached to aromatics and peak B at δ 3.1 ascribed to a proton of the terminal ethynyl group. The molecular weight

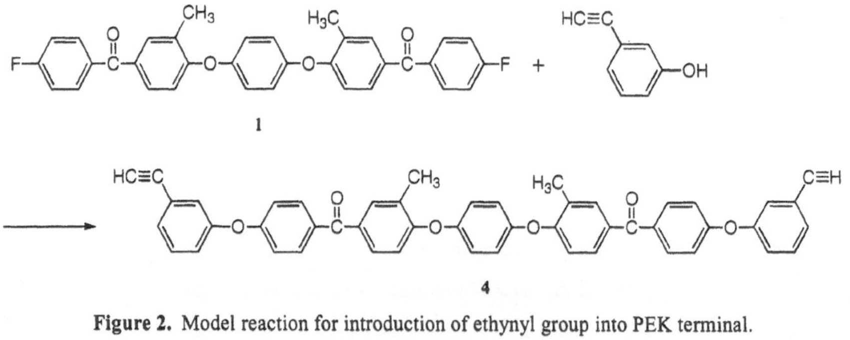

Figure 2. Model reaction for introduction of ethynyl group into PEK terminal.

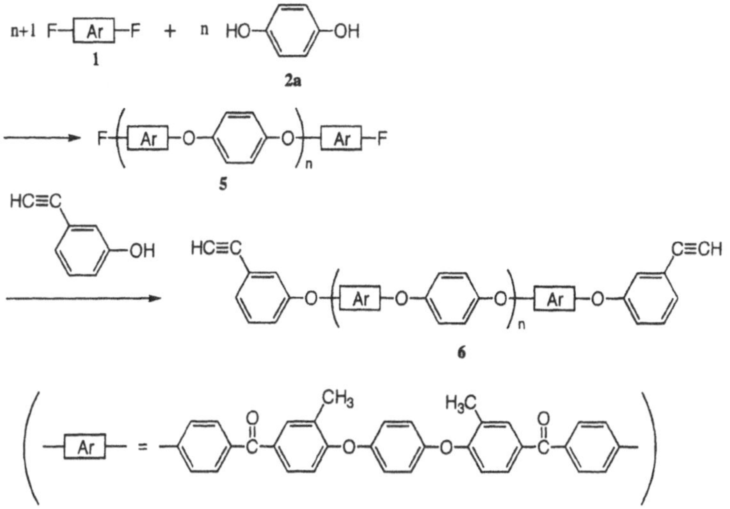

Figure 3. Synthesis of acetylene-terminated poly(aryl ether ketone) **6**.

was calculated from the area ratio of the both peaks, which value was very close to the threoritical value. Furthermore, the molecular weight determined by GPC analysis well agreed with the calculated value or that by NMR analysis. These data support that the acetylene group is introduced at both chain ends, and imply that the molecular weight could be controlled by the feed ratio. **6** was soluble in chloroform, DMAc and NMP, but insoluble in acetone and methanol. This behavior is similar with that of **3a**, having the same polymer unit without the terminal group. The molecular weight of **6** hardly affected the solubility. **4** was soluble in the above solvents as well as in acetone.

Table 3. Synthesis and solubility of ether-ketone compound **4** and poly(aryl ether ketone) **6** bearing acetylene groups at chain ends

| | | | | | | Polymer | | | | |
| | | | | | | | Solubility[d] | | | |
$[2a]_0/[1]_0$	Structure	Yield (%)	M_n[a]	M_n[b]	M_n[c]	Chloroform	DMAc	NMP	Acetone	Methanol
---	4	99	742	-----	-----	+	+	+	+	±
0.82	6a	95	3000	4300	4400	+	+	+	-	-
0.95	6b	87	12000	12000	12000	+	+	+	-	-
0.97	6c	95	24000	28000	20000	+	+	+	-	-

a) Caluculated from the feed ratio.

b) Determined by GPC.

c) Determined by ^{1}H NMR.

d) +: soluble; ±: swelling; -: insoluble.

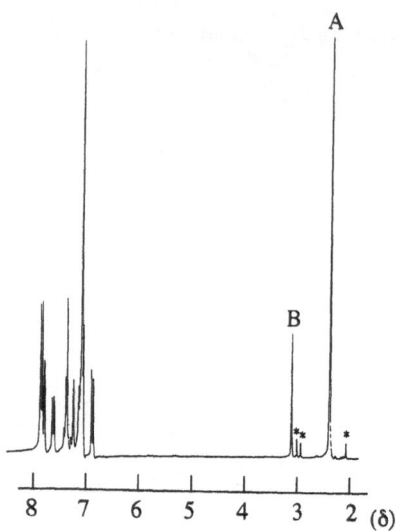

Figure 4. ^{1}H NMR spectrum of **6a**.

Curing of Acetylene-Terminated PEK

In order to study the thermal behaviors and the curing of PEK **6**, DSC analysis was carried out under nitrogen. Figure 5 showed a DSC trace of polymer **6b**. The first scan showed an exothermic peak at 316 °C, on the other hand, no exothermic peak was observed in the second scan. The product after the curing became insoluble in the organic solvents solubilizing non-cured PEK **6**. These data indicate that the terminal ethynyl groups are thermally reacted with each other to give the crosslinked polymer. **4** was used for comparison of the curing behavior between an acetylene-terminated ether-ketone compound and polymer. Thermal properties of **4** and **6** before and after curing are summarized in Table 4. The exothermic peak due to the thermal crosslinking was observed in the first scan of all the

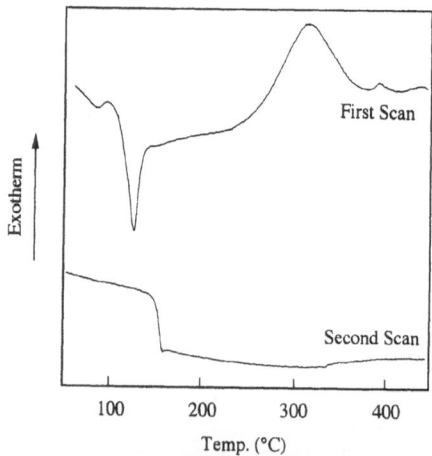

Figure 5. DSC traces of **6b** of the first and second scans.

Table 4. Thermal properties of **4** and **6** before and after curing

Product	Before Curing		After Curing	
	$T_m^{a)}$ (°C)	$T_c^{b)}$ (°C)	$T_g^{c)}$ (°C)	$T_d^{d)}$ (°C)
4	---	259	---	445
6a	117	316	151	432
6b	131	318	152	423
6c	128	320	150	430

a) Melting point (endothermal onset by DSC).

b) Exotherm peak temperature.

c) Glass transition temperature of the cured material.

d) Onset temperature of the initiation of thermal decomposition.

samples examined, and the peak of **6** was higher than that of **4**. The crosslinked polymer showed Tg around 150 °C and the Tg value scarcely affected the molecular weight of PEK **6**. Non-crosslinkable PEK **3a** having the same polymer unit showed Tg of 148 °C, which value is almost the same as the cured polymer of **6**. On the other hand, Tg of the cured material from the ether-ketone compound was not observed, probably because of the high crosslinking density. From TGA analysis, the crosslinked polymers were found to be stable up to 400 °C and their initial decomposition temperature was ca. 430 °C. The stability properties of the cured material was similar with those of the non-crosslinkable PEK, as shown in Table 2.

The thermal reaction of the ethynyl group by the curing procedure was confirmed by FT-IR spectroscopy (Figure 6). In the chart of **6a** (before curing), two characteristic absorbances at 3290 cm^{-1} ascribed to the hydrogen-carbon bond of ethynyl group and at 2110

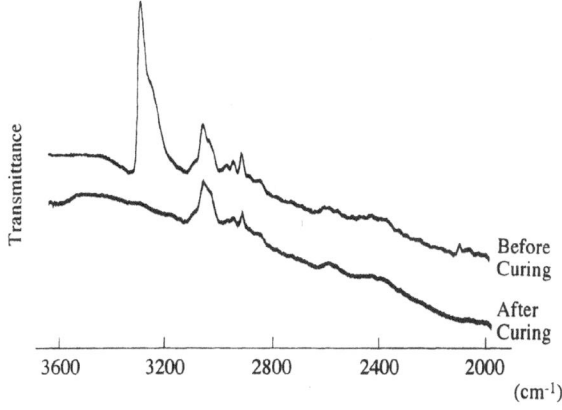

Figure 6. FT-IR spectra of **6a** before and after curing.

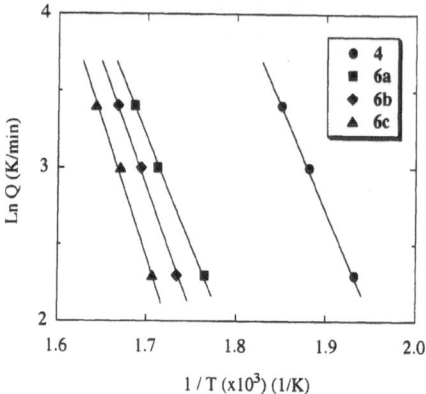

Figure 7. Repcirocal of temperature versus logarithm of heating ratio.

cm⁻¹ due to the carbon-carbon triple bond were observed. After curing,these two peaks completely disappeared, indicating that the ethynyl group is quantitatively reacted during the curing.

In order to evaluate the quantitative reactivity of the acetylene group of 4 and 6, activation energy in the curing treatment has been determined by DSC measurement. The energy is calculated according to Thomas's equation:

$$LnQ = -Ea/RT + B$$

where Q, Ea, T, R, B are heating rate (K/min), activation energy (cal), exotherm peak temperature (K), gas constant (1.987 calK⁻¹mol⁻¹), and constant, respectively. Figure 7 shows relationships between repcirocal of the temperature and logarithm of the heating ratio. In all samples, the linear relationships were shown. The activation energy calculated from the slope was as follows: **4**, 2.6 kcal; **6a**, 28.4 kcal; **6b**, 32.6 kcal; **6c**, 37.6 kcal. The activation energy of **4** was much lower than that of **6**, indicating big difference of the reactivity of the acetylene group between **4** and **6**. The higher the molecular weight of PEK **6**, the larger the activation energy.

Application for Electronic Device

As a possible application, a matrix resin of PEK **6** with conductive carbon black has been prepared, and evaluated as a novel carbon-film resistor. As a matrix resin of the resistor, a commercial phenol resin is often used because of cheap price and of mild curing condition. However, the resistor from the phenol resin normally shows no good moisture resistance (stability) owing to its higher hygroscopic property. In order to improve the moisture stability, development of a matrix polymer having lower hygroscopic nature is desired. Acetylene-terminated PEK**6** has no polar functional group, and hence, the cured polymer is expected to show high stability toward moisture.

At first, the carbon black was added in 50 weight % NMP solution of **6a** to give the resistive ink containing 8.3 vol. % of the carbon black. The carbon-film resistor (Figure 8) was obtained by printing the ink on the alumina substrate, followed by curing at 250 °C for 1 h. For comparison, the carbon-film resistor using a commercial phenol (Novorak-type) resin as a matrix polymer was also prepared under the similar curing conditions. Environmental characteristics are shown in Figures 9 and 10. The test of a long-term heat resistance

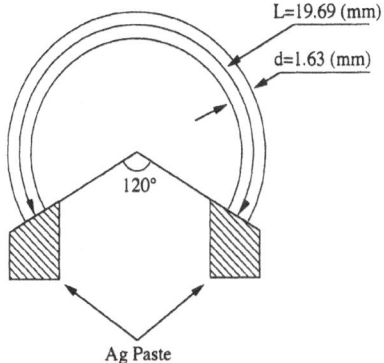

Figure 8. Pinging pattern of carbon-film resistor.

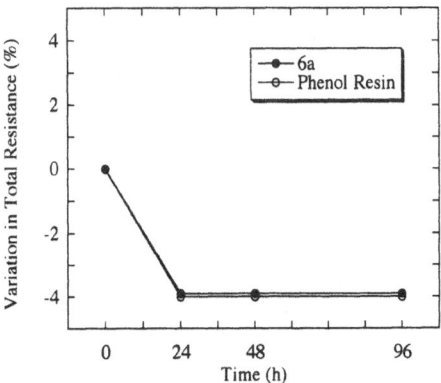

Figure 9. A variation in total resistance in the test of long-term heat stability.

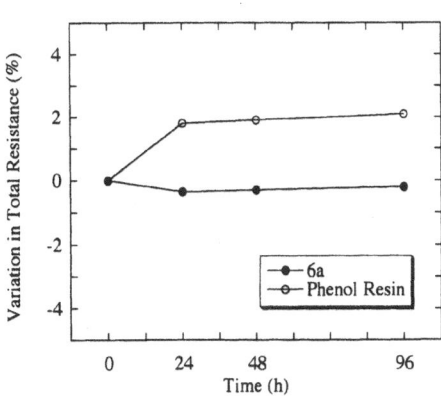

Figure 10. A variation in total resistance in the test of moisture stability.

was performed as follows. The resistor sample was kept in a test chamber at 85 ± 3 °C for appropriate period. Then, the total resistance was measured and the variation in the total resistance was determined. In the test of moisture stability, the sample was allowed to stand in a test chamber of 95 % humidity at 60 ± 3 °C. The resistance value of the resistor obtained from PEK **6a** decreased after 24 h at 85 °C, and afterward kept almost constant. The resistance behavior of the sample from **6a** in the test of the long-term heat stability was almost the same as that from the phenol resin. On the other hand, the resistor from **6a** showed good moisture stability; the resistance scarcely changed in the high humidity condition during the test. The variation from **6a** was smaller than that from the phenol resin, indicating that the resistor from **6a** shows higher moisture stability than that from the phenol resin.

CONCLUSION

Fully aromatic poly(ether-ketone) **3** containing more than two alkyl substituents per one unit was synthesized. The resulting polymer showed high solubility toward organic solvents. Synthesis of acetylene-terminated PEK **6** was performed by the introduction of ethynyl group into the fluorine-terminated PEK. The curing of the polymer produced insoluble materials of high thermal stability. PEK **6** was used as a matrix polymer of carbon-film resistor. The resistor showed a very good moisture stability, implying the possibility of PEK **6** as a novel matrix polymer of the carbon-film resistor with high moisture resistance. Evaluations of **6** as coating and adhesive materials are now under way in our laboratory.

REFERENCES

Bennett, G. S., and Farris, R. J., 1994, The synthesis and characterization of amine-termianted poly(aryl ether ketone)s as a function of side group and molecular weight, *J. Polym. Sci., Polym. Chem. Ed.*, 32:73-87.

Hergenrother, P. M., 1986, Acetylene-terminated prepolymers, *Encyclopedia of Polymer Science and Engineering*, 2nd ed., John Wiley & Sons, New York, Vol. 1, pp61-86.

Hergenrother, P. M., Bryant, R. G., Jensen, B. J., and Havens, J., .1994, Phenylethynyl-terminated imide oligomers and polymers therefrom, *J. Polym. Sci., Polym. Chem. Ed.*, 32:3061-3067.

May, R., 1986, Polyetheretherketones, *Encyclopedia of Polymer Science and Engineering*, 2nd ed., John Wiley & Sons, New York, Vol. 12, pp313-320.

Núñez, F. M., de Abajo, J., Mercier, R., Sillion, B., 1992, Acetylene-terminated ether-ketone oligomers, *Polymer*, 33:3286-3291.

Percec, V., and Auman, B. C., 1984, Functional polymers and sequential copolymers by phase transfer catalyst, 2, synthesis and characterization of aromatic poly(ether-sulfone)s containing vinylbenzyl and ethynyl chain ends, *Makromol. Chem.*, 185:1867-1880.

Risse, W., and Sogah, D. Y., 1990, Synthesis of soluble high molecular weight poly(aryl ether ketone) containing bulky substituents, *Macromolecules*, 23:4029-4033.

Ueda, M., Seno, Y., Haneda, Y., Yoneda, M., and Sugiyama, J., 1994, Synthesis of *tert*-butyl-substituted poly(ether ketone) by nickel-catalyzed coupling polymerization of aromatic dichloride, *J. Polym. Sci., Polym. Chem. Ed.*, 32:675-681.

Wang, F., Roovers, J., and Toporowski, P. M., 1993, Synthesis and molecular weight characterization of narrow molecular weight distribution fractions of methyl-substituted poly(aryl ether ether ketone), *Macromolecules*, 26:3826-3832.

INDEX